Praise for *The Light Ages*

"Compulsive, brilliantly clear, and superbly well written, *The Light Ages* is more than just a very good book on medieval science: it's a charismatic evocation of another world. Seb Falk uses the monk John of Westwyk to weld us into the medieval ways of imagining as well as thinking. And there are surprises galore for everyone, no matter how knowledgeable they may think they are. I can't recommend it highly enough."
—Ian Mortimer, author of *The Time Traveler's Guide to Medieval England*

"An expansive survey of Eastern polymaths, squabbling theorists, political schemers, and optimistic overreachers." —*The New Yorker*

"Like a fictional scientist cloning dinosaurs from wisps of DNA, Seb Falk takes barely surviving fragments of evidence about an almost forgotten astronomer in a storm-chilled, clifftop cell to conjure the vast, teeming world of scientific research, practice, and invention in the Late Middle Ages. . . . Profoundly scholarly, wonderfully lucid, and grippingly vivid, *The Light Ages* will awe the pedants and delight the public."
—Felipe Fernández-Armesto, author of *Out of Our Minds*

"Falk doesn't only paint the Middle Ages as a time of intellectual sophistication; he allows room for some of its more outlandish ideas as well."
—Mary Wellesley, *Guardian*

"As fascinating as it is exquisitely written. . . . [T]he range of mathematics, astronomy, and engineering (the mechanical clock is a true medieval marvel) is impressive. More impressive still is the elegance with which [Seb Falk] tells the tale." —Tom Whipple, *Times* (UK)

"The phrase 'medieval science' is often understood as a contradiction. Here, the historian and Cambridge Universi— a lucid explanation of why it is not."

"Thanks to men like Westwyck, the Dark Ages were anything but dark; Falk's book is a lucid and eloquent reproof to anyone who says otherwise."
—Mathew Lyons, *Prospect*

"[*The Light Ages*] lluminates not just the visionaries of the past but also the troubled state of anti-intellectualism in the modern world. . . . [I]t also draws an engaging portrait of a time of expanding horizons."
—Chris Allnutt, *Financial Times*

"[A] brilliant study of medieval astronomy and learning. . . . [R]iveting."
—Tom Hodgkinson, *Spectator*

"Seb Falk has framed a fascinating book around his personal quest to understand how scientific thinking flourished. *The Light Ages* reveals the intellectual sophistication that flourished against a backdrop of ritual and liturgy. It offers for most of us a novel perspective on a 'dark' historical era and should fascinate a wide readership."
—Lord Martin Rees, author of *On the Future*

"Seb Falk lays out the wonders of medieval science. . . . The mechanical clock, spectacles, advances in navigation, a grasp of tides and currents— these were among the achievements of the Middle Ages." —*Economist*

"Long before the word 'scientist' was coined, John of Westwyk devised a precision instrument to explore the universe and our place in it. Falk re-creates the schooling of this ordinary (if gadget-obsessed) medieval monk in loving detail. There's a world of science on every page."
—Nancy Marie Brown, author of *The Abacus and the Cross*

"A book to change historical presuppositions." —*Tablet*

"Astronomy, a science in which medieval thinkers excelled, is a good vehicle . . . to explain the arithmetic, map-making or medicine that went with it. If anyone can make it clear, it is Seb Falk . . . with his carefully constructed narrative and prose as plain as a well-forged . . . astrolabe."
—Christopher Howse, *Daily Telegraph*

"Seb Falk, in his magnificent book *The Light Ages*, [lets] us inhabit, for a spell of seven finely crafted chapters, the vibrant mind of a 14th-century Benedictine monk, John Westwyk. . . . [A]s if John Westwyk and Seb Falk, separated in time but not in spirit, were joining hands while guiding us along; or as if *The Light Ages* were Mr. Falk's own clever astrolabe, seeking to make that shimmering light in the distance look, as well it should, wonderfully close and luminously real."

—Christoph Irmscher, *Wall Street Journal*

"*The Light Ages* is unambiguously and successfully an antidote to the cliché of the 'Dark Ages' as a millennium of stagnation and regression. It does not achieve this by inviting us simply to marvel at inventions that confound our expectations. . . . Rather, Falk's approach is to explain the things we share with our medieval forebears and the things we differ on: to reveal, in other words, how they saw the universe."

—Kate Wiles, *Literary Review*

"If you think the term 'medieval science' is a contradiction then you should read this hugely enlightening and important book."

—Jim Al-Khalili, author of *The World According to Physics*

"Falk offers a sense of the international nature of medieval scholarship, debunking the image of isolated, repressive monastic communities and highlighting the influence of both Muslim and Jewish innovators."

—Meilan Solly, *Smithsonian*

"Seb Falk is undaunted. . . . This is a book teeming with interesting themes and lines of enquiry Seb Falk is right. It is time for us moderns to show a little humility and admire the light emanating from the Dark Ages."

—Michael Duggan, *Irish Examiner*

"Irresistible. . . . By the time readers are finished with this book, they'll know their way around an astrolabe with the familiarity of a 14th-century alchemist. And Falk is such an agreeably wonkish enthusiast. . . . [A] revelation."

—Steve Donoghue, *Christian Science Monitor*

"The rewards are rich: Falk gives us a genuine understanding of how people long-dead saw their world and a respect for their considerable scientific acumen."

—James Hannam, *Catholic Herald*

"How blissful it was to leave the 21st century behind and immerse myself in this debut. . . . [W]e embark on a wondrous voyage of discovery: navigating by the stars, multiplying Roman numerals, curing disease and telling the time with an astrolabe, the medieval answer to a smartphone."

—*Bookseller*

The Light Ages

The Light Ages

The Surprising Story of Medieval Science

SEB FALK

W. W. NORTON & COMPANY
Independent Publishers Since 1923

Copyright © 2020 by Seb Falk
First American Edition 2020

Originally published in the UK under the title THE LIGHT AGES:
A Medieval Journey of Discovery

For information about permission to reproduce selections from this book, write to
Permissions, W. W. Norton & Company, Inc., 500 Fifth Avenue, New York, NY 10110

For information about special discounts for bulk purchases, please contact
W. W. Norton Special Sales at specialsales@wwnorton.com or 800-233-4830

Manufacturing by LSC Communications, Harrisonburg
Production manager: Erin Reilly

Library of Congress Cataloging-in-Publication Data

Names: Falk, Seb, author.
Title: The light ages : the surprising story of medieval science / Seb Falk.
Description: First American edition. | New York, NY : W.W. Norton & Company, 2020. |
Includes bibliographical references and index.
Identifiers: LCCN 2020028315 | ISBN 9781324002932 (hardcover) |
ISBN 9781324002949 (epub)
Subjects: LCSH: Science, Medieval—History.
Classification: LCC Q124.97 .F35 2020 | DDC 509.4/0902—dc23
LC record available at https://lccn.loc.gov/2020028315

ISBN 978-0-393-86840-1 pbk.

W. W. Norton & Company, Inc., 500 Fifth Avenue, New York, N.Y. 10110
www.wwnorton.com

W. W. Norton & Company Ltd., 15 Carlisle Street, London W1D 3BS

2 3 4 5 6 7 8 9 0

To Fergus and Vivian

Bot for men sein, and soth it is,
That who that al of wisdom writ
It dulleth ofte a mannes wit
To him that schal it aldai rede,
For thilke cause, if that ye rede,
I wolde go the middel weie
And wryte a bok betwen the tweie,
Somwhat of lust, somewhat of lore,
That of the lasse or of the more
Som man mai lyke of that I wryte.
                                    – John Gower

Contents

List of Illustrations

The Light Ages

Prologue

The Mystery Manuscript

Derek Price, they said, was 'not socially house-trained'.[1] It was not easy to fit in, in 1950s Cambridge. Coming from a lower-middle-class Jewish family did not help, and he had no wartime medals to point to, only his training at the undistinguished Southwest Essex Technical College. He had become interested in the history of science while teaching mathematics in colonial Singapore, and sent letters seeking a job as a lecturer. The professors told him to enrol as a student.[2] From the day he arrived at Christ's College, whose alumni included Charles Darwin and the Queen's cousin, war hero Lord Mountbatten, Price was desperate to prove himself.

One chilly morning in December 1951, he got his chance. A few months after starting his research on the history of scientific instruments, he had an appointment to visit the medieval library of Peterhouse, Cambridge's oldest college. There was just one manuscript there that interested him – number 75. It contained – so its nineteenth-century cataloguer had hesitantly guessed – 'directions for making an astrolabe (?)'.[3] It was, as Price later recalled, 'a rather dull volume . . . and had probably hardly been opened in the last five hundred years it had been in the library'.

> As I opened it, the shock was considerable. The instrument pictured there was quite unlike an astrolabe – or anything else immediately recognizable. The manuscript itself was beautifully clear and legible, although full of erasures and corrections exactly like an author's draft after polishing (which indeed it almost certainly is) and, above all, nearly every page was dated 1392 and written in Middle English instead of Latin . . .

The significance of the date was this: the most important medieval text on an instrument, Chaucer's well-known *Treatise on the Astrolabe*, was written in 1391 ... The conclusion was inescapable that this text must have had something to do with Chaucer. It was an exciting chase.[4]

The chase got hotter when Price spotted the beginning of a word: 'chauc'. The rest was buried in the manuscript's tight nineteenth-century binding, but Price quickly persuaded the Peterhouse librarian to have the binding cut apart. On the day when the disbound leaves returned from the conservators, Price and two distinguished professors were ejected from the hushed library for whooping with raucous delight.[5] The full word was indeed revealed to be 'chaucer'. This 'dull' manuscript was a draft instruction manual for a completely unknown scientific instrument. And it was seemingly written by the hand of Geoffrey Chaucer, the greatest English writer before Shakespeare.

With his characteristic single-minded energy, and equally characteristic disregard for the cautious norms of Cambridge scholarship, Price rushed to publicise his discovery. 'Chaucer Holograph Found in Library', trumpeted the university newspaper *Varsity*, above a collage of the manuscript and Price, with wavy hair and thick-framed glasses, looking a little younger than his twenty-eight years (image 0.1). *The Times* of London, a few days later, was more hesitant. 'Possible Chaucer Manuscript: Discovery at Cambridge' ran its headline. The story quickly spread across the world, in newspapers from Copenhagen to Chennai.[6] But was Price right? Or was the *Times*'s hesitation justified? And why did it matter?

The shock was not simply that a new work by the famous *Canterbury Tales* author had been discovered, but that this was a scientific treatise. 'Was Chaucer a Scientist Too?' ran the incredulous headline in the Indian newspaper *The Hindu*. Never mind that historians – including Price himself – were already well aware that Chaucer had written another scientific-instrument manual, the *Treatise on the Astrolabe*. In the 1950s, just like today, the general view was that the phrase 'medieval science' was a contradiction in terms.

It is often supposed that science began with the Renaissance. In his multimillion-selling book *Cosmos* (1980), the superstar of popular

Photo: Maggi
Mr. Derek Price and part of the manuscript he has discovered.

0.1. Image of Derek Price and Peterhouse manuscript 75, published in *Varsity* on 23 February 1952. The collaged image placed Price's head over the crucial 'chaucer'.

science, Carl Sagan, drew a timeline featuring a range of famous names and events in the history of science. After a smattering of ancient figures such as Pythagoras and Plato, around the year 400 he marked 'onset of "Dark Ages"'. A wide blank space takes us almost to 1500, where we find 'Columbus, Leonardo'. 'The millennium gap in the middle of the diagram', Sagan lamented, 'represents a poignant lost opportunity for the human species.'[7] Sagan never claimed to be a historian – so maybe, one might suggest, he should have left the subject alone – but many who do claim that label have given their readers the same misleading impression. Bookshops are full of titles like 'The Invention of Science' that place its birth – in Europe at least – in a time of revolutionary ferment around 1600 that followed the discovery of the New World and the invention of the printing press.[8] Even university courses in history of science often begin in that period. One fairly recent book is entitled *Science: A History.*

Though excellent, it begins in 1543, and its first part is named 'Out of the Dark Ages'.[9] The medieval reality, however, is a Light Age of scientific interest and inquiry.

Curiously, the concept of the Dark Ages itself comes from the medieval world. Early Christians had written of the pagan darkness before the birth of Jesus. Humanist scholars in fourteenth-century Italy took that old Christian metaphor and turned it on its head. They described the darkness of a supposed cultural decline, between the fall of the Roman empire around 400 and their own Renaissance revival of classical learning. For scholars keen to divide human history into easy chunks, it was both convenient and evocative. It gave them an enemy to define themselves against. That became particularly appealing where the Protestant Reformation took hold, and earlier centuries could be mocked as enslaved to Roman Catholic superstition. Introducing a selection of English literature in 1605, the Anglican antiquarian William Camden dismissed the Middle Ages as 'overcast with darke clouds, or rather thicke fogges of ignorance'.[10] The idea of the Dark Ages peaked in the eighteenth century: in his monumental *History of the Decline and Fall of the Roman Empire* Edward Gibbon described 'the darkness of the middle ages', implicitly contrasting them with his own Age of Enlightenment.[11]* But as historians developed a new appreciation for the brilliance of medieval culture and learning, the term 'Dark Ages' began a steady decline. It lingered longer in the English-speaking world, where it served as a shorthand for Britain before the 1066 watershed of the Norman Conquest. Even there, though, it could not last, and historians now prefer the less pejorative term 'Early Middle Ages'.

---

* Indeed, even during the 'Dark Ages' scholars were using very similar metaphors to contrast their own period with earlier stagnation. In his prologue to Einhard's life of Charlemagne (742–814), the German monk Walahfrid Strabo wrote that 'Charlemagne was able to offer to the cultureless and, I might say, almost completely unenlightened territory of the realm which God had entrusted to him, a new enthusiasm for all human knowledge. In its earlier state of barbarousness, his kingdom had been hardly touched at all by any such zeal, but now it opened its eyes to God's illumination. In our own time the thirst for knowledge is disappearing again: the light of wisdom is less and less sought after and is now becoming rare again in most men's minds.' Einhard the Frank, *The Life of Charlemagne*, tr. L. Thorpe (London, 1970), 23.

Yet the spectre of the Dark Ages still lurks behind mentions of the medieval world, and especially its scientific achievements. The word 'medieval' is routinely used to sum up the barbarous crimes of terrorist groups. Politicians, journalists or judges brandish it metaphorically to condemn torture or female genital mutilation, dismiss an investigation as a 'witch-hunt' (though witchcraft trials belong firmly to the Early Modern period), even to bemoan poor cellphone coverage.[12] A resurgence of slightly different usage followed the appearance of the phrase 'get medieval on your ass' in the ever-quotable 1994 movie *Pulp Fiction*. When his post as Chief Strategist to President Donald Trump came under threat in August 2017, Steve Bannon reportedly threatened to 'go medieval on enemies of Trump and his populist agenda'. Bannon's words aroused annoyance and amusement on social media. The historian and television presenter Dan Snow jokingly asked his Twitter followers if Bannon might 'Raise a small, unreliable army of ungovernable nobles & poorly equipped, conscripted peasants and immediately get dysentery?' He followed that tweet with 'Lack the most basic understanding of scientific method, embrace quackery & astrology and depend on an imagined deity to bring you victory?'[13]

That second tweet from Snow, however light-hearted, reminds us how negative stereotypes of medieval science have lingered. It is understandable. Our eyes are drawn to the most striking objects, and our minds to simple summaries. In an age when the world's tallest building was Lincoln Cathedral, who would doubt the immense power of religious faith? But belief in God never prevented people from seeking to understand the world around them. Loyalty to texts and traditions never meant the rejection of new ideas. Channelling money and creative energies into religious art and architecture never restricted the range of medieval people's interests. The relationship between faith and the study of nature was – and remains – a complex one, as we shall see throughout this book. Disputed ideas have occasionally caused conflict, of course. But to imagine 'science' and 'religion' as two separate, inevitably antagonistic opponents, or to suggest that such closed-mindedness as does exist has always been on the side of religion, is far too simplistic. The Middle Ages were much more than battles and black boils.

A more detailed picture requires a wider range of sources. The most commonly reproduced medieval images are the jewels of imagination and craftsmanship: exquisite books of hours; tapestries of mythical beasts; painstaking calligraphy. Most scientific writings were not so beautiful, any more than research results published in scientific journals today are immediately appealing to the casual reader. When Derek Price first leafed through Peterhouse manuscript 75, he would most likely have opened it at one of the many handwritten mathematical tables that filled it. It would have looked something like image 0.2. No unicorns there.

The manuscripts and instruments you will find featured in this book are mostly not the precious art objects displayed in exhibitions of library treasures, sumptuously decorated with gold leaf. Medieval science books do survive, in large numbers, but they are usually not the sort of books whose decorated motifs adorn banknotes and postage stamps, symbols of national pride, as hard for scholars to get their hands on as the Crown Jewels. As Price found, they have sometimes been neglected by historians, and may be in poor condition. Still, librarians and archivists work tirelessly – and usually without recognition – to preserve them, so are invariably pleased to help someone study them. I have rarely been refused access, and it is always mildly surprising to me that no one ever wants to check I have washed my hands. (You almost never use gloves to handle a medieval manuscript.) But manuscripts like Peterhouse 75 are no less remarkable, no less important, than the ones that glint in display cases. In this book we will read the sometimes scrappy texts, but we will also handle bits of brass instruments, decipher sketched diagrams. They are the surviving witnesses to the forgotten world of medieval science. We will examine them not only for their contents but also to find out how they were made, kept and used, read, bound, borrowed and sold, decorated and discarded.

What was 'medieval science'? The phrase itself is controversial. We think we know what science means: science is what scientists do. They undergo standardised training, obtain internationally recognised professional qualifications, and use universally accepted methods, in purpose-designed spaces, to obtain reliable answers to

0.2: Table of mean motions of the planet Mars, in Peterhouse, Cambridge MS 75.I.

questions that have themselves been posed in a standard way. Medieval science was not much like that. True, science today did evolve from knowledge-gathering activities stretching back to the Middle Ages and much earlier, and those activities investigated natural phenomena very similar to what scientists investigate today. Medieval people sought to build understanding of why things in nature behave as they do and used their understanding to make future predictions. But they were not scientists, and their science included activities that would not be considered science today. If we study medieval science looking only for precursors and forerunners of the way we do things now, we will inevitably find it failing to be quite like us – especially if we measure it against an idealised 'scientific method' that even some modern sciences do not live up to.

So should we not use the word 'science' at all, since we will only be disappointed when we fail to find it? That is what some historians have demanded. Medieval investigation of nature, they argue, was driven so strongly by the belief that that nature was created by God, and was directed so single-mindedly towards understanding the divine mind behind creation, that it was an entirely different enterprise – one usually given the name 'natural philosophy'. For medieval people, study of the world – that is, the whole created cosmos – was a route to moral and spiritual wisdom. As Isaac Newton – hardly himself medieval, but standing on the shoulders of several medieval giants – wrote in an afterword to his monumental *Principia mathematica*, 'thus much concerning God; to discourse of whom from the appearances of things, does certainly belong to Natural Philosophy'.[14]

Newton wrote that in Latin, still the universal language of learning in the 1700s. Latin had a word *scientia*, which gave rise to the English word 'science' – but it is an unsatisfying translation. *Scientia* in the Middle Ages could mean knowledge or learning in a general sense, or a way of thinking. Or it could refer to any organised branch of knowledge, from mathematics to theology. It did not have the restricted definition that 'science' has in modern English. Nonetheless, I use the word 'science' throughout this book, since its meanings are flexible and accommodating. (I do not use the word 'scientist', which was coined in the nineteenth century, since it conveys too precise an image of a modern professional, something medieval

8

philosophers, astronomers and physicians certainly were not.) You will, I hope, recognise the family resemblances between the activities described in this book and their descendants in modern science. Yet much has changed in motivations, methods and language, and we must temper our expectations accordingly. Historians of war can observe that the Crusades were fought in entirely different ways and for different reasons from modern conflicts but still have no hesitation in recognising them as war. We can do the same for science.

Disparaging the 'Dark Ages', as we have already seen, has always been about making ourselves seem better by comparison. But we should not award points for being like us. Viewing the past as an imperfectly developed version of the present day can lull us into complacency about the state of our own knowledge, allowing us to ignore what we still do not know or cannot do, as well as how fragile the structures and status of science are. The measure of medieval ideas should never be 'how closely do they match our superior modern ways?', but rather, 'how important were they in their time?', and 'what impact did they have?' Understanding the history of scientific ideas in their proper context – seeing the science first through the eyes of the people who made it – allows us to appreciate that science does not progress in a constant straight line. Progress there undoubtedly has been, but it has not been a series of 'Eureka' moments by great men. Progress can be slow and gradual. Scientific understanding has sometimes hit a dead end, or taken a step sideways, or backwards. And it still can.

If understanding the past of science starts by understanding it on its own terms, then we must learn to appreciate the motivations of the men – and some women – who practised science and who produced the scientific ideas, books and equipment described here. This means getting inside the heads of individuals. But first we need to know something about who those individuals were. That was why it mattered so much whether the astronomical instrument described in Peterhouse manuscript 75 was invented by the satirical Southwark poet Geoffrey Chaucer, or by someone else entirely.

Derek Price did not live to see the question resolved. His research into the mysterious scientific instrument, which he had given the

Middle English name *Equatorie of the Planetis* ('Computer of the Planets'), was warmly received when he published the final results in 1955, but as he himself admitted, he had found nothing more than an accumulation of 'pointers' to Chaucer's authorship. Chaucer scholars remained wary. As literary experts, they admitted they were uncomfortable judging a scientific text, and there was also a sense that involvement with mundane scientific prose was a stain on Chaucer's poetic reputation. The matter was especially sensitive since the *Equatorie* text had a whiff of astrology, 'in which', Chaucer had disingenuously protested, 'my spirit has no faith'. Price's claims were never widely accepted, and the *Equatorie* was included in only one anthology of Chaucer's works.[15]

Price himself moved on. Disenchanted by the closed academic culture of the United Kingdom, he crossed the Atlantic Ocean in 1957, symbolically adopting his mother's Jewish family name, de Solla, on arrival in the USA. He still dreamed of bringing his family back to Britain, but those dreams were soon dashed when he applied for a job at Cambridge University. He had 'a touch of genius', according to one of his references – but in a private letter the same referee admitted, 'I don't believe he will get [the job], for purely personal reasons.' Sure enough, it went to the other candidate. Price was not happy. 'I had always thought they were fair and moral, but one cannot pretend it was either,' he railed. 'Perhaps they do not like my personality or the colour of my eyes. They might at least tell me so. The more I think of the casual insult that Cambridge have handed me in exchange for six years hard and not unproductive work, the more I despair of Britain. It seems I must resign myself to living for ten years in this country while my children grow up as Americans.'[16] Despite this distressing setback, his hard work was rewarded just a few months later when he was asked to found a department of history of science at Yale University. He remained a professor there until his death in 1983, enjoying the prestige of advising governments on science policy and cultivating a reputation as a pipe-smoking 'scientific detective'.[17]

While Price became one of the most important historians of his generation, shaping research in the new field of science studies, the *Equatorie* manuscript remained a mystery. A series of Chaucer experts took different sides in the debate, using arsenals of analytical tools. Could the handwriting be proved to be Chaucer's? In any case, was the scribe of this manuscript really the original author (or translator) of the text? Did the writing style and vocabulary match the poet's, and was the astronomy in the manuscript in keeping with his interests and abilities? Most scholars came to accept that Chaucer's authorship would never be proved – a huge number of manuscripts from the Middle Ages are simply anonymous. But as long as Manuscript 75 could not be definitely assigned to someone else, the case stayed open. Recently, however, it slammed firmly shut.

Kari Anne Rand, an elegant, softly spoken Norwegian scholar, had long taken an interest in the *Equatorie*. She had researched it at the University of Oslo in the 1980s and published an academic book about it in 1993. She had shown that the manuscript was certainly a draft roughly written out by its author, using a London-like dialect similar to Chaucer's, but it was impossible to say any more than that. When twenty years passed and no one had made any new breakthroughs, Rand became impatient. She decided to take up the trail again herself. She searched the libraries of Europe for other instrument manuals from the same timeframe, until she found one whose handwriting perfectly matched Peterhouse 75.[18] A gift for the library of Tynemouth Priory in the far north-east of England, this matching manuscript had been produced at Tynemouth's prosperous mother monastery, St Albans, in about 1380. Turning to the first page, Rand found that its donor – and scribe – had written his name there: 'Dompnus Johannes de Westwyke'. Brother John of Westwick. Not Chaucer. A monk.

Brother John is the perfect guide to the story of medieval science. We do not know much about him – but that is precisely what makes him so suitable. It is entirely appropriate that a book about medieval science should centre on an almost unknown figure. Too many histories are narratives of 'Great Men' – that is one reason why historians were so keen to ascribe the *Equatorie of the Planetis* to a famous name. A true story of science should not be a parade of famous names but should represent the ideas and achievements of the nameless majority of scientifically minded people. Our guide is no household name but an ordinary monk (one of the 2 per cent of Englishmen in holy orders) who lived and died in the late fourteenth century. A man born in a rural manor, educated in England's grandest abbey, exiled to a clifftop priory. A crusader, inventor, astrologer. John of Westwyk was, in many respects, not all that unusual. But following the life of an ordinary scientific monk gives us a true picture of medieval belief and thought. It was an era of modest anonymity. Far from trumpeting his achievements, John Westwyk did not even write his name in his most important and original work, the *Equatorie* discovered by Derek Price. Like most monks who never rose to positions of power within their orders, he left few traces in archives. From those traces, however, we can attempt a reconstruction of the kind of unexceptional life that is so often forgotten by historians. In such a tale of an unknown figure, there will inevitably be gaps. But piecing together what we can know about his life and science allows us to experience the wonder of an age of selfless scholarship.

On Westwyk's journey through medieval science we will meet a fascinating cast, none of them household names. The Spanish Jew-turned-Christian who taught a Lotharingian monk about eclipses in Worcestershire; the clock-building English abbot with leprosy; the French craftsman-turned-spy; the Persian polymath who founded the world's most advanced observatory. Medieval science was an international endeavour, just as science is today. Religious belief spurred scientific investigation, but deeply devout people had no problem with adopting theories from other faiths. We should not underestimate the immense global variety of scientific ideas during a period lasting close to a thousand years, but watching how one individual knew what he knew will help us understand the ways medieval thinkers built on each other's work and influenced other scholars working in different languages thousands of miles away.

What Westwyk knew, above all, was the central science of the Middle Ages – astronomy. As the political poet – and friend of Chaucer – John Gower wrote:

> The science of Astronomy
> I thinke for to specify
> Withoute which, to telle plain,
> All other science is in vain.[19]

Astronomy was the first mathematical science; the models and formulae of modern science could not exist without it. It was of obvious interest to devout scholars attempting to read the mind of God through Creation, as the regular motions of the heavens demonstrated His perfection. It also had immense practical significance, influencing timekeeping and the calendar, geography and architecture, navigation and medicine. As a student of astronomy and user of instruments, John Westwyk represents this meeting of theory and practice well. This book will get you doing science with him, learning the science as and when he learned it. From counting to 9,999 on your fingers to casting a horoscope or curing dysentery, understanding something of how medieval science was not just thought but really *done* – not just admiring astrolabes but weighing the brass in your hands – is essential to appreciating its achievements.

'The past is a foreign country', wrote L. P. Hartley, just as Derek Price was struggling to settle in alien Cambridge.[20] So I invite you to accompany me on a journey to the fourteenth century, to share in the scientific life of an unknown monk.

A NOTE ON TRANSLATION, TRANSLITERATION AND NAMES

In sharing samples of medieval scientific ideas, and painting portraits of the people behind them, I have tried to balance conflicting priorities. Making the material understandable runs the risk of making it too modern. Putting names into familiar English (or Latin) forms, such as 'Thomas Aquinas' and 'Albertus Magnus', may make it easier for you to find further reading about them; but 'Tommaso d'Aquino' and 'Albert von Lauingen' could more clearly demonstrate the multiculturalism and multilingualism of medieval science. Faced with such conflicting priorities, I am unashamedly inconsistent. Sometimes I translate medieval English, sometimes I allow you to appreciate its musicality. Quotations in other languages are translated (by me, unless otherwise stated), but if they are poems you will find the original rich rhythms and rhymes alongside. I try to give names in something like the forms their owners would have used, but sometimes I felt it best to give two versions. (The index should help you resolve any uncertainty.) Names – and other words – from languages that did not use the Roman alphabet, such as Arabic, are generally Romanised in a simple form. I hope that, like me, you will enjoy the sensation of sounding out an unfamiliar word and rolling it across your tongue – as Chaucer himself clearly did.

I

Westwyk and Westwick

How do you reconstruct a lost life? How can we hang a history of science on the story of a single fourteenth-century monk when we can't be sure when or where he was born, what his background was, or how he came to donate a manuscript made at St Albans to the priory at Tynemouth? John Westwyk left us two precious books of astronomy (plus some sketches and notes in at least two others), but of his own biography we have little more than his name.

We begin with that name. It is a decent handle to grab – the measureless majority of medieval people are now nameless to us – but still a name by itself may not seem like much. And with the name John, it isn't. John was by far the most common name for a man in fourteenth-century England. In 1380, the year John Westwyk left St Albans to begin an international odyssey, the Benedictine monastery had fifty-eight members. Twenty-three of them were called John.[1]

'Westwyk', however, does tell us something. Like the surnames of almost all the monks, it is a toponym, telling us where its bearer was from. These monks were resisting fourteenth-century fashion: outside the cloister, occupational surnames like Tyler or Smith were becoming popular alternatives to the surnames based on birthplaces. Within a few generations families would start to pass their surnames down, so that what looks like a toponym might tell us only where a person's ancestors were from. But before 1400, Johannes de Westwyke was reliably John from Westwick.[2]

Westwick hardly exists now. In fact, it barely existed in John's day. Then, as now, it was commonly known as Gorham – for reasons important to our story, as we'll shortly see. Now, it lies in the

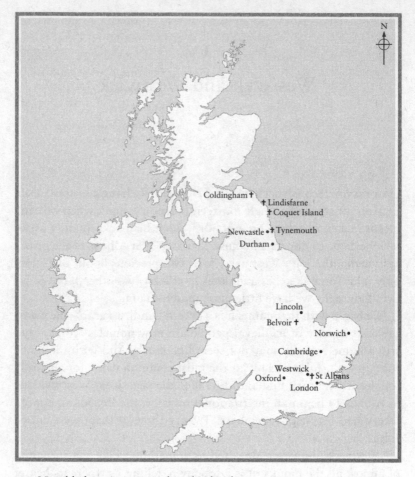

1.1. Notable locations named in this book.

Gorhambury Estate, seat of the earls of Verulam, still a rural manor with a faint feudal flavour. Then, its acres of woodland and meadow had just been bought by the wealthy abbey of St Albans. Fifteen miles north of London, where the Vale of St Albans rises to meet the chalk hills of the Chilterns, the River Ver runs through rolling country-side, with well-drained mixed-clay loam providing good arable soil. On the river's southern bank was Westwick, a manor dotted with farmhouses and fishponds. The theatre of the ruined Roman city of

Verulamium lay on its eastern edge, offering residents a reminder of the glories of the past – and useful building materials. From there the horizon was crowned by the imposing bulk of the great Norman abbey church, built in the 1090s using red Roman bricks salvaged from Verulamium, and larger even than the cathedral constructed at Canterbury in the same decade. Our journey of medieval scientific learning starts appropriately with that panorama of agriculture, classical heritage and faith.

It takes less than an hour to walk down from the centre of old Westwick to the Roman city, across the River Ver and up to the abbey. John was not the only monastic entrant to make that journey; in his time, the chamberlain of St Albans, responsible for the monks' clothing, bedding and washing facilities, was one William of Westwyk.[3] Monasteries were never isolated from their surroundings. Although the individual monks had taken a vow of stability, committing themselves to remaining within the cloistered enclosure, a monastery, especially one as large and centrally located as St Albans, was deeply involved in its local community – at least as important to the local economy, life and culture as a university is today. The monks received constant reminders of this relationship. At the heart of the abbey church was the shrine of St Alban, visited day and night by pilgrims who left gifts in the niches of the richly painted marble tomb. Towards the end of John Westwyk's life, the community began an imposing addition to the saint's chapel: a two-storey watching loft, where abbey tenants took turns to supervise the pilgrims praying and making offerings to the martyr. The wooden structure, the only one of its kind left in Britain, is decorated with a carved frieze showing daily life at the end of the fourteenth century. The brothers tending to the shrine saw vivid scenes of harvesting and hunting, herding and milking. From a sow and her litter, to a man mowing rye, to the squirrel with a nut, a whole year's rural life is depicted in unmistakeable detail (image 1.2).[4]

This mattered to the monks. The produce of their estates did not just put food on the refectory table; it put clothes on their backs, books and scientific instruments in their library and stone in their ever-expanding cloisters. The business of land management fills a substantial proportion of the surviving abbey records, and even when

1.2. Detail from the St Albans watching loft. A dog leads a pig by the ear; a couple use bellows to stoke a winter fire.

writing national history – a subject in which they were pioneers – such issues were never far from the monks' minds. Brother Matthew Paris, the greatest chronicler of late-medieval England, paused in his account of the disputes between the Pope and King Henry III to copy out a charter dated 1258. In that charter the abbot allocated five men from Westwick (and manure from St Albans) to work on the nearby manor of Kingsbury. Their task was to boost the supplies of bread and beer for the monks and their guests.[5]

Growing up on the manor before his admission to the monastery aged about twenty, John Westwyk would have witnessed the business of land management. Westwick's hay meadows and woodland, its mill, fishponds, pigsties and cow-houses, are all recorded in surviving surveys of the manor.[6] Positions of power within the cloister were dominated by the land-holding and merchant classes, but as someone who never rose to such prominence John was more likely the son of a mid-ranking peasant, a *valettus* or yeoman. These made up the largest group of recruits from the monastery's manors. Poorer peasants (bondmen or villeins) were rarely allowed to take monastic vows, though with the abbot's permission – and payment of a small registration fee – they could attend the St Albans school. The abbey needed a supply of literate clerks, but not as much as it needed an agricultural workforce, particularly after the Black Death came to St Albans in 1349. It hit Hertfordshire hard, leading to a dearth of farmworkers. In the years immediately following the plague, the abbey and its local satellites had to buy almost all of the corn they needed from outside, and the shortage of labour meant that wages skyrocketed in the succeeding decades. Landowners could not object to their villeins being educated, but they certainly objected to ambitious labourers leaving the land to pursue a career in the Church or some other profession. So they sought to control their education and career options.[7]

As the son of a yeoman, John Westwyk would have been steeped in farming as a business, a science and a way of life. It is the science that concerns us most, as we seek to understand how John went from peasant, to monk, to astronomical-instrument designer. At a fundamental level, farming was inseparable from astronomy. The seasonal cycles of agricultural labour depicted on the St Albans watching-loft were the rhythms of John's childhood: sowing and reaping, farrowing and slaughtering, working and feasting, all dictated by changing conditions in the fields. The signs of those changing conditions could be read in the heavens.

All human cultures mark the passing of time by the differences they observe in the world around them. Our choice of which differences to mark depends firstly on what we can observe and secondly on what is important in our lives. How we mark the differences – the shapes of our calendars and our rituals – depends on the connections we make between those two things. In the agricultural society of pre-modern Europe, where higher latitudes make the seasons easily observable, it was natural to monitor the solar cycle. Conversely, among the largely nomadic peoples of Arabia, for whom seasonal changes were less significant, the lunar calendar was a more sensible choice. That did not make it inevitable that Islam would use a lunar calendar and Roman Christianity a solar one, but political and religious decisions were made from options limited by geography and lifestyle, filtered through tradition.[8]

If the young John Westwyk was up at first light on the feast of St Luke, 18 October, watching through the chill autumn mist, he could see the Sun rise directly behind the squat Norman tower of St Albans abbey church. He, or his father, could take this as their signal to scatter the year's seeds of winter wheat, as recommended for the month of October in a Middle English poem written a few decades later:

Januar	By thys fyre I warme my handys	By this fire I warm my hands
Februar	And with my spade I delfe my landys	And with my spade I dig my lands
Marche	Here I sette my thinge to sprynge	Here I start the work of spring
Aprilis	And here I here the fowlis synge	And here I hear the fowls sing
Maii	I am as lyght as byrde in bowe	I am as light as a bird on a bough
Junij	And I wede my corne well i now	And I weed my corn well enough
Julij	With my sythe my mede I mawe	With my scythe I mow my meadow
Auguste	And here I shere my corne full lowe	And here I shear my corn fully low
September	With my flayll I erne my brede	With my flail I earn my bread
October	And here I sawe my whete so rede	And here I sow my wheat so red
November	At Martynes masse I kylle my swyne	At Martinmas I kill my swine
December	And at Christes masse I drynke redde wyne.[9]	And at Christmas I drink red wine.

Standing in the fields of Westwick, where the Chilterns slope down to the River Ver, with each new day of the autumn John would see the Sun rise a little further south along the horizon, until the winter solstice, when for a week it rose in the same place, two hand-breadths to the right of the abbey. Then it would begin to move back. Successive Suns would crest the horizon ever more towards the north, rising behind the abbey once again a little before St Scholastica's Day in February and continuing until mid-June, when the Sun came up over the river, just by the mill where the nuns of St Mary de Pré ground their malt and oats.[10] Through the year it covered almost a quarter of the horizon, passing each spot twice and constantly moving back and forth, except for its week-long pauses at the solstices (the Latin *solstitium* means 'Sun standing still').

Such are the gradual changes that have marked the solar year ever since humans first formed settled communities. This is folk astronomy: not a precision science of careful measurement and finely tuned models but an accumulation of ancient wisdom. Even so, it shared some basic principles with the scholarly astronomy that John Westwyk would later learn. It made predictions; it divided space and time according to observations performed over many years; and above all, it was founded on the common-sense understanding that while things on Earth changed constantly, growing and decaying in ways beyond the comprehension of mankind, the movements of the heavens were in a constant, endlessly repeating cycle. It is this understanding that allowed Stonehenge to be constructed in perfect alignment with the midsummer sunrise and – more important to its builders – the mid-winter sunset. Folk astronomy is by definition not written down, but ancient calendars like Stonehenge are monumental evidence of its significance. What knowledge could be more important than the knowledge that, when the Sun went down on the darkest day of the year, it would indeed return with strengthening brightness?[11]

Along with the gradual movement of the sunrise – and sunset – along the horizon, John could observe two further changes in the Sun and its light. The quantities of light and darkness within a day changed; and so did the lengths of shadows (which were easier to measure than the corresponding height of the Sun in the sky). Those two changes were duplicated within each year: just as the Sun rose behind the abbey church twice, so every day of the year had its twin, when the hours of daylight were the same and the midday shadows were the same length, as the Sun culminated at the same height above the horizon.

These symmetries are recorded in a manuscript written at St Albans abbey in the 1150s, and still carefully preserved and used there in John Westwyk's day. During that decade a scribe with a gorgeous calligraphic hand made copies of a tract on the Trinity by the Church Father Hilary of Poitiers, the epistles of St Paul, and a liturgical compendium with initials intricately picked out in red, blue, green and gold.[12] More importantly for us, he also copied a practical manual of scientific agriculture from the latter days of the Western Roman Empire. It was called *The Work of Farming*, by Palladius, a high-ranking Roman who nonetheless saw himself as a down-to-earth farmer. Palladius went through the farming year month by month, dispensing pithy advice on the times to plant and pick, how to assess soil quality, where to buy bees, and why ceramic piping was better than lead (he was well aware that the latter was toxic). At the end of each month-chapter he gave the length of the shadows for each hour of the day, pointing out that the months come in symmetrical pairs: 'August matches May', he noted, 'by the comparable course of the Sun.'[13]

Palladius gives one set of hourly shadow-lengths for each month. They range from two feet at noon in June or July, to twenty-nine feet in the first and last hour of the day in January and December. He listed twelve sunlit hours for every day, summer and winter. This meant that the length of each hour varied during the course of the year. In the summer, each of the twelve daytime hours would be rather longer than each of the twelve night-time hours; but in the winter the situation was reversed, and the twelve daytime hours would fly by. These 12 + 12 *unequal hours*, invented in ancient Egypt and used by Jesus, were still common in medieval Europe. It made sense when there was far more work to do in the summertime fields, and the monks of St Albans were well accustomed to adapting their canonical hours of prayer to the passing seasons.[14] It was only in John Westwyk's century that the *equal hours* with which we are all familiar, and which astronomers had preferred for centuries, came into common use. As we shall see in the next chapter, this was not because the monks or civil authorities found the unequal hours cumbersome or confusing. The change was simply spurred by the spread of mechanical clocks that beat out regular time without reference to the shifting seasons.

The St Albans monks, who could read their copy of Palladius while hearing the steady chimes of the abbey's monumental clock, understood all this. They saw the shadows shorten as the springtime Sun moved across the equator to warm the northern hemisphere; John Westwyk himself wrote out a table of this changing solar declination.[15] The monks also realised that the shadow-lengths Palladius gave were not precise: the shadows at a given time of day never stayed a constant length for an entire month, and Palladius did not specify the height of the object casting the shadow, or say where in the world his measurements were correct. We can calculate that he was working with a shadow-casting gnomon five feet long – convenient for a farmer roughly that height, who could estimate the time by looking at his own shadow – and at the latitude of northern Italy or southern France, which is where Palladius was from.[16] And astronomers in Westwyk's day made tables with shadow-lengths calculated to the nearest sixtieth of a foot for each day.[17] But such precision did not matter to the farmers of Westwick. They could observe simply that the shadow of their own body, or of a sundial-stick planted in the ground, gradually decreased in length each morning and increased by the same amount in the evening, and did the same in the morning and evening of the year.

Nor did that precision matter much to the monks: they read their Palladius with a bigger picture in mind. Beyond its practical usefulness for monks who managed substantial agricultural estates, the regular predictability of the shadows demonstrated that the universe was well ordered. Sundials made according to such calculations had a symbolic function, much as they do today. They were not there to tell the time so much as to show that time could be told. The lengthening and shortening shadows, just like the labours of the months depicted on the St Albans watching-loft, reminded the monks of the regular patterns of their lives in a divinely ordained world.

No wonder *The Work of Farming* was so popular. Copies were made in monasteries from Canterbury to Coventry. A few decades after Westwyk's time, Humphrey, Duke of Gloucester, the youngest son of King Henry IV and a major patron of St Albans abbey, had it translated into English, which widened its popularity still further. The production of an accessible vernacular version of a work of agricultural science had obvious practical value to a wealthy landowner, but it was clearly more than that. Humphrey's anonymous translator turned Palladius' prose manual into poetry. For a prince proud of his classical learning, the translation burnished his literary and humanist credentials. On top of that, its messages of prudent estate management could easily be read as metaphors for stable royal governance.[18] Unconstrained by narrow disciplinary divisions, medieval writers felt no need to distinguish astronomy from agronomy, politics from poetry. So when Palladius advised farmers to pick beans before sunrise, and to wash and chill them to keep them safe from bugs, a medieval translator might relish the chance to demonstrate his literary creativity:

Now benes in decresyng of the moone	Now beans when the Moon is waning,
Er day and er she rise, uppluckéd sone,	Quickly picked before dawn, before the Sun picks up,
Made clene, and sette up wel refrigerate,	Cleaned and set well chilled, they
From grobbis save wol kepe up their estate.[19]	will keep their [fresh] condition, free from bugs.

We have already seen folk astronomy; now this is literary astronomy. Scientific – if not overly complex – content was blended with traditional wisdom and put to poetic purposes. The flexibility of Middle English spelling even allowed the translator a poetic pun on 'soon' and 'Sun'.

Of course, each reader could choose to focus more on the poetry or the practicality, and to extend their learning in different directions. Several copies of Palladius' manual contain annotations showing its readers had also read the *Georgics*, by the Roman poet Virgil.[20] Written in the first century BCE, the *Georgics* was popular in medieval England, partly for agricultural insights like this autumnal advice:

Libra die somnique pares ubi fecerit horas	When Libra balances the hours of day and sleep
et medium luci atque umbris iam dividit orbem,	And splits the world, half light and half in shade
exercete, viri, tauros, serite hordea campis	Then work your oxen, men, sow barley in your fields
usque sub extremum brumae intractabilis imbrem.	Till winter's rain sets in and work must fade.[21]

The *Georgics* is both more literary and more astronomical than Palladius' work. Laid out in Latin hexameter, it went beyond talk of lengthening shadows to include a substantial amount of star lore. Virgil did not just write of the changing seasons; he made clear that it is the constellation Libra which 'balances the hours of day and sleep' at the September equinox. He suggested that farmers sow their wheat in November – later than the English poem we saw earlier, because the soil in Virgil's Italy was drier in the autumn – but he also advised them, more specifically, to 'let the Pleiades set in the morning, and the Cretan star of burning Corona depart, before you commit seeds to the furrows'.[22] Many stars rise and set each day, but here Virgil was not talking about that daily rotation of the heavens. He was referring to the annual rotation which made some stars disappear from view for months at a time.

How did that annual rotation of the heavens affect the stars? If you stand at the North Pole and look up, you will have the North

A. Observer at North Pole

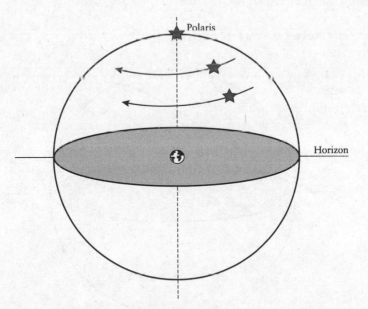

1.3. The stars as viewed from the North Pole (A, *above*), the equator (B, *over*) and the village of Westwick (C, *over*).

B. Observer at equator

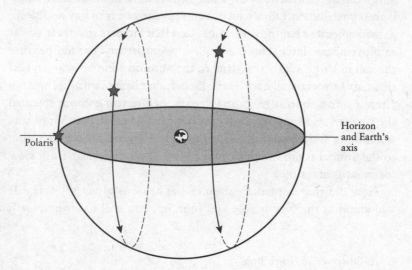

C. Observer at 51°45' N (Westwick)

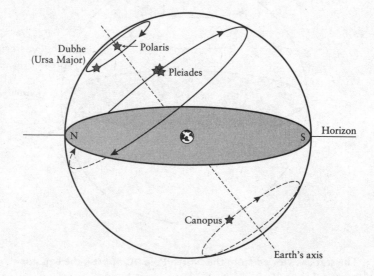

Star (Polaris) directly overhead. As the Earth rotates daily on an axis running from its centre almost exactly to Polaris, all the other northern hemisphere stars will move around Polaris in horizontal circles. None rise or set (see image 1.3a). If, on the other hand, you stand at the equator (image 1.3b) and look towards the north, Polaris – and with it the Earth's axis – will be exactly on the horizon. All the stars – of both the northern and southern hemispheres – will rise and set vertically; none will stay up all night. Wherever you are in the world, the stars circle the pole. The altitude of the pole above the horizon tells you your latitude. So if, like John Westwyk – and Virgil before him – you observe the sky from a middle latitude in the northern hemisphere (image 1.3c), stars at or near the north celestial pole, including Polaris and some well-known constellations like Ursa Major (the Plough) will never set, while some in the southern hemisphere, such as Canopus, the second-brightest star of all, never rise. And some stars, such as the much-mythologised seven sisters of the Pleiades, rise and set. They will always rise and set at some time each day, but that may be during daylight hours or at night. It changes with the seasons, since it simply depends on whether the stars are on the same side of the Earth as the Sun. That varies through the year as the Earth (as we might say) completes its annual revolution.

When Virgil wrote of the Pleiades setting in the morning, his readers knew that he meant the first morning setting – the first date, in the late autumn, when the stars could be seen falling beyond the western horizon just before the Sun's rising glow arrived to blot out their light. We are preserving the legacy of this seasonal astronomy when we call the hottest part of summer the 'dog days'. Ancient astronomers marked this season by the first appearance, shortly before sunrise in late July, of the Dog Star, Sirius, the brightest star in the sky.[23]

Medieval stargazers understood this too – the fact that they experienced the annual cycle in terms of the Sun's apparent revolution around the Earth, rather than the Earth around the Sun, made no difference to observations of relative positions. As they looked up at the heavens unpolluted by streetlights, they awaited the seasonal reappearance of familiar stars with eager anticipation. If John Westwyk was awake before dawn, shivering in the cold of a clear night, he might look hopefully for the warming rays of the Sun, in the direction where he knew it would rise. There, before the expanding dawn blotted them out, he could see constellations rising as the heavens turned. They were slightly different every day. Those stars which were the final heralds of daybreak above the abbey church on St Luke's Day in October could not have been seen at all a few weeks earlier, because they were too close to the Sun. Watching the stars near sunrise and sunset, it was easy to imagine the Sun moving steadily on a narrow annual path through the stars of the zodiac, while the stars themselves stayed resolutely fixed in relation to each other. Although they were invisible to him at that season, John would have known that behind the Sun on the feast of Luke were the trapezoid of four faint stars which form the constellation Libra.

So the Sun was in Libra; but on that day in mid-October the weighing-scales of that constellation no longer balanced the hours of day and sleep: the autumn equinox was weeks in the past. Fourteen centuries earlier, when Virgil had written his *Georgics*, the idea of Libra balancing the hours of day and night could be understood almost literally, as the Sun entered that constellation on the September equinox. But as the centuries passed the constellations drifted slowly eastward. While the Sun glided in its steady annual circle against the background of the stars, that background itself was creeping fractionally forward. The rate of this fractional drift was one degree in about seventy-two years; not enough to be noticeable by any individual in one lifetime, but certainly noticed by ancient and medieval astronomers who drew on their predecessors' observations. This phenomenon became known as the *precession of the equinoxes*, for it seemed that the equal day and night of the equinoxes occurred just a little too early each time.

Astronomers worked hard to refine their models for this motion of the *fixed stars*. They were called 'fixed' because the constellations kept the same constant shape, quite unlike those few 'erratic stars' which moved steadily through the firmament and were called by the Greek word for 'wanderers': *planetes*. The immediate solution to the slow drift of the constellations was to separate the visible groups of stars from the named positions they had once occupied in the sky – albeit, confusingly, without giving new names either to those groups (*constellations*) or their old positions (*signs*). So when night and day were of equal lengths in mid-September in the late fourteenth century, John Westwyk would have observed that the Sun was in front of the stellar stick-figure – the *constellation* – of Virgo; but astronomers of his day also knew that the 30-degree segment of sky running eastwards from the Sun's position on the equinox was the *sign* of Libra.

John would have observed the Pleiades' morning setting in mid-November, around the time of the feast of Martinmas. He might, like the Middle English poet who marked each month's labour, have taken that astronomical event as his signal to slaughter a pig: 'At Martynes masse I kylle my swyne.' The St Albans monks made a similar association. The same scribe who made their copy of Palladius' *Work of Farming* also produced an astronomical calendar with vividly decorated initials (images 1.4, 1.5, plate section). The KL of each month – which stands for Kalends, the first day of the month – became a frame for a scene symbolising the agricultural work typical of the season. The docile pig which the October farmhand drove into the woods to forage for acorns was brutally culled by a bearded yeoman in November.

This was universally understood, but none of it was officially organised. On the other side of the world, things were different. In 1280, the official astronomers of China's Yuan dynasty promulgated the Season-Granting System (*shoushi li*). The Mongols who had conquered China took very seriously the imperial responsibility to bestow an accurate calendar on the people, and they gave them far more than a simple list of days. The state Astrological Commission instructed dozens of official planners and mathematicians to produce a compilation of astronomical data. It was designed to assist with the planting and harvesting of crops as well as with state rituals and astrological decision-making, though it had little impact outside the literate elite.[24]

If agriculture at St Albans lacked that level of astronomical precision, the monks certainly linked their lives to the fortunes of the nearby farms. Few meant as much to them as the abbey's old manor of Westwick. The manor had been given away by an over-generous abbot, Geoffrey de Gorham, as dowry for his sister's marriage in 1130. The fact that it was renamed Gorham only fired later monks' continuing outrage at the loss. Yet the St Albans chronicler Matthew Paris, so often brutal in his judgements, spared Geoffrey the worst of his criticism. Matthew recognised the abbot's many achievements: expanding the abbey's revenues and buildings; founding a hospital and nunnery.[25]

Geoffrey really concerns us because he was – almost – the first named master of the abbey school, where John Westwyk would first have learned the sciences of number. Geoffrey never took up that appointment. Clearly an educator of some repute – though not yet a monk – he had been headhunted from the northern French town of Gorron as St Albans sought to reform its schooling. But he took longer to arrive than expected and the monks found a substitute. Geoffrey made ends meet by teaching at nearby Dunstable. While waiting for the St Albans post that had been promised to him, he organised the performance of a miracle play telling the story of St Catherine (a favourite saint of the Norman royal family).[26] He had no costumes for the performance, so borrowed some splendid robes belonging to the choir of St Albans. The play was a tremendous success. However, the following night his house burned to the ground, taking all his books and the precious borrowed robes with it. Geoffrey paid his debt to the monastery with the only thing he had left: his life. He offered himself as a sacrifice to God and St Alban, taking the monastic habit and quickly rising to the abbacy. Matthew Paris remarks drily that, as abbot, Geoffrey was particularly careful in looking after the monastery's choral robes.[27]

The development of the St Albans school was vital in widening access to the monastery. By the 1370s, when John Westwyk took his vows, the admission process was an exacting one. A collection of letter templates from his time includes a formal notification sent to the sponsor of a would-be monk who had failed a probationary period, which reveals that references were required. And at such a prestigious house as St Albans, a novice certainly had to be literate. Literacy was not as rare in medieval England as is often assumed – around half the population had a basic level, sufficient to read a familiar prayer. But John had to attain a higher standard to be accepted into the monastery. Attendance at the St Albans school did not guarantee his entry, but it was a likely first step.[28]

Large abbeys often had their own schools, but at St Albans the abbey and town school were one and the same. The monk-chronicler Matthew Paris boasted that 'there could scarcely be found in England any school better, or more productive, or more useful, or more full of scholars'.[29] Located just beyond the monastery walls, it was open to fee-paying scholars from outside the cloistered community but was controlled by the abbey. Sixteen places were reserved for Poor Scholars, who paid no fees. They were looked after by the almoner (the brother responsible for all the abbey's charitable activities) and boarded in the abbey almonry. According to the rules laid down in 1339, the Poor Scholars had to shave a tonsure on their heads and say the daily office of Matins. They were admitted 'for five years at most, since this time is enough for them to become proficient in grammar'.[30]

It was, then, a *schola grammaticalis* – a grammar school in both name and mission. The only textbook we know of was Priscian's sixth-century classic *Institutes of Grammar*, and the exams consisted of tests in writing and composition. The school's goal was to prepare its pupils for their monastic profession, which required above all the ability to read and sing the liturgy. Even so, passage from the school to the monastery was not automatic. One alleged failure had been Nicholas Brakespear, later to become Pope Adrian IV (1154–9). That uncertainty of destination, as well as the fact that masters were permitted to admit extra fee-paying pupils to build their income, created demand for a more rounded education.

So while John Westwyk did not receive a thorough scientific training at the St Albans grammar school, he probably did get at least a grounding in arithmetic. This would have included basic numeracy and understanding of the elementary functions of addition and subtraction, halving and doubling. No beginner's primer survives from the medieval period, but the mathematical texts that were popular in monasteries start from the assumption that their readers had already attained that elementary level.

These basic mathematical functions were done using Roman numerals. John Westwyk was born in the middle of a very gradual transition from that system to the Hindu-Arabic decimals we use today. The digits 0 to 9 were popularised in the Latin west from the twelfth century onwards.[31] That was the great period of Christian scientific translation, when scholars in Spain and southern Italy worked feverishly to make Latin versions of the most important works of Arabic and Greek learning. The new numerals greatly eased calculations in the advanced astronomy and mathematics that were spreading northwards from the shores of the Mediterranean. A key figure in their popularisation was the Italian mathematician Leonardo of Pisa, better known as Fibonacci. But the English monks who eagerly adopted and taught them were well aware that they had their true origin much further east, in India by way of Islam:

> Algorizmi said: since I had seen that the Indians set up IX symbols in their universal numbering . . . I wanted to reveal what might be done with them; something which – God willing – should be easier for learners.[32]

This opening sentence of a guide to the new numerals was copied at the East Anglian monastery of Bury St Edmunds in the thirteenth century. The Bury brother wrote it out in Latin, but he was surely well aware that 'Algorizmi' had first composed the treatise in Arabic. The author of that original Arabic version – now alas lost – was the ninth-century polymath Muhammad ibn Musa al-Khwarizmi. Originally from central Asia, al-Khwarizmi encountered, and took forward, the Indian arithmetic while working for the intellectually prodigious Abbasid Caliphate in Baghdad. In sharing these skills, four hundred years later and 2,500 miles to the north-west, the Benedictine scribe was punctilious in recording their Arabic and Indian heritage.

The new numbers arrived in Europe packaged with relatively advanced treatises in arithmetical theory. The medieval Latin translators called this 'algorismus', in honour of al-Khwarizmi; this is the source of our modern word 'algorithm'.[33] It was clear what the benefit of the new numerals was for the complex calculations used in sophisticated arithmetic and geometry, but less clear what the benefit of switching to a new system would be for everyday use. Although all the numbers that survive in Westwyk's handwriting – the extensive astronomical and trigonometrical tables he produced later in life – use what we often (mistakenly) call 'Arabic' numerals, he would certainly have had his first maths lessons with the Roman system.

The crucial difference between the Roman and Hindu-Arabic

numerals is that the latter have a place-value capacity built in. The meaning of a digit depends on its place on the page or tablet. In the number 21 the digit 1 signifies 'one', but in the number 12 it means 'ten'. This is not the case in Roman numerals, where I is always 'one' and X is always 'ten', whether they appear in CIX or XIII. Our decimal system is just one possible form of place-value notation. Although the numbers 1 to 9 and the placeholder 0 come from fifth- or sixth-century India, the concept of place-value notation goes back much further, to the numbers of the Babylonians, invented some time before 2100 BCE. That system, partially inherited from the Sumerians and passed in turn to ancient Egypt, Greece and India, was base-60 or *sexagesimal* (from the Latin for sixtieth). Understanding this sexagesimal system is an important starting point for any study of medieval mathematics and astronomy.

The Babylonians wrote units from 1 to 59 in characteristic cuneiform wedge shapes. (Those numbers were built up by repetition of smaller strokes, as the system evolved from a non-place-value system, but they were read as complete units.) Above 60, they used the same number-signs one place to the left. So, for example, they wrote our '70' as 110 – and for greater clarity, we can add a comma, making 1,10. The number to the left of the comma is a multiple of 60. An additional comma would indicate a further power of 60. So the number 2,21,40 contains three sexagesimal powers: the 2 represents 2 x 3600, the 21 represents 21 x 60, and the 40 represents 40 x 1. Thus 2,21,40 is equivalent to (2 x 3600) + (21 x 60) + (40 x 1) = 8500. The system might feel unwieldy, but the Babylonians needed only fourteen different symbols to make it work – far fewer than the twenty-six symbols in the modern English alphabet.*

It may seem strange to combine a base-60 element with a base-10 (decimal) element, but this is what we do whenever we write the time in hours, minutes and seconds. Or if sailors give their position in degrees, minutes and seconds (though most now use decimals of minutes instead of seconds), it is because we have stuck with the sexagesimal system we inherited from the Babylonian pioneers in the

---

* The ancient Greeks, and following them the early medieval Arabs, used their alphabet of letters for numbers too: an extended alphabet of twenty-seven letters served to represent numbers 1 to 9, 10 to 90 and 100 to 900.

sciences of space and time. John Westwyk would use the same sexagesimal system when he came to compute precise planetary positions.

Just as today we sometimes write numbers using words (like 'ten' or 'twenty'), sometimes Roman numerals and sometimes the Hindu-Arabic – and yet we speak the numbers the same way regardless – so did Westwyk and his fourteenth-century monastic colleagues. Even after they had learned Hindu-Arabic numerals and sexagesimal notation, they continued to use the Roman system too. They could appreciate the value of the decimal-sexagesimal system for difficult mathematical exercises, especially involving fractions, and for broader scientific applications, especially the most important mathematical science, astronomy, which divided the sky into degrees and minutes. But the universal clarity and familiarity of Roman numerals ensured their continued popularity outside academic settings. When, in around 1440, a friar at Warrington in the north-west of England sat down to translate some instructions for making a sundial, he converted the Hindu-Arabic numerals in the original Latin text into Roman numerals for his Middle English translation.[34] No doubt his readers appreciated his consideration.

In 1396 the monks of St Albans finally reversed the two-hundred-year outrage of Geoffrey de Gorham's marriage gift. John Westwyk was born on a manor held by the Earl of Oxford, but Earl Robert, favourite of Richard II, forfeited the estate when he was tried for treason under the Merciless Parliament in 1388. Eight years later the St Albans abbot sealed the deal and bought back Westwick-Gorham for 900 marks. Such a substantial sum required everyone to chip in, and the abbey chronicler recorded the names of the monks and other benefactors who contributed to the purchase. He carefully noted the amount each had given – using Roman numerals:

> Item: received by gift of various brethren and others, in aid of the purchase of the manor of Westwick, as follows. By gift of master Nicholas of Redclif, Archdeacon, xl marks. By gift of master Roger Henrede, Sacrist, vi li [*librae*, pounds] xiii s [*solidi*, shillings] iiii den [*denarii*, pence]. By gift of Thomas Sydon, the Abbot's servant, vi li xiii s iiii d . . .

The list continues through fifteen donors, finishing:

By gift of Robert Trunch, xi s & viii den.
Total: L li II s VIII den.[35]

These Roman numerals are used with something approaching a place-value system: pounds, shillings and pence. (The Babylonians' general place-value system, too, had evolved from these kinds of units that were specific to the thing being measured.) There were twelve pence in a shilling and twenty shillings in a pound. To make matters more complicated, money was also sometimes counted in marks, with one mark being two-thirds of a pound, or 13s 4d. So while Nicholas of Redclif had given 40 marks, Roger Henrede and Thomas Sydon probably did not think of their donation as the seemingly random number £6 13s 4d but rather as a round 10 marks. The abbey chronicler added up all those marks, pounds, shillings and pence, giving the correct total (in Roman numerals) of £50 2s 8d.

If that seems an impressive feat of arithmetic, consider that before the currency reforms of the 1960s and 1970s, children across the former British empire had to learn to add and subtract the twelfths and twentieths of pounds, shillings and pence. (Most of the rest of the world decimalised their currencies in the nineteenth century.) In the same way, adding and subtracting Roman numerals is easy with just a little practice. To begin with, it helps to think of X, ten, as a I with a line through it representing a tallied group of ten strokes. V, five, is a X halved horizontally. A basic addition such as VII + XVIII can easily be accomplished by writing the numbers together and re-sorting them: VIIXVIII becomes XVVVIIIII, which is easily simplified to XXV.

In practice, of course, such sums can be done in your head. For tougher calculations, the Roman numerals could be converted into a more flexible format. In his enormously influential eighth-century textbook *On Time-Reckoning*, the Northumbrian monk Bede – not only 'Venerable', he was also an outstanding polymath – introduced his readers to two options: the Greek alphabetic system and what he called 'that very useful and easy skill of flexing the fingers'.[36]

How did monks like Bede do decimal arithmetic on their hands? Hold your hands up with your palms facing away from you and your

1.6. Finger-counting positions, in Bede's *On Time-Reckoning*.

thumbs together (image 1.6). Start on the left, with the three outside fingers of your left hand. Those three fingers, bent fully, partially or not at all, combine to represent the units from one to nine. That is why the technical term for integers was *digiti*, the Latin for fingers – and hence our numerical 'digits' and digital technology.[37] Next, multiples of ten were shown using different shapes made by bending the left thumb and forefinger over one another (the Latin for tens was *articuli*, which also meant knuckles). The hundreds column is the right thumb and forefinger, and thousands are on the last three fingers of your right hand. Numbers from 0 to 9,999 can be thus shown using two hands. Because the fingers provide four distinct place values – as it were, columns of thousands, hundreds, tens and units – adding and subtracting large numbers column by column becomes easy, and even basic multiplication is possible.

There were two reasons to use your left hand for the smallest numbers. First, it meant that the number could be read correctly by someone facing you, from their left to right. These hand gestures were about communication as much as counting. They could be used in the marketplace, where noise or language could prevent conversation, or in the monastery, where silence was often required. Bede even suggested that the numbers could be used as an alphanumeric code to pass messages in dangerous situations. The second reason for starting with the left hand was so, if the calculation involved only numbers below one hundred, your right hand was free to take notes, point or demonstrate. Bede's perfectly practical system came directly from the classroom, where monks also learned to use their hands to help them memorise musical theory and locate days and dates in the cycles of the Sun and Moon.

It was fine to use your digits to work with digits, but for more complex arithmetic it was easier to calculate with *calculi* – pebbles. As John Westwyk learned to work with numbers he would quickly have become adept at using an abacus or counting-board. A simple arrangement of small stones on a board with lines created a decimal place-value layout. Some versions added intermediate positions for groups of five, fifty, five hundred, and so on, reducing the number of pebbles required. In other versions the counters themselves were numbered from one to nine, and the abacus was then simply a frame providing columns for units, tens, hundreds, and so on. Monks also drew abacus frames in their manuscript books, laying out columns – often decorated like the colonnades of their cloisters – for the placement of counting stones. When not covered with those *calculi*, the spaces between columns could be filled with the text of their arithmetic lessons.[38]

The use of the abacus remained popular into the early modern period, even as other increasingly sophisticated techniques became widespread. In *The Philosophical Pearl*, a bestselling textbook written by a Carthusian monk, which went through twelve editions in the sixteenth century, the section on arithmetic began with a woodcut showing the two approaches to the subject (image 1.7). On the left is Boethius, the late-Roman pioneer of the liberal arts. Another polymath – the omnivorous all-rounder is a common character in medieval sciences – Boethius wrote works on logic and music as well as arithmetic; but he is most famous for his meditation on the human condition, *The Consolation of Philosophy*, so influential across the centuries that it was translated into English by Alfred the Great, Geoffrey Chaucer and Elizabeth I.[39] In it Boethius, like many astronomers before and after him, contemplated the vastness of the universe, the cosmic insignificance of Earth, and the cold distance of the stars. His presence here reminded readers that mathematics was about more than abstract quantities.

On the right of the woodcut is a figure of equal repute: Pythagoras. The great Greek philosopher here uses a counting-board to display

1.7. *The Image of Arithmetic*. Frontispiece to Book IV of Gregor Reisch's *Margarita Philosophica* (1503). Illustration by Alban Graf.

the numbers 1241 and 82. The line furthest from him is thousands, next are hundreds, and so on – but note the 50 in the gap between the lines of tens and hundreds. Boethius, on the other hand, displays the Hindu-Arabic numerals of *algorismus* and their potential for showing fractions. Between them is Lady Arithmetic, her dress decorated with increasing powers of 2 and 3. Although the pen-and-paper possibilities of the Hindu-Arabic numerals were to win out in the long term (driven in large part by the spread of sophisticated banking and accountancy), the sheer versatility of counting-boards or abacuses ensured their continued popularity well into the modern age. In practised hands, they could be as effective as an electronic calculator. A dramatic public competition held in Tokyo in 1946 saw a Japanese abacist beat an American calculator-operator in a series of mathematical challenges, demonstrating both superior speed and accuracy.[40]

For less expert users, a counting-board could serve simply to record intermediate stages in calculations. Medieval mathematicians knew many techniques for simplifying sums, often breaking them up into series of sections that could be worked out mentally, or with basic operations of the abacus. John Westwyk would undoubtedly have learned some such techniques. One, which has a variety of names, including the Russian Peasant Method and the Egyptian Method, was invented independently in several places and may well have been taught at the St Albans grammar school. It turns large and difficult multiplications and divisions into a series of halvings and doublings. The popularity of this method helps explain why the earliest textbooks on arithmetic using the new Hindu-Arabic numerals, including the one copied by that Bury Benedictine, teach how to halve and double them as separate procedures between addition and multiplication.

The beauty of doubling and halving is that you do not need to know a separate process: you only need to know how to add a number to itself. Let us say you want to multiply 43 by 13. Write the numbers side by side, and start doubling the larger one and halving the smaller one (ignoring any remainders). In a few moments you will have:

43	13
86	6 (ignoring the remainder)
172	3
344	1 (ignoring the remainder again)

When you cannot halve any more, strike out the rows where you have an even number in the halving column (in this case, 86 : 6), and add up what's left in the doubling column. So 43 x 13 = 43 + 172 + 344 = 559. With a little practice, this can be done very quickly – and since it uses mental arithmetic, it is no more difficult with Roman numerals than with Hindu-Arabic. It works because it depends on the fact that any number can be made up of powers of 2. So 43 x 13 = 43 x (1 + 4 + 8).*

It works just as well for division. Say you want to divide 729 by 34 (or DCCXXIX ÷ XXXIV). Simply begin by doubling 34 until you can't go any further without passing 729:

XXXIV	(1)	
LXVIII	(2)	
CXXXVI	(4)	
CCLXXII	(8)	
DXLIV	(16)	(doubling 544 will clearly take you past 729)

Now, starting from the last line, add together the largest numbers you can to get as close as you can to 729 (this takes a little practice). When you have done that, the respective row numbers you have used will add up to your answer. Here:

DXLIV (row 16) + CXXXVI (row 4) + XXXIV (row 1)
   = DCCXIV (714)
So 729 ÷ 34 = 16 + 4 + 1 = 21 (remainder 15)

Again, this can be done using mental arithmetic, but if John

* The reason you only keep the rows with an odd number in the halving column is because that is where you will 'lose' a remainder, which needs to be added back in at the end. If you are multiplying by a number which is a power of 2 (e.g. 8), there will be no such remainders: you will strike out all but the last line, since multiplying by 8 is a simple series of doublings.

Westwyk did have to resort to his counting-board, he would have found that the Roman numerals corresponded perfectly to its columns, making direct transcription of the answer very easy. Transcribing from the counting-board into the Hindu-Arabic numerals would require a moment's extra thought.

These techniques became very easy with practice. Knowing them, and having the option to resort to a counting-board if necessary, most monks had no need to reject the methods that had served their predecessors perfectly well in favour of the new *algorismus*. If their work or interests did require them to make frequent multiplications, or to use fractions, monks might prefer to draw up a reference table of Roman-numeral multiplication, rather than learning a whole new arithmetic.[41] Such reference tables and counting-boards are the tools that John Westwyk would have had at his disposal as he began his education at the St Albans grammar school. They would serve him well until his growing interest in astronomy forced him to get to grips with the new numerals and multiplication techniques of Hindu-Arabic algorismus.

To progress from the school to full membership of the monastery required John Westwyk to undergo up to ten years of training. From the day he first put on his novice's outfit (for which he had to pay the large sum of £5, effectively an entrance fee to the abbey) he was drilled in the rules and customs of monastic life. Its basic principles had changed little since St Benedict had written the Rule for his order almost a thousand years earlier.[42]

This was a good time to seek stability and certainty. While the Black Death had been devastating, its political and social effects had been softened by the stability of the fifty-year reign of Edward III. But Edward died in 1377, a year after his eldest son and heir, and the throne passed to his ten-year-old grandson, Richard II. Richard inherited conflict in France and Castile, Ireland and Scotland, as well as the demographic and economic consequences of the plague. The country badly needed strong leadership. The grievances of poor people against unfair taxation and working conditions would soon erupt in widespread rioting in the 'Peasants' Revolt' of 1381; a little later the poet John Gower would write:

. . . for now upon this tyde	. . . for now at this time
Men se the world on every syde	men see the world on every side
In sondry wyse so diversed,	changed in so many ways
That it wel nyh stant al reversed	that it well-nigh stands all
	reversed.[43]

It must have seemed sensible to be safely enclosed in a monastery (though the 1381 rebels' assault on St Albans abbey would prove nowhere was truly safe). In an era of uncertainty, John Westwyk could find the relief of routine in both the monastic liturgy and the study of astronomy.

Chanting the psalms every day, John would sing of the fingers of God, who had set the Moon and stars in place (Psalm 8); who had set them to govern the night while the Sun governs the day (Psalm 136); who numbers the stars and calls them by their names (Psalm 147). Gradually memorising the texts, his memory would be jogged by the initial images painted in many psalters which summarise the contents of each psalm as a vivid aide-mémoire.[44] The artists often chose to include the stars. Their magnitude and permanence were the perfect reminder of the power of God.

As a child, John Westwyk could observe how important it was for farmers to understand the cycles of the Sun and the skies. As he grew older, contemplation of the stars gave meaning to the vast cosmos; a glimpse into the mind of God. Measurement and mathematical analysis could only heighten his sense of a world that was precisely designed and obedient to God's laws. In later chapters we will follow his exploration of that world. First, though, we shall see how the science of astronomy ruled the daily lives of the monks.

2

The Reckoning of Time

John's walk from Westwick to St Albans crossed the River Ver by the old royal fishpond. As he ascended the curving slope of Fishpool Street towards the square of Romeland, where the town's fair was held, an imposing new fortification came into view above him (image 2.1). The massive gatehouse of St Albans abbey had been built a few years earlier as a symbol of the monastery's authority over the town. Passing through its arch, there was no escaping the fact that Westwyk was entering an institution of great power.

St Albans epitomised the wealth – and tendency to corruption – of late-medieval monasteries. Among the travelling pilgrims in *The Canterbury Tales* is an immensely fat, greasy-faced monk. Chaucer layers on details of his clothes lined with expensive squirrel fur, his gold jewellery and his love of hunting. When his turn comes to tell his tale, the Host asks this monk: 'What shall I call you, my lord? Don John, or Don Thomas, or Don Albon?'[1] John, as we noted in the last chapter, was by far the most popular medieval name, given to a third of Englishmen at this time. Thomases were also common (including the influential abbot of St Albans when Chaucer was writing); but people named Alban are unheard of in the fourteenth century, and Chaucer can only have been making fun of the reputation of John Westwyk's house. Yet we should not take this sharp satire as a straightforward factual description of the monastic life in the Middle Ages. In the eight hundred years since its foundation, the Benedictine Order had, unsurprisingly, become a victim of its astounding success, attracting individuals less dedicated to a life of humility than St Benedict envisaged, but a succession of abbots and popes had recognised the rot and worked doggedly to reform it. As a result, if

2.1. St Albans Abbey gatehouse (1365).

by the fourteenth century the Benedictines lacked the austerity of the Cistercians, whose monasteries perched on mountaintops and lurked in remote valleys, if they lacked the commitment to preaching among the people of the Dominican and Franciscan friars, the order had at least reached an equilibrium of moderate commitment – and dedication to learning – that made Benedict's severe but sweet Rule accessible and attractive to converts from a cross-section of society, as well as to important patrons.[2]

So when John Westwyk entered his local monastery, he was entering one of the most powerful organisations in the country. Its location a day's walk from London on the main north-west road gave it wealth and influence. Its dedication to St Alban, England's first martyr, and its supposed foundation in the eighth century, gave it prestige. Its rebuilding by an energetic and well-connected abbot shortly after the Norman Conquest gave it a spectacular new church. This was supplemented in succeeding centuries by a complex of courts and cloisters, the headquarters of a network of associated parish churches and priories, hospitals and schools spread from southern England to the Scottish border. The leader of this organisation – and President of the English Benedictines – was Thomas de la Mare. He had become abbot in 1349, after the Black Death had killed the previous abbot and forty-seven of his monks, and had overseen a remarkable recovery. He had cultivated a close relationship with the court, ensuring royal support at a time of financial and political uncertainty. And apart from the commanding gatehouse, he built a new scriptorium for the monastery, where the scientific and philosophical books brought from the thriving university at Oxford could be studied and copied.[3]

Compared with the tribulations of the outside world, the St Albans cloister must have seemed like paradise. That was, of course, what was intended: a space whose inhabitants coexisted in harmony and in contemplation of the divine. Nevertheless, harmonious coexistence could not be accomplished without norms. Even where men born to different social stations lived together as equals, like the Apostles, each one had his role to play, his tasks to fulfil. There was no freedom of uninterrupted meditation: the monks' life was structured incredibly tightly. In this chapter we shall see how, through this tight structure, religion took support from science – and, in turn, spurred its progress.

Upon exchanging his yeoman's garb for the black habit of a Benedictine, John Westwyk's immediate task was to master the Rule written by the order's founder around 540. The first topic Benedict had covered after introducing the core monkish qualities of obedience and humility was the regular routine of the liturgy, and so the first thing the novices had to learn – after the Rule itself – were the patterns of the daily offices. From Nocturns and Lauds through the numbered hours of Prime, Terce, Sext and Nones (with additional Masses), to Vespers and Compline, each had its canon of antiphons, psalms, prayers, readings and responses.[4] All were carefully performed and precisely timed.

Since this *horarium* of offices occupied some ten or eleven hours of the day, beginning around 2 a.m. – depending on the season – and finishing shortly before 7 p.m., ensuring attendance could be challenging. Brothers in positions of authority were excused from the routine offices to allow them time to accomplish their administrative duties (which is how Chaucer's richly dressed monk could be outside the confines of the cloister), but attendance for everyone else was strictly enforced. Shortly after becoming abbot of St Albans, Thomas de la Mare issued a new rule-book. Apart from tightening up the dress codes, he decreed that any monk missing the midnight office of Nocturns (later renamed Matins) would forfeit the privilege of eating meat the following day. If that was a fish day, the offender would get no fish or dairy products.[5]

Keeping time in the monastery was a weighty responsibility. It fell to the sacrist. As well as looking after the supplies of candles, bread and wine for Communion, and accounting for all the furnishings and utensils of the church, he also had to maintain and ring the bells which alerted the community to the start of offices.[6] Fortunately for the sacrist, these onerous responsibilities could be delegated to more junior monks.

The instructions for one such junior monk in central France survive in his tiny pocket-book (just four by three inches, smaller than two credit cards side by side). In neat eleventh-century handwriting, alongside some poems set to musical notation, we find a set of Latin instructions, which begin like this:

> On Christmas Day, when you see Gemini lying almost over the dormitory, and the sign of ORION above the chapel of All Saints, prepare to stir the signal bell.
>
> On [the feast of] the Lord's CIRCUMCISION, when you see the bright star [Arcturus] which is in the knee of ARCTOPHYLAX, over the space between the first and second windows of the dormitory, just above the rooftop, then go to light the lamps.
>
> On the feasts of St LOMER and St AGNES, [do it] when you observe the scales which VIRGO is said to hold, that is, two bright stars, raised high over the space which is between the sixth and seventh dormitory windows.
>
> . . .
>
> And on the feast of St VINCENT, when you see them just rising above the fifth window, near the roof, and – note this carefully – to observe them you must move back a little from the usual place towards the juniper

bush, on the path to the well, so you can see and count the windows.[7]

These instructions give an admirably simple list of the stars that would be visible at the correct time of night. The use of the stars to regulate the times of prayer goes back to the earliest monastic instruction books. In the late sixth century the Bishop of Tours wrote a guide to *The Course of the Stars* which put classical astronomy to explicitly religious use.[8] He drew sketches of selected constellations, noting at what time of year they could first be seen rising just before dawn (image 2.2). For each month he identified a constellation that would help you get up in time to sing the Nocturns with the crowing cock, and explained how many psalms could be recited before dawn. Yet by the later Middle Ages such star-gazing was becoming harder. Ever grander monastic buildings blocked their inhabitants' view of the horizon. That is why our eleventh-century monk had to back away from his dormitory, down towards the juniper bush, which many monasteries grew for medicinal purposes.

Even then, these observations using buildings as instruments were not particularly precise. Churches could themselves be positioned in harmony with the heavens: several twelfth-century Cistercian abbeys were, similarly to Stonehenge, carefully aligned to the sunset. Whilst those ancient standing stones point towards midwinter, the White Monks often chose to make their abbeys line up with the dying rays of Michaelmas (29 September).[9] But such orientation was more a matter of symbolic annual commemoration than daily measurement. Only much later, in the seventeenth century, did enterprising astronomers lay meridian lines in the floor of some Italian and French churches, making them the best solar observatories of their day.[10] In any case, the Benedictines did not orient their churches astronomically. The nave of St Albans abbey is oddly aligned, almost 25 degrees south of the ideal east–west line, but that was simply a matter of practical convenience to the builders, who sought to minimise the effect of the slope on which it stands.

In such a large and wealthy monastery, rather than sending a novice out into the night, the monks could use an alarm clock to drag them from their standard-issue white night-caps and woollen blankets.[11] The earliest such devices were water-driven. Monastic

2.2. Constellations explained by Gregory, Bishop of Tours, in *De Cursu Stellarum* (*The Course of the Stars*). The stars shown may be tentatively identified as: the two bright stars of Canis Minor (unnamed in the text); Sirius, with its four neighbours in Canis Major (which Gregory calls 'Quinio'), and Ursa Major ('which the common people call The Wain').

chronicles rarely share much detail of how they worked, but we know for sure there was an alarm of this kind at the East Anglian monastery of Bury St Edmunds – since the monks used its water to extinguish a fire that threatened the martyr's shrine in 1198.[12]

The oldest surviving Latin description of a water alarm is found in an eleventh-century manuscript from the monastery of Santa Maria de Ripoll, in the foothills of the Pyrenees.[13] Unlike today's clocks, it had no face. Instead, as water flowed from a container, a float sank, gradually priming an alarm consisting of a rod with a few bells hanging from it. This mechanism had to be reset each time it was used, which meant that the sacrist still needed to know, or guess, the time and the length of the night when he refilled it.

Estimating elapsed time is notoriously hard to do. For shorter periods, from seconds up to a few hours, the monks often marked time by how long it took to perform prayers, psalms or even whole offices. Medieval people also commonly estimated *lengths* of time in appropriately spatial ways, such as how far someone could walk in that duration.[14] But by John Westwyk's day there were several instruments that could assist the sacrist's guesswork. Foremost among them was the astrolabe. The monks were well aware of its properties – in fact, the Ripoll manuscript contains an early Latin manual, *On the Uses of the Astrolabe*, which highlights its value 'to find the truth of each hour of the day, in summer or in winter, without doubt in your method'. 'This seems ideally suited', its anonymous author continues, 'for celebrating the divine office hour by hour, and especially for science. Everything proceeds more pleasantly and smoothly when the Lord's services are appropriately carried out at the appointed times under the rule of the just Judge.'[15]

Ripoll was among a handful of monasteries which had led the way in bringing the knowledge of astronomical instruments from Arabic works in the eleventh century and developing it for a Latin Christian audience. Whether this house in the hills of Catalonia received the science from Moorish Spain ahead of monasteries in northern France and southern Germany, as geographical intuition would suggest, is hotly debated by historians.[16] Either way, a pivotal figure in this reception of Islamic sciences was a monk of the monastery of Reichenau, set on a small island in Lake Constance on the Swiss-German border, known to history as Hermannus Contractus – Hermann the Lame.

Hermann was born into a noble family in 1013. The precise cause of his disability is not known, though according to a legend found in one manuscript, he was mauled in childhood by 'his father's bear' while playing in the woods around his castle. (Some historians sceptical of this legend have tried to use descriptions of his symptoms to diagnose him with a particular neurological condition, but such retrospective diagnosis is at best somewhat speculative.)[17] In any case, his infirmity did not prevent him from mastering the liberal arts, including writing both history and hymns. He also designed a multiplication table to simplify the tricky process of computing complex fractions.[18] But his greatest achievements were in astronomy. Starting from a revised version of the *Uses of the Astrolabe* text found in the Ripoll manuscript (and several others), Hermann completed the treatise with essential instructions for making an astrolabe.[19]

Hermann was one among many monks working on this astronomical material in the early eleventh century, but from an early stage his fame outshone his contemporaries'. Thus it was that when Matthew Paris, the St Albans chronicler, came to compose a book of astrological prediction around 1250, he illustrated it with a picture of Hermann holding an astrolabe, alongside the great Greek geometer Euclid (image 2.3).[20]

The astrolabe, however, was a complex and astronomically advanced device. (You will learn how to use it in Chapter 4.) It was expensive, too. Such instruments were certainly present in monasteries: we find them listed among the books in library catalogues, as if there is no difference between instruction manuals and the objects they describe.[21] But it is likely that most monks – who at least in principle had to renounce the ownership of private property – would have used some simpler instruments to observe the stars and tell the time.

We can see the first of these in the image below, in Euclid's left hand. He is not holding a telescope – first made in 1608 – but a simple sighting tube, also known as a dioptra. Normally fixed on a stand, sighting tubes were used to observe the motions of the heavens. Euclid himself, for instance, instructed his students in the fourth century BCE to aim their tubes at the constellation of Cancer when it

2.3. Euclid and Hermann the Lame with astronomical instruments, drawn by the St Albans historian Matthew Paris *c.*1250.

was rising, and then quickly move to the other end of the instrument to see Capricorn setting, thus demonstrating that the two constellations are diametrically opposed in the heavens. The tube could be fitted with a protractor, allowing you to measure altitudes. Or, if two tubes were fixed together, with the first aimed at the Pole Star, and the second turning around it at a constant angle, the daily rotation of any star around the pole could be observed. You could also see the Pole Star itself moving slightly, since it was not precisely on the axis of the spheres' rotation. Such heavenly rotation was often described as the rising and setting of the *celestial equator*: the great circle in the sky – effectively a projection outwards of the Earth's equator – that brings the Sun and stars above the horizon. The equator turns at a constant rate, since it is at right angles to the Earth's axis.

This same heavenly rotation is what makes a sundial work – and we can be sure that monks had those, in all shapes and sizes. The most common kind in John Westwyk's day was the cylinder dial, sometimes called a shepherd's dial. Cylinder dials were known to the Romans, but the earliest surviving manual on their construction was written in the eleventh century as a companion piece to Hermann the Lame's works on the astrolabe. (Historians always assumed that Hermann himself had written it, but recent research suggests otherwise.)[22] That first manual calls the cylinder a 'travellers' dial', but that name was inappropriate, since it worked only at a single latitude. Whatever the name, the design consisted of a cylinder with hour-lines running down it. A gnomon (pointer) rotated around the cylinder and had to be set by the user for the correct day of the year. A diagram showing the whole cylinder, as if unwrapped and laid flat, survives in a fourteenth-century manuscript from the wealthy Augustinian priory of Merton, near London (image 2.4). This Merton manuscript contains a fascinating collection of scientific texts; we will meet it a few more times as our story continues.

One of Geoffrey Chaucer's Canterbury pilgrims tells a saucy (and somewhat disturbing) tale about a monk – named John, like so many of his brothers – who seduces a cash-strapped woman by giving her money he has borrowed from her parsimonious husband. They meet in a garden early in the morning. John 'embraces her hard, and kissed her often' and sends her away, saying, 'let us dine as soon as that ye may, for by my cylinder it is prime of day'.[23] The monk is fond of smutty jokes, and 'by my cylinder' is undoubtedly sexual innuendo, but it is also clearly a reference to his use of a sundial to observe the prayer times – even if Chaucer gives no indication that this monk, excused from his monastery for the business of managing its estates, will do any actual praying. In any case, prime was far too early to dine, as St Benedict had dictated that the first (and often only) meal of the day must be no earlier than midday.[24]

What, though, did 'prime' mean? It may seem a simple question, but that is only because the hours that govern our lives are generally accepted. We forget all too easily that they are not natural, merely conventional. Prime marked the first hour of the day, but that could be a single point in time or an extended duration – or, indeed, the religious office assigned to that period. Even the concepts of hour and day require clarification. There was no agreement about when the day began. One scholar named Robertus Anglicus (Robert the

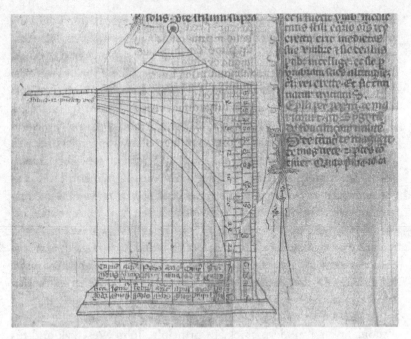

2.4. Wraparound diagram for a cylinder dial, early fourteenth century.
This manuscript was once at the priory of Merton, near London.

Englishman), writing in southern France in the 1270s, complained
of the confusion:

> Some people start the day from sunrise, like many Latins; some, speak-
> ing sloppily, from the first appearance of sunlight or the day at dawn.
> But some start it at noon, like astronomers; for the astronomers say
> that Thursday begins at noon of the same day ... Anyway, some
> people begin the day at midnight, like the Chaldeans; others, like the
> Jews, from sunset.[25]

The most 'natural' time to start the day, Robert concludes, is
sunrise – which he defines precisely as when the centre of the Sun
can be seen. But as he himself had shown, his conclusion was far
from universal.

As for the hours, when we met Palladius and his shadow in Chapter 1, we saw that unequal hours – the system with exactly twelve

hours from sunrise to sunset every single day, no matter how long the Sun stayed up – were slowly superseded in John Westwyk's century by the equal hours we use today. This was an uneasy transition, and nowhere more so than in the monasteries, where the old unequal hours allowed the patterns of prayer to vary smoothly with the shifting seasons. Sundials, including cylinder dials, could be engraved with lines either for the unequal canonical hours or for the equal hours. The six curved lines marked on the unwrapped cylinder in image 2.4 show the six unequal hours of sunlight on either side of noon, but many surviving cylinder dials (mostly made in later centuries) have lines for equal hours.[26] Meanwhile, if you wanted to equip a water clock like the Ripoll alarm with unequal-hour markings, you would need multiple seasonal sets, since the flow of water would hardly vary through the seasons (except when it froze). Equal hours could be marked more simply, but the sacrist would still need to judge the seasonally appropriate time for the monks' rest when he refilled the water tank.

So it is hardly surprising that 'prime' does not translate easily to a moment on a modern clock. In principle, John Westwyk and his monastic brothers would gather in their vast, dimly lit abbey church soon after daybreak to sing the monastic office of prime around 6 a.m. But according to that poet who translated Palladius' agricultural manual for the Duke of Gloucester, the six unequal morning hours were half-prime (7), prime (8), half-undron (9), undron (10), midday (at 11!), and noon (12).[27] In the *Canterbury Tales*, on the other hand, a humorously strutting rooster called Chauntecleer, who kept strict astronomical time through the changing seasons, knew instinctively that on 3 May, when the Sun had risen 41 degrees above the horizon, it was prime – which works out at nine o'clock in equal hours. Chaucer, well practised in such calculations, was adamant that Chauntecleer could not be wrong:

Wel sikerer was his crowyng in his logge	More accurate was his crowing on his perch
Than is a clokke or an abbey orlogge.[28]	Than a clock or timepiece of an abbey church.

However John Westwyk defined 'prime' – and you can be sure

that in each situation people telling the time knew what they were referring to, just as today we may wish to clarify the time zone before telephoning someone in a different country – he could mark it on the world's most advanced astronomical clock, set on a raised platform in the abbey church of St Albans. John must have passed it several times a day, steadily ticking on his right-hand side as he entered the cavernous church through the cloister door. It was the pride of the monastery. The offer to build it probably helped its inventor, Richard of Wallingford (image 2.5, plate section), to be elected abbot in 1327, though its spiralling costs meant that it was still incomplete at his death – from leprosy – in 1336.[29]

The mechanical clock was surely the most significant invention of the Middle Ages. Imagine our lives today without timekeeping. At the clockwork revolution around 1300, the possibility dawned of reliable machines that could keep universally agreed time in equal hours: all our GPS systems and online-delivery slots stem from this moment. That it was an invention whose time had come is obvious from how many people were trying to make it work, and how quickly it spread when they succeeded. Around 1230 one French engineer drew a device that would 'make an angel keep its finger pointing towards the Sun' as it moved across the sky. He also sketched a perpetual-motion machine, which was supposed to make use of the special properties of mercury.[30] Other experimenters discussed the possibility that magnets might provide perpetual motion. Meanwhile, in Spain in the 1270s, artificers working for King Alfonso X (known as Alfonso the Wise) designed a clock mechanism that used the slow flow of mercury to temper the driving force of a falling weight. And Robert, the Englishman whom we heard complaining about the varied starts to the day, wrote hopefully in 1271 that 'clockmakers are trying to make a wheel that will move precisely with the motion of the celestial equator'. Such a wheel, he explained, would be turned by a hanging weight but, he noted sadly, 'they cannot quite accomplish their task'.[31]

They soon proved him wrong. Just two years later there was a clock at Norwich Cathedral priory, surely mechanical, and records survive from the following decade of clocks in Dunstable, Exeter, London, Westminster and Oxford.[32] Of all these clocks, not a single fragment survives. Again and again we will see the irresistible medieval drive to tinker, to redesign, to incrementally improve or upgrade technology. When that happened, the attraction of reusing or recycling components – and the limitations of storage space – left little material evidence. Historians are dependent on descriptions, drawings and financial records.

So it is in financial records that we find the first complete upgrade, in Norwich. In the early 1320s the cathedral employed three full-time clockmakers, as well as additional craftsmen, over a three-year period, to make an entirely new clock. The two principal horologists, Roger and Laurence of Stoke, were well rewarded for their services: each received a new fur robe every year, in addition to their weekly wages, and the abbey also covered Roger's medical expenses. Other workers were offered the right to dine at the abbot's table. The project did not, however, go entirely to plan. The first attempt to engrave the dial, which weighed a massive eighty-seven pounds, failed, and the Norwich sacrist was unable to recover all the money he had advanced the contractor in London. Two new London craftsmen were employed, but these two also abandoned the job. For the third attempt, Roger of Stoke personally rode down from Norwich to London to oversee the work; this time it was satisfactorily completed. In total the clock cost £52 9s 6½d, more than 10 per cent of the cathedral's enormous annual income.[33] The competence that Roger and Laurence (who may have been father and son) showed in bringing such an important project to completion is surely what made Richard of Wallingford hire them to work on his own, far more complex, clock for St Albans.

What made a clock a clock? We saw with the water-driven alarm at Ripoll that a basic timekeeping device did not need a face or a dial. Many early mechanical clocks marked the hours simply by ringing a bell, and the word 'clock' derives from the medieval Latin for 'bell': *clocca*, like the modern French *cloche* and the German *Glocke*. Instead, what defines the mechanical clock – and excludes most of the water-based devices which had been used worldwide and developed over millennia – is its reliable, self-regulating driving mechanism. (I say 'most' because water-based clockwork mechanisms had been used to power astronomical devices in China for over three hundred years. These were not fully mechanical, relying in part on a constant flow of water, and do not seem to have spread beyond a few outstanding examples. Nevertheless, they remind us how often an invention that at first appears revolutionary turns out on closer inspection to be hard to distinguish from a long history of incremental improvements; in this case, the ever more creative Chinese uses of water to power astronomical clockwork.)[34]

The heart of the self-regulating driving mechanism was the escapement. This component rationed the continuous energy produced by the falling weight, transmitting it in regular parcels to the timekeeper. The earliest clockmakers mostly made their mechanical escapements in the form of a 'crown' wheel, named for its saw-teeth, which alternately pushed two plates fixed on opposite sides of a rod (known as the 'verge'), so it oscillated back and forth. Since the wheel could turn only one tooth at a time, the speed of the clock was governed by how long it took for the verge to rotate first one way and then the other. Richard of Wallingford used a slightly different version, called a strob, which comprised two wheels fixed together (image 2.6). Each wheel had pins sticking out from its rim in alternating positions, pushing different sides of a single pallet attached to the verge, which thus oscillated.[35] It is not clear which version was invented first. But either way, the escapement was not what made the St Albans clock so special.

Only in the last fifty years have historians realised quite how remarkable the St Albans clock was. After its disappearance when the abbey was dissolved in the sixteenth century, the clock was known only from the rather vague descriptions of monks and visitors to St Albans, none of whom understood its workings. The first suggestion that its design might be reconstructed was made by the 'scientific detective' Derek Price. As his work on the *Equatorie* was nearing completion, Price visited a Cambridge library to see a manuscript made at St Albans. It was written a little after John Westwyk's day, but its scribe was obviously inspired by the glories of his generation, writing out an epitaph for abbot Thomas de la Mare, 'a shining Sun of English monks'. Price discovered a short treatise, just six neatly written pages, entitled 'Instructions to divide the wheels for an astronomical clock for the motions of the planets'.[36] He announced his discovery in a short article in a specialist magazine, suggesting it

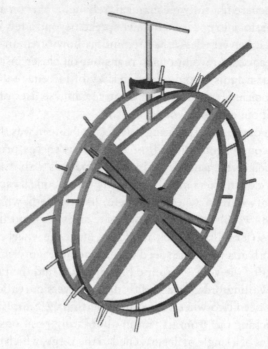

2.6. The strob escapement, which governed the St Albans clock. As the double wheel turns, the pins alternately engage with the half-moon-shaped pallet, turning the verge (bar) at the top.

could be a fragment of a description of Wallingford's clock, but without more text – or any diagrams – he had no way to substantiate his speculation or make sense of the mechanism. A decade later, however, Price was proved right when an almost complete description of the clock was discovered by another historian, a courteous, bespectacled mathematician-turned-philosopher named John North.[37]

Amongst the manuscript collection bequeathed to Oxford University by the seventeenth-century astrologer Elias Ashmole, North found a small, thick book. Its 201 leaves of parchment, bound in undecorated leather-backed wooden covers, contained an almost complete collection of Richard of Wallingford's scientific works. On several pages it was labelled as the property of John Loukyn, who was a lay brother at St Albans in John Westwyk's day. Lay brothers were typically excused from some of the religious duties of other monks, but in return performed some expert service for the monastery. Loukyn was a sub-sacrist, which required him to assist with the sacrist's duties of stock-keeping and maintenance. His manuscript included remarkably detailed machine drawings of Wallingford's clock. It seems likely he was responsible for keeping it ticking along smoothly, two generations after the designer's death. By this time, care of the clock was probably a standard part of any sub-sacrist's job.[38]

Captivated by the manuscript's elaborate diagrams (image 2.7), North dedicated much of his career to studying and publicising Wallingford's work. The result of North's research is that Wallingford – a complex character whom we shall get to know better in Chapter 4 – has been recognised as the greatest English astronomer of the later Middle Ages (though he is still hardly the household name he should be). Wallingford's achievements signal the important role of monks in the story of science and remind us how religion and science went hand in hand. They also help explain why St Albans was such a centre for scientific study in the decades following his death, right up to and beyond the time of John Westwyk.

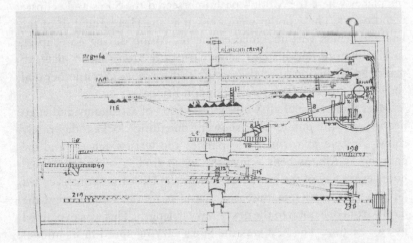

2.7. Drawing of part of the gearing for the St Albans astronomical clock. Note, near the top, the wheel carrying the lunar nodes, with the dragon's head on the right and marked with 177 teeth (ΙΛΛ) on the left.

Wallingford's clock made all of the most vital astronomical questions manifest for the monks in their stalls down beneath its position high in the abbey's south transept. Like most clocks, it struck the equal hours on a bell. Unlike the others, which rang the bell only once at each hour, it chimed multiple times on the hour, from once at one o'clock to twenty-four times at the end of the day. Such hour-striking is so familiar to us – albeit only from one to twelve – that it may seem trivial, but it required a clever piece of technology, which Wallingford invented: a barrel with pegs that released the striking mechanism and stopped it after the correct number of strokes (image 2.8). The same principles of hour-striking, as well as a strob-type escapement, were used in designs sketched out 150 years later by Leonardo da Vinci, so it seems Wallingford's ideas spread widely.[39]

However, it was not the bell that made Wallingford's clock such a marvel, but its dials. The unequal hours were forged into an iron web of fixed curves, so the seasonal time could be read by the rotation of a celestial plate behind them (image 2.9, plate section). The phases of the Moon were graphically represented by a ball, half black and half white, which rotated behind a window and was illuminated just the right amount at all times. A further display showed the lunar nodes: the places where the Moon's path intersected the Sun's and eclipses could occur. These nodes were known as the Head and Tail of the Dragon (which was imagined eating the Moon during an eclipse), so Wallingford had his craftsmen carve a dragon-shaped plate, to make the prediction of eclipses graphically obvious. Yet another dial showed the time of high tide at London Bridge.

Most impressive of all Wallingford's feats of astronomy and engineering was the pointer that showed something clocks today rarely show: the *true* solar time. Our phones and wristwatches – and the chiming St Albans bells – give the *mean* time: twenty-four equal hours every day of the year. That is why the UK's time zone is called Greenwich Mean Time. But, as medieval astronomers well knew, the days – from one noon to the next – vary in length. This was explained by two factors. The first was the Sun's varying speed in its annual journey across the background of fixed stars.* The second factor was

---

* The change in the Sun's speed at different times of year, on its circular path through

2.8. Hour-striking mechanism from Richard of Wallingford's clock (1:4 scale reconstruction).

that the Sun's annual path through the wider zodiac band of constellations is set at an oblique angle to the celestial equator (image 2.10).

The Sun always follows the same precise path on its yearly circuit of the constellations. That path is called the *ecliptic*, because an eclipse can occur if the Moon comes to that line when it is new or full. The angle between the two fundamental circles – ecliptic and equator – is about 23½ degrees. Remember that the turning of the equator, rising and setting, is how we measure time. When the Sun, on its annual ecliptic path, crosses the equator, which it does at the equinoxes, it will be angled most sharply to the equator. That sharp crossing angle means it moves less along the equator, giving a shorter day at the equinoxes. Fast-forward three months, and the Sun is at its farthest from the equator, touching the tropics of Cancer and Capricorn on the summer and winter solstices before turning back towards the equator (as John Westwyk had observed across the River Ver at sunrise). At these solstices the ecliptic runs parallel to the equator, and so the day is longer.

the constellations, is understood today to be a result of the Earth's elliptical orbit.

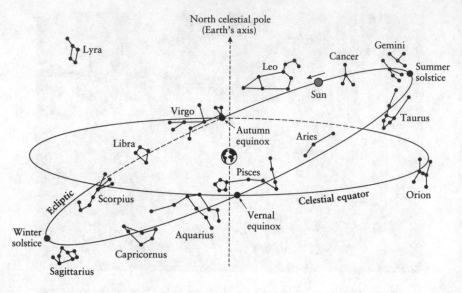

2.10. The ecliptic with the zodiac constellations, angled at about 23½ degrees to the equator.

The differences in the day's length – known today as the *equation of time* – can be up to thirty seconds per day, and these can accumulate to throw clock time off by a quarter of an hour at certain points in the year. If you have ever wondered why the earliest sunset is several days before the winter solstice, and why mornings in the northern hemisphere keep getting darker until early January, this is why. It was all well understood in the Middle Ages: the challenge for Richard of Wallingford was to reproduce it on his clock. This he did with a quite astounding feat of design and craftsmanship: an oval gear with 331 precisely spaced iron teeth. Thus, to the slow eleven-second beat of the clock, the monks could watch the stars turning clockwise, and see the Sun carried with them while it tracked slowly in the opposite direction on its annual journey around the ecliptic.

The mechanical clock was an obvious symbol of authority, of its maker's mastery of the motions of the world machine. It is not surprising that clocks – albeit far simpler than the St Albans design – were rapidly commissioned by civic authorities, in regions as far afield as

Rouen and Strasbourg, Bern and Prague.[40] Clocks have had symbolic power ever since. When it was proposed to silence the iconic 'Big Ben' clock of the Palace of Westminster during four years of maintenance starting in 2017, many members of the British parliament, who legislate to the sound of its hourly chimes, howled in protest. They intuitively understood the impact of such silence, especially at a time of considerable national unease.[41] For the monks of St Albans, the clock was less a symbol of authority than of the perfectly designed cycles of Creation. They watched its dials turn as surely as the seasons, seeing the Sun traverse the equator and head north as spring turned into summer.

As the seasons changed, so did the rhythms of John Westwyk's life in the monastery. All the divisions of time were interrelated. For instance, each of the twenty-four hours was astrologically assigned to one of the seven planets – including the Sun and Moon, for they too wandered among the stars. And each planet gave its name to one day of the week. The cycle of seven planets ran inwards in the order accepted since classical times, which was based on the length of their orbit: Saturn, the longest, then Jupiter, Mars, the Sun, Venus, Mercury, and the Moon. On a Sunday, the first hour was ruled by the Sun. The second hour was then ruled by the next planet in the inward sequence, Venus; the third hour was ruled by Mercury; and the fourth by the Moon, which was considered the innermost planet. The sequence then immediately restarted at the outermost planet – Saturn – followed by Jupiter, then Mars. After those seven, the eighth hour of the day would again be governed by the Sun. So would the fifteenth hour, and the twenty-second. That just left two more hours, assigned to Venus and Mercury in turn, so that the following day began with the Moon – Monday. Each day was thus named for the third planet inwards after the previous day: Mars after the Moon, Mercury after Mars, and so on. This is why the Sun's day still follows Saturn's in modern English, and why, in most Romance languages, we see the midweek sequence of Mars (*martes* in Spanish), Mercury (*miércoles*), Jupiter (*jueves*) and Venus (*viernes*). We cannot be sure quite why the ancients chose a seven-day week, but the imperfect fit of seven days into twenty-four planetary hours explains why the days are in this order.

Sunday, the Lord's Day, meant an earlier start, to make time for extra readings and hymns in the midnight office. The early start clearly caused problems even for St Benedict's own monks. His Rule stresses the need to take 'special care' to rise on time so that the extended liturgy could be performed in full. The monk who fails to wake his brothers on time, Benedict adds, must make amends to God through prayer.[42] On the other hand, Sundays also meant feasting. The normally strict dietary rules, which restricted the monks to one main meal a day in winter (albeit supplemented by snacks) were relaxed, and meat was permitted. This was just the most basic variation in a calendar which, by John Westwyk's era, had become rich in multilayered complexity. The different seasons meant different prayer-times and meal-times – a longer day but an extra main meal in summer; a single, later main meal in the penitential season of Lent. On top of this, each monastery had its particular celebrations, commemorating patron saints, founders, earlier abbots and other notable forebears. Within the communal calendar, individual monks could benefit from occasional bloodletting. This was originally a health measure put in place by a St Albans abbot who had trained at the renowned Italian medical school of Salerno, but by Westwyk's time it meant a two-day vacation from most monastic responsibilities.[43] And of course there were the greatest feasts of the Christian Church. Some of these, like Christmas, were fixed on the same day every year; others, like Easter, were partly dependent on the cycles of the Moon and so shifted their position in the solar calendar.

The solar calendar was complicated enough by itself. Although the days of each month were sometimes numbered as they are today, it made more sense to recall the cycle of fixed holy days. Rather than saying that today is X days after the start of the month, as we do, medieval people were more likely to note that it was a day or two before or after a certain saint's day. John Westwyk must have known some version of a 365-syllable mnemonic for this purpose. Like all the best mnemonics, these were tailored to the precise information being learned, and could even be personalised by individuals, but the first two lines John learned would have gone something like this:

Cisio janus epi lucianus & hil, fe mau mar sul
Pris wul fab ag vin, pete paulus iul agne battil.[44]

At first sight, this is complete nonsense. On closer inspection, its thirty-one syllables contain the names of the most relevant saints commemorated in January – relevant, that is, to the house in the north-west of England where this version was written. The basic idea had travelled from eleventh-century Germany, as had the first five syllables, which gave the mnemonic its name, *Cisiojanus*: *cisio* for the circumcision of Christ, celebrated on 1 January, and *janus* for the name of the month.[45] Syllables 6 and 7 – *epi* – were also a constant fixture, referring to the feast of Epiphany on 6 January. Beyond that, monks were free to insert whichever saints and celebrations gave structure to their local year, provided that they kept the correct number of syllables in each month and, of course, the correct total of 365. You can see that this version, written around 1400, highlights the feasts of Saint Hilary, who still gives his name to the spring term in some universities and legal systems (13 January); Wulfstan, the eleventh-century Anglo-Saxon bishop of Worcester who worked to manage and mitigate the upheavals of the Norman Conquest (19 January); and Balthild, the English slave who became queen of Burgundy in 648 (30 January).

Three hundred and sixty-five syllables of sometimes obscure saints might sound a lot for Westwyk to memorise, but it was only part of what novices had to master. Before they could be fully ordained, each new entrant into the monastery was required to learn Benedict's Rule and the complete psalter of 150 psalms. On top of this, he would have by heart the entire liturgy: versicles and responses; hymns, canticles and antiphons.[46] Such feats, astounding to us now, were commonplace for medieval scholars, who developed an arsenal of memorisation techniques, from simple rhymes to entire imaginary castles; many of these medieval techniques are still used today in foreign-language courses and competitive memory sports.[47] It is easy to see how memorisation was a crucial tool of learning in an era when the production of texts was a more laborious and expensive business than it is today. What is less intuitive to us now, but more important to understand, is that memorisation – which we may belittle as 'rote learning' – was a deeply creative activity. It was central to meditation as well as the production of new ideas, which required a solid foundation of layers of prior thought.[48]

As well as learning the liturgy, the novices had to develop the singing skills to do justice to the sacred music. John Westwyk would have performed it several times each day, his notes bouncing off the newly carved nave-screen and floating to the high vaults at the heart of the church. Yet music was a science too. It made important advances in the Middle Ages, both in terms of the theory of mathematical relations underlying harmonies (a development to which Hermann the Lame made a significant contribution) and in techniques for writing and communicating new ideas. The later Middle Ages saw the growth of rich polyphonic harmonies, and monasteries competed to honour God with the most ornate music. Some abbots resorted to hiring professional singers, but Thomas de la Mare was determined that the St Albans monks themselves would perform the music. A guide to musical notation, written for the St Albans novices in Westwyk's time, still survives. There were some concessions: the monks did not have to memorise the new music, which was much more complex than the traditional plainchant. To help them read the notes, they were allowed to bring candles into the choir. Older monks deplored this innovation, which they said would rot the novices' memories.[49] New technologies for learning have always had their detractors.

It is hardly surprising that more complex knowledge is more likely to be written down. We duly find that the monks' surviving books contain many calendars that lay out the intricate structure of their year. These often appear where they would be most useful: as part of the psalters that give the complete set of psalms to be chanted in the choir. Since mature monks would have known the psalms inside out, these psalters were mainly studied by novices. The twelfth-century St Albans calendar we met in the previous chapter had more than just images of docile pigs being fed and slaughtered: it contained the basic information a young monk like John Westwyk would need to learn not only the fixed but also the movable feasts of the year. If we are to understand how the monks divided up their year, we must learn, as John did, to read their coded calendars.

The January page of the St Albans calendar is shown in image 2.11. It is divided into five columns, of which the last and widest is the list of feast days. Some of these will already be familiar from our memorisation of Cisiojanus, like Saints Hilary, Felix, Maurus and Marcellus – *hil fe mau mar* – on consecutive days in the middle of the month. Immediately to the left of these saints, we find a double column with days of the month. They are not numbered from 1 to 31, as we are used to, but according to the Roman system of Kalends, Nones and Ides. The Kalends was the first day of the month – and the reason each month in this calendar begins with the large illuminated initials KL. The Latin word *kalendarium* had originally referred to financial records with deadlines, but by the later Middle Ages it had taken on the same sense of a layout of the year as the modern English word 'calendar'. The Ides ('IDUS' in the Latin calendar) was a day in the middle of the month that, in early Roman calendars charting the lunar year, marked the full Moon. This was the thirteenth or fifteenth day of the month. Nine days before that, counting inclusively (so on the fifth or seventh), was Nones, which can be seen in the St Albans calendar with a fat stylised NO stretching across the double column. After Nones we see a countdown to Ides, from the VIIIth day before Ides (which in January was Epiphany) to II Id' and then Ides itself. Nones was preceded by a similar countdown, and after Ides we find a countdown to the Kalends of the next month. Here it begins '[ante diem] XIX Kalendas Februarias': the nineteenth day before the Kalends of February. Because we are counting inclusively, the last day of the month was always II Kl'. Thus the legacy of ancient Rome maintained its influence, every day of the medieval year.

To the left of that enumeration of days, we find a repeating sequence of letters A to G. Starting with A on 1 January, these seven *ferial letters* give a permanent calendar of weekdays, allowing the monks to find the date of every Sunday – or any other day – in any year. If you know that the first Sunday of the year will fall on 1 January, all the A's in the calendar will be Sundays. Or, if the first Sunday is 3 January, the Sunday letter will be C. In that year, every D in the calendar will be a Monday, every E a Tuesday, and so on. By now you will not be surprised to hear that medieval astronomers had mnemonics for this too. *Altitonans Dominus Divina Gerens Bonus Extat Gratuito*

2.11. January page from a St Albans calendar, mid-twelfth century.

Coeli Fert Aurea Dona Fideli ('The good Lord is ruling, thundering on high; He freely brings divine golden gifts of heaven for the faithful') is one of them. Its twelve initial letters give the Sunday letter for the first day of each month.[50] Memorising that verse – which survives in the same Merton Priory manuscript where we found a cylinder dial, among many others – told brothers that if 1 January (A for *Altitonans*) was a Sunday, 1 October (A for *Aurea*) would be too; 1 February and 1 November (both D's) would be Wednesdays, and so on.[51]

That comfortable sequence concealed some uncertainty, for the year was not always 365 days. Since the decree of Julius Caesar in 46 BCE that the Roman empire should move from a quasi-lunar year of 355 days to something approximating the solar cycle, the 365 days had been supplemented by an extra day every four years. This leap day was added by repeating 24 February, the sixth day before the Kalends of March, which is why in many European languages the word for a leap year is something like 'twice-six': *bissexto* in Portuguese; *bissextile* in French. Why did the Romans choose that particular day? According to the Northumbrian monk Bede, whose eighth-century textbook *On Time-Reckoning* was the essential manual of calendrical sciences for much of the Middle Ages, it was out of devotion to the god Terminus, whose feast was the previous day.[52] The Christian calendar thus patiently preserved its pagan roots – and each leap year required a second Sunday letter.

Getting the length of the solar year right was hard enough. Matching it up with lunar cycles was twice as hard. But that was the critical task of Latin astronomers. They needed to align festivals of the Roman calendar under which Christianity had been founded with those of the Jewish calendar in which their religion had its roots. The attempts by astronomers to chart the solar and lunar cycles precisely, and to improve the Christian calendar that depended on them, gave rise to a complete discipline of calendrical astronomy: *Computus*. Computus was the essential science for any medieval monk to learn. Its practical application ruled their lives, through the patterns of their prayers.

Beneath the St Albans calendar lie layers of careful luni-solar astronomy, regulated by cycles hundreds of years long. On the surface we see only its simplest results, in that leftmost column of unevenly spaced Roman numerals. These are the Golden Numbers, which give the date of every new Moon in a given year. To find the golden number for any year, add 1 to it, then divide by 19. Discard the whole-number result: the remainder is the Golden Number. For example, for the year 1377:

1377 + 1 = 1378
1378 ÷ 19 = 72, remainder 10
  [i.e., 72 x 19 = 1368; 1378 − 1368 = 10]

The new Moons in the year 1377 are thereby assigned to days in the calendar with Roman numeral X. For January, we can see this is the day after Ides − or the nineteenth day before the Kalends of February − or the feast of St Felix: 14 January.

Since there were roughly twenty-nine and a half days between new Moons, and January and February together have fifty-nine days, the Golden Numbers were the same in March as in January. So in 1377, it was 14 March. To go from new Moon on 14 March to the next full Moon, we add thirteen days; that will take us beyond the spring equinox on 21 March. To find the next Sunday, we need the Sunday letter for 1377 − D − and the calendar tells us that the next D is 29 March. That day, the first Sunday after the first full Moon after the equinox, will be Easter Sunday, the most important day in the Christian calendar.

How did this work? All John Westwyk needed was to know the Golden Number and Sunday letter for the year, and he had all the Christian festivals from Shrovetide to Pentecost mapped out. We now know how to calculate the Golden Number (which in any case was just one more than the previous year's). They were based on the eleven-day difference between the solar year of 365 days and twelve lunar months of twenty-nine and a half days each − 354 in total − which is why Golden Number IX comes eleven days after X in image 2.11. The Sunday letter ran in a similarly straightforward sequence over twenty-eight years, that is, the seven days of the week multiplied by the four years of the leap cycle.

To make it easier, reference tables were commonplace. Many

survive from medieval monasteries, laying out the architecture of the church calendar for decades at a time. As monks drew them up, copied and adapted them, they tried out different methods, or moved from Roman numerals to the Hindu-Arabic. The system John West-wyk used was a streamlined version of calculations that had been refined over centuries – though Golden Numbers so impressed their users that legends arose about their marvellous origins. They were called 'golden', according to some medieval writers, because the Romans had painted them in gold lettering. Others demurred, claiming it was because their value was greater than gold.[53]

The computus rested on a combination of impressive astronomy, historical compromises and practical assumptions. The supporting astronomy was laid out in admirably clear treatises like Bede's *On Time-Reckoning*, and copied into elaborate monastic manuscripts.[54] The compromises, bashed out between bishops in Alexandria and Rome (with important contributions from Ireland), and largely final-ised by the end of the seventh century, were the resolution of a conflict within early Christianity about when to commemorate the crucifix-ion and resurrection of Jesus Christ. Those events had taken place during the Jewish springtime festival of Passover, which began at the full Moon in the Hebrew lunar month of Nisan. The first part of the compromise was to celebrate the Lord's Resurrection on the Lord's Day: Sunday. The second was that the Easter full Moon would be the first one after (or on) the spring equinox. The third was that Easter could not be celebrated on the same day as that full Moon. So if that Paschal full Moon happened on a Sunday, Easter would need to be the following one. And the fourth, following intense historical debate across Christendom, was the definitive standardisation in 525 of the number of years since the birth of Jesus – the Common Era (or annus Domini) we still use today.[55]

The practical assumptions were, first, that the spring equinox could be fixed on 21 March; and second, that the solar and lunar cycles could be precisely meshed within an overall cycle. These assumptions were based on astronomy, and from the beginning astronomers were aware of their inherent limitations, but for a long time they were will-ing to put up with them. From ancient Greece they adopted a cycle which placed exactly 235 lunar months in 19 solar years, and they accepted the Julian calendar year of 365¼ days, though they knew that as a measure of time between spring equinoxes it was too long. In fact, as far back as the second century BCE astronomers had observed that the time between spring equinoxes (the *tropical year*) was different from the time it took the Sun to return to the same star (the *sidereal year*). The difference between the two caused the *precession of the equinoxes*, that slow drift of constellations we encountered in the previous chapter. The Sun takes a little longer than 365¼ days to complete its annual journey through the stars, but it takes a little less than 365¼ days to return to its position over the equator on its spring journey northwards.

Precisely how much less was the object of increasingly refined calculation. From the point of view of the calendar, it was this difference between the tropical and Julian years that mattered, since it meant that the observed equinox gradually crept earlier and earlier in the year. More urgent for computists was the problem that, as Bede put it, 'the Moon sometimes appears older than its computed age'.[56] It had long been realised that the 19-year cycle was a little longer than 235 Moons; this was resolved by having the lunar calendar jump over one day at the end of the cycle. But that was not enough: the mounting discrepancy in the years before each lunar leap, and the fact that the precise moment of the new Moon could happen at any time of day, meant that the Moon was sometimes two days old when the tables said it should be new. Since any peasant could observe the phase of the Moon – as Hermann the Lame dismissively noted – this caused considerable embarrassment to the computists.[57]

They reacted by rejecting the old calendrical assumptions. In the eleventh century concerned monks like Hermann began to produce ever more accurate lunar tables, using careful observation of eclipses to make the timing of new and full Moon as precise as possible. They broke down the barrier between the easy practical calendars with their purely notional cycles, and the continually refined astronomical models based on careful calculation and observation. This was, as before, an international endeavour. Bede's manual had been read at the court of Charlemagne, and Irish computistical texts had influenced Swiss monasteries; now the manuscript writings of Hermann and his successors were enthusiastically copied out all over Europe.

The mathematical techniques multiplied in the twelfth century, as Christians across the continent competed to create computistical solutions. It was not just men: one particularly condensed system devised by Abbess Herrad in around 1180 for the nuns in her Hohenburg convent reduced the entire 532-year Easter cycle to a short series of tables full of cryptic letters, dots and lines.[58]

One key instance of twelfth-century transmission featured Walcher, the prior of Great Malvern. Nestled at the foot of the verdant Malvern Hills in the west of England, the priory had been founded under bishop Wulfstan (the one commemorated in that Cisiojanus mnemonic) as part of the Norman reformation of English monasteries that also saw the rebuilding of St Albans abbey. Walcher was its second prior. He came from Lotharingia, the region of small territories between modern-day France and Germany, and he brought with him the interests in observational astronomy and calendar calculation that were widespread there.

Science and innovation were not, to be sure, unknown to the western monasteries. At Malmesbury Abbey, forty miles south and around eighty years earlier, a young monk named Eilmer had carried out an experimental flight. Inspired by the myth of Daedalus, he fastened wings to his hands and feet and leapt from a tall tower. According to the abbey chronicle, he flew more than two hundred metres, before a gust of wind caused him to fall and break his legs. He was lame for the rest of his life but survived well into old age. So if we can believe the somewhat disapproving chronicler – who was one of the most reliable historians of his age – Eilmer piloted an experimental glider, not wholly without success, almost five hundred years before Leonardo da Vinci sketched a similar flying machine.[59]

But back to Prior Walcher. Concerned that inaccurate tables were undermining the effectiveness of medical interventions that depended on astrology, he observed several lunar eclipses in the years around 1100, using an astrolabe to determine their precise mid-point.[60] His observations contradicted the standard models, which assumed that the Moon's motion was constant. But Walcher had no new model to replace them – until, that is, he met Pedro Alfonso.

Pedro, whose birth name was Moses, had converted from Judaism to Christianity in his native Spain. His hometown of Huesca had only recently been captured from the Arab Banu Hud rulers of Zaragoza. This was a time of significant, if sometimes uneasy, cultural interchange, and Pedro took full advantage of the access he had to scholarship from the Islamic world. As well as astronomy, he wrote a widely copied defence of Christianity against Islam and Judaism, and – equally popular – a collection of moral fables based on Arabic and Hebrew sources. One copy of this collection calls him physician to King Henry I of England.[61] Although that is improbable, he was certainly in England when he met Walcher around the year 1120.

Walcher later recalled Pedro's revelatory teaching about the mean and true motions of the Moon, including the cycle of the lunar nodes – the Head and Tail of the Dragon that Richard of Wallingford would later display on his monumental clock. Pedro was not able to explain the theories as precisely as he might have liked because, Walcher lamented, 'he had left his books across the sea'.[62] Even so, this was an important moment – and not just in the personal lives of a Lotharingian monk and an Aragonese Jew working together in medieval Worcestershire. For copies of Walcher's and Pedro's works soon spread from this hub in the west of England through the Benedictine networks. In surviving monastic manuscripts – including one linked to St Albans – we find them written out with that old guide to the *Uses of the Astrolabe*, as well as the earliest attempts to adapt the Indian-influenced astronomical tables of al-Khwarizmi for the Christian calendar.[63] The practices and needs of computus clearly supplied a space for Greco-Arabic sciences to spread and develop.

And develop they did. Far from the stereotype of a stagnant scientific environment which did no more than preserve the ideas of the ancients, computists in the twelfth and thirteenth centuries continued to refine their astronomical models, with ever more accurate estimates of the solar and lunar cycles. Scholars became more outspoken in their criticism of the increasingly unrealistic ecclesiastical calendar. In the 1260s the Franciscan friar and proponent of empirical science Roger Bacon wrote, at the Pope's request, a series of wide-ranging tracts on educational reform. In the third of these he condemned 'the corruption of the calendar'. It was, he thundered, 'intolerable to any wise person, horrible to any astronomer, and ridiculous to any computist'.[64] Bacon accepted that only the Pope could change the centralised calendar but urged him to take action on this score. For Bacon, reform of the sciences was an essential part of defending Christendom against existential internal and external threats.

Later popes did admit the problem and even commissioned leading astronomers to propose ways of redesigning the calendar. One proposal, a system of 'New Golden Numbers' designed for the papacy by two French astronomers, survives in the *Très Riches Heures* of Jean, Duke of Berry, a sumptuous illuminated book of hours whose artistry is justly celebrated but whose astronomical contents are often ignored.[65] But although such proposals would certainly have improved the accuracy of the calendar, they were not implemented. Science is one thing, policy is quite another, and the practical problems with the implementation of a new calendar – not least the need to revise or scrap thousands of painstakingly produced books across Europe – outweighed the political will to carry it out. So the overestimates in the Julian year continued to accumulate, and as late as 1532 the French satirist François Rabelais could begin his novel *Pantagruel* in a year when 'the month of March was missing from Lent, and mid-August was in May . . . because of the irregular bissextiles, when the Sun stumbled like a sinner to the left, and the Moon went five fathoms out of her way'.[66] It would be another half-century before Pope Gregory XIII took the decisive step of reforming the calendar, cancelling the leap day in three out of every four centennial years (the years divisible by 100 but not by 400). He decreed that ten days in early October 1582 be skipped, so that the next spring equinox would fall on 21 March, as it had when the Christian calendar had first been agreed. By then the Reformation had taken many countries outside the influence of the Catholic Church, and most – including Britain and her colonies – stubbornly persisted with the old calendar until the eighteenth century.

In the meantime, though, medieval astronomers could incorporate new theories of mean and true motions to enhance their calendars and accurately predict eclipses, while maintaining the core content of the Christian year. In John Westwyk's day, the very latest thing was the *Kalendarium* of Nicholas of Lynn. This Oxford friar, according to later legend, sailed from his Norfolk birthplace to explore the North Pole. He also composed an astrological calendar valid for seventy-six years – four nineteen-year cycles – in 1386.[67] Each month covered at least four pages. The first contained much of the same information we saw in the twelfth-century St Albans calendar: the Golden Numbers, the ferial letters, the wide column of the month's key feasts. Monks who copied this calendar could customise a feast day with their own favourite local saints – removing Wulfstan, or adding Alban on 22 June. By now such calendars had switched from Roman to the Hindu-Arabic numerals, so Nicholas included a column numbering the days of the month from 1 to 31, just as we do. On the rest of that first page, and three or four thereafter, he supplied a wealth of monthly astronomical data. He computed it with meticulous precision for his location at Oxford: 51 degrees and 50 minutes north of the equator – defined, as we saw, by observing the pole at that altitude above the horizon.

The monks took great pains in their reproduction of Friar Nicholas's calendar: margins filled with decorative flourishes and floral fronds, key words picked out in blue, red and gold. They proudly laid out an admirable array of tables: the daily position of the Sun on its ecliptic journey, the length of each day from sunrise to sunset, and accurate dates – and even times – of both new and full Moons. Nicholas also provided tables showing the lengths of shadows at each hour of the day. These were similar to the shadow-lengths of Palladius we encountered in Chapter 1, but while that Roman writer had given only a single set of figures for the whole month, Nicholas calculated a fresh row for each day. His data were in the equal clock hours that had by then become commonplace. They were more precise, with the shadows calculated to the nearest sixtieth of a foot and solar altitudes given to a minute of arc (a sixtieth of a degree). And unlike Palladius' five-foot Roman farmer, Nicholas made it clear that the person casting this shadow was to be six feet tall. He supplemented his calendar with tables predicting the solar and lunar eclipses for the next seventy-six years, together with pictures illustrating the extent of each eclipse. Further tables had astrological functions – including a table showing the week of seven planets, ruling each of the twenty-four hours in turn. Almost as an afterthought, towards the end of the collection he added a small correction table showing how far the accumulated error in the Julian calendar would take the computed Sun away from its true position.

*

Even if the ecclesiastical calendar did not change to reflect the progress of medieval astronomy, the mathematical techniques of computus remained an important part of monastic learning. Young monks in the fourteenth century made use of popular textbooks, some of which included tools to use your hand as a mnemonic. Like the finger-counting methods we have already met, this *manual computus* went back as far as Bede.[68] It was updated in the thirteenth century, so each knuckle now represented a Golden Number or Sunday letter (image 2.12). Monks could use the same manual methods for a raft of mental tasks, from multiplication to mastering music theory. Medieval learning did not necessarily mean laborious reading, reciting and writing. It could be impressively varied, including board games like

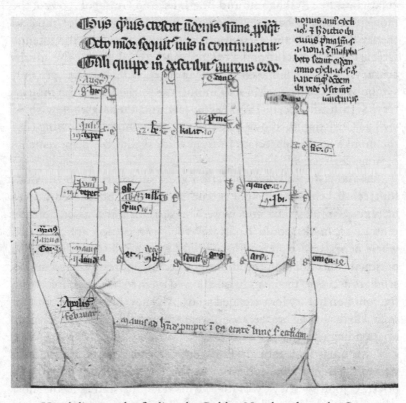

2.12. Hand diagram for finding the Golden Number, from the *Computus manualis* (*c.*1281) by Baldwin of Mardochio.

77

Number Battle, where players competed to master multiplication and arithmetical theory.[69]

By John Westwyk's day the monks of St Albans had more time and better facilities for learning than they had ever had. The previous abbot had rearranged the schedule of offices, bringing the daily Mass earlier and exempting students from some services so they had an unbroken stretch of time in the morning for reading.[70] He gave them a special space to study, conveniently located close to the cloister. Later, the cloister itself was decorated to reflect the breadth of monastic learning, with windows depicting leading figures of the liberal arts. The windows included classical philosophers and poets, of course, but also medical thinkers, mathematicians like Pythagoras and Boethius, and Guido of Arezzo, the monk thought to have designed the hand mnemonic for musical theory. Geometry and astronomy were represented by the totemic Greek masters Euclid and Ptolemy, and astrology by the ninth-century Persian Albumasar (Abu Ma'shar). Palladius was present, symbolising the agriculture so important to the life of the monastery. Significant recent thinkers in law and theology – Jewish as well as Christian theology – had their own windows, showing that the monks could appreciate both new ideas and the achievements of non-Christians.[71]

The monastery's book collection grew substantially in the mid-fourteenth century, as abbots enthusiastically purchased copies of texts both classical and new. The scriptorium was completely rebuilt, so books could be more efficiently copied and repaired where necessary.[72] Some of these books were marked for the abbot's personal collection, but the abbot granted the more advanced monk-scholars access to them, and also allowed them to borrow books from the common library for extended study. And new books were brought to St Albans by the brothers who were granted the immense privilege of attending university.

First things first, though. Only in exceptional cases were monks sent to the Benedictine college at Oxford before full ordination as a priest. If John Westwyk was to take his scientific studies to the most advanced level, he first had to complete that long process of ordination.[73] After a probationary year in which the Rule was repeatedly read to him in full and he was challenged to change his mind, he committed himself for life in the ceremony of profession. In the presence of all the brothers, he made the three-fold vow of obedience, fidelity to monastic life, and 'stability', or permanent commitment to the cloister.[74] He wrote it out in his own hand.

From his profession, John then began further training for the priesthood. Progression through three lower ranks of acolyte, subdeacon and deacon, before final ordination as a priest, usually took about three years. But it could be much quicker. For example, five St Albans novices a little younger than John Westwyk were ordained both acolyte and sub-deacon in St Paul's Cathedral on the very same day in March 1382. One of the five, Thomas Boville, proceeded to the third rank of deacon just six months later. Another was made a full priest by the Bishop of London the following St Valentine's Day.[75] St Albans ordinations were conducted by a variety of friendly bishops, after local son Pope Adrian IV granted the monastery a special exemption from episcopal supervision.[76] In Westwyk's lifetime it was generally the Bishop of London, but the Archbishop of Canterbury Simon of Sudbury, who had held that bishopric before his promotion to the primacy, also continued to ordain some St Albans monks on his regular visits to the capital. Bishops kept good records of their ordinations, carefully recording names, dates, status and locations of the ceremonies, but their registers have not all survived the centuries. For no St Albans brother do we have a record of all four ordinations. For most – including John Westwyk – we have none at all. The ordinations from the years 1368–79, for example, are completely lost. So we can only say that it was probably in that decade that Westwyk, aged in his early twenties, made his profession and was ordained. He had already left St Albans by 1380, when that list of monks containing so many Johns was compiled.

Ordination as a priest did not mean the end of learning. Careful reading was an integral part of the monastic life, and while this was centred on Scripture and theological studies, the monks also read widely in 'grammar' – in the widest sense that encompasses philology and even philosophy – and history. And since the monks' methods of meditative reading emphasised moving from fixed objective experience to subjective inner contemplation, they could use scientific description of the world and the comforting handle of the regular calendar as concrete starting points for meditative prayer. There were plenty of scientific books in the library at St Albans – especially the works of Richard of Wallingford, as we shall see in Chapter 4. But for access to the latest and greatest works in medieval scholarship, a scientifically minded monk like John Westwyk must have begged for the chance to study at university. We cannot know whether he got that chance – here, too, records are fragmentary – but the expertise he demonstrated later in life makes it more than possible. Either way, the paramount importance of universities in the story of medieval science means we must now make our way across the Chiltern Hills to Oxford.

3

Universitas

Richard of Missenden was a St Albans monk several years younger than John Westwyk. Like Thomas Boville, whom we briefly encountered at the end of the previous chapter, he was ordained deacon at a ceremony in St Paul's Cathedral on 20 September 1382.[1] Unlike Thomas, none of his other ordinations appear in the surviving bishops' registers. But we do know a fair amount about him from other sources. The St Albans chronicle tells us that when new abbots were elected in 1396 and 1401 Richard was the abbey's sub-cellarer, helping to manage the monks' supplies of food and drink and maintaining the kitchen equipment. We know from the same chronicle that he contributed funds to build a water-mill on one of the abbey's estates around that time. We know he was prior of Beadlow, twenty miles north of St Albans, when the monks abandoned that poverty-stricken priory in the autumn of 1428. He was then put in charge of Redbourn, the little house an hour's walk up the old Roman road, where small groups of brothers could retreat for their periodic bloodletting vacations. And we know that the following May, he represented the abbey in a legal dispute over the payment of tithes. The witness statements from that case were copied into the almoner's record-book, and each witness began by sketching the outline of his life story. It was there that the sixty-seven-year-old Richard Missenden recalled his five years at the University of Oxford.[2]

Richard said only that he went there after four years at St Albans, aged twenty-five or twenty-six. But that is more than we know about most of the monk-students. Since the earliest days of the university they had been required to register with a master, but no medieval matriculation roll has survived.[3] We can say a great deal about how and what they studied, but when it comes to the identities of the individual monk-students we are reduced to searching for scattered references. We only know, for instance, that John Heyworth, the archdeacon in Westwyk's time, had once been a student because a later monk, contemplating his blank marble tomb, recalled that he was a Bachelor of Canon Law.[4]

It is not surprising, then, that we cannot say with certainty whether John Westwyk went to university – though his use of high-level texts makes it more than possible.[5] What we can say is that many St Albans monks did. The Benedictines were slower than some other groups – notably the Franciscan and Dominican friars – to realise what a university education could offer their members. But by Westwyk's time they had joined in with gusto. In an attempt to drive up the standards of monastic learning, in 1336 the Pope called on monasteries to send one in twenty of their monks for a higher education. St Albans was among the most enthusiastic: as many as 15–20 per cent of the brothers went.[6] Many of them left the university before taking a degree.[7] But while they were there, they played their part in one of the most important institutions in the history of Western science.

The universities did not appear out of nowhere but evolved after centuries of gradual development in monastic and cathedral schools, catalysed by the flood of translations of Arabic and Greek philosophical and scientific works in the twelfth century. We have already stepped into the monastic schools through the work of Bede, Hermann and others. The monks in those schools studied the seven liberal arts.

Based on ancient Greek ideas of broad foundational training, the seven arts were laid out as a curriculum in the later Roman empire.[8] They were 'liberal' because they were suitable for a free or noble person; and the word 'art' did not denote the narrow range of aesthetic activities it does today, but was any skill worth studying. Early in the fifth century, a writer in the Roman outpost of Carthage named Martianus Capella wrote a vivid allegory that personified the arts as seven bridesmaids to Learning. They were strikingly dressed and symbolically equipped, with knives to prune children's wayward pronunciation, or surveying tools to measure the globe.[9] Boethius took up the idea in the 520s, and the seven were soon divided into two groups. There was the *trivium* of verbal sciences: grammar, rhetoric and logic; and the *quadrivium* of mathematical sciences: arithmetic, geometry, music and astronomy. They were further popularised in the following century by Isidore, the long-serving bishop of Seville. Isidore gave a summary of the by-then-standard liberal arts at the opening of his encyclopaedia, *The Etymologies*. An ambitious attempt to summarise all human knowledge, *The Etymologies* was probably the most popular and influential book after the Bible throughout the Middle Ages.[10]

These seven standardised arts became the basis of studies at the schools which grew up around cathedrals like Chartres and abbeys like St Victor in Paris. Herrad, abbess of Hohenburg, whom we saw in the previous chapter devising computistical tools for her nuns, included a guide to the arts in her compendious *Garden of Delights* (*c.*1180). There she drew the seven maids in a circle, as if under the arches of a cloister (image 3.1).* At the centre was Lady Philosophy, with the wise teachers Socrates and Plato seated just below her. Outside the circle were the 'poets or wizards', whose work, Herrad warned, was impure and valueless.

Like all the best educational schemes, the liberal arts were sufficiently clearly defined to be useful but sufficiently flexible to be accommodated to students' changing needs. Scholars enthusiastically redefined, prioritised and subdivided them. The Spanish *converso* Pedro Alfonso, for example, in his book of moral fables, *A Scholar's Guide*, was sure of only six: logic, arithmetic, geometry, physics, music and astronomy. One possible candidate for the seventh spot, he suggested, was 'the science of natural things' – but he admitted its claim faced two contrasting rivals: necromancy and grammar.[11] Later, in a letter sent to the philosophers of Paris around 1120, he highlighted the importance of astronomy above all – 'more useful, more enjoyable and more important than the other arts'.[12] He urged them to abandon outdated texts and to learn through practice (*experimentum*) – ideally with himself as teacher.

We do not know whether Pedro's sales pitch won him any new

---

* The unique twelfth-century manuscript was burned in the siege of Strasbourg during the Franco-Prussian War in 1870 but has been partially reconstructed from nineteenth-century copies.

3.1. The Liberal Arts, in the *Hortus Deliciarum* by Herrad of Landsberg, abbess of Hohenburg (made *c*.1180, destroyed 1870).

customers. But the schoolmasters of Paris were by no means resistant to the expansion of education he was suggesting. They were well aware that many questions about the created world could not be answered by the Bible. So they were willing to learn not only from other texts but also by looking around themselves. As one of the twelfth-century monk-teachers at St Victor's wrote: 'This whole visible world is like a book written by the finger of God ... to make manifest the wisdom of God's mysterious workings.' According to this influential ancient metaphor, the book of nature sat alongside the book of Scripture. It was not only legitimate to study both of these two 'books'; it was an integral part of praising God.[13]

At the end of the twelfth century these 'books' were read in entirely new settings: the universities. As towns and cities expanded, their newly wealthy citizens demanded educational opportunities. The cathedral and monastic schools could not meet such demand, but a number of independent masters were well placed to fill the gaps. The most gifted masters attracted crowds of students. They were keen not just to study the latest philosophy but also to take advantage of opportunities for skilled administrators, lawyers and theologians, both in the Church and in government.[14] Following the example of trade guilds, which were multiplying in the prosperous European cities, groups of masters and students began to unionise. In this way, they won recognition and protection from civic authorities. The Latin word *universitas* simply describes these unions of students and masters, regardless of any buildings or formal courses. As they came together, they gradually formed Europe's first universities.

In Bologna, whose law schools had developed an international reputation, it was the students who asserted their rights, winning a charter from the Holy Roman Emperor in 1158. If they were unhappy with their treatment by local authorities, they could vote with their feet – and some did, setting up a rival university in Padua in 1222. In Paris, on the other hand, it was the masters who formed a guild to resist regulation by the bishop and the chancellor of Notre-Dame (which ran the largest cathedral school). The establishment of a university at Oxford towards the end of the twelfth century is more mysterious – as late as the 1180s nearby Northampton was a greater centre of study – but masters may have been attracted to this middling market town by its role as a local legal centre. They probably also appreciated that the bishop who might claim authority over education was more than a hundred miles away, in Lincoln.[15] Oxford university was well enough established by 1209 to survive a four-year abandonment: both masters and students walked out when the townspeople executed two students whose housemate had committed murder. Some of them settled at the equally insignificant marshland town of Cambridge instead.

With growing support from Church and state authorities, universities sprang up all over Europe. By 1500, they had educated as many as a million students.[16] Yet from their beginnings, the universities had distinct identities reflecting their organic origins. That included their subject specialisms, which might be inclined towards one of the three disciplines of higher learning. Bologna, we have just seen, was a centre for law, while Padua and Montpellier rapidly acquired a reputation for medical training, especially as the pioneering school of Salerno declined in importance.[17] Paris, true to its outgrowth from the cathedral schools, specialised in the third and greatest of the higher subjects: theology. Oxford, too, focused on theology, but there the lower faculty of liberal arts was more influential than in other universities. That helped it to attract leading masters of the trivium and, especially, the mathematical quadrivium.[18]

The arrival of the translations changed everything. We have already seen that scholars in monasteries like Ripoll and Reichenau were enthusiastically reading Arabic texts on the astrolabe, as well as the new Hindu-Arabic numerals, around the year 1000. At the end of the next century this trickle of texts turned into a torrent of translations. A cohort of energetic linguists made Latin versions of scientific works from Greek, Hebrew and Arabic. They were helped by growing trade links between northern Italian cities like Pisa and Venice and the eastern Mediterranean, where the Crusades had turned old pilgrimage routes into highways for the traffic in goods and knowledge from the libraries of Antioch and Byzantium. There was important contact, too, in southern Italy, where a Tunisian translator known as Constantine the African brought a whole library of medical books first to Salerno and later to the abbey founded by St Benedict at Monte Cassino. In the twelfth century many translators moved to Spain, where territorial conflicts between Christians and Muslims provided new opportunities for cultural exchange. The energetic marketing of Pedro Alfonso, who in his letter to the Paris masters praised Constantine the African and boasted of his own translation of al-Khwarizmi, may also have turned attention to Iberia.[19]

By the middle of the century, Toledo (taken by Castile in 1085) was the most important centre for translation. Later, powerful patrons including the King of Castile would sponsor systematic translation schemes, but in the twelfth century Toledo was overrun by individual scholars, motivated by their hunger for new scientific writings. The most prolific of these was Gerard, who travelled from the Italian city of Cremona in around 1140, searching for Ptolemy's great second-century astronomical compendium, the *Almagest*. The translators were not just getting Greek works: they could also mine a rich seam of Islamic and Jewish scholarship, which had built considerably on the science of the ancient Greeks – as well as Indian sources – since the first great translation movement in ninth-century Baghdad. And the translators did not always work from written texts: they also Lati-nised the science taught by their Arabic masters, transcribed tables, reworked diagrams and sketched scientific instruments. Gerard of Cremona himself, according to a short biography written by his stu-dents, 'saw the abundance of books in Arabic on every subject and, regretting the poverty of the Latins in these things, learned the Arabic language, in order to be able to translate'.[20] Over the next forty years he translated more than seventy scholarly works, by luminaries such as Euclid and Ptolemy, al-Khwarizmi and his Abbasid contemporary, the pioneer of optics Ya'qub ibn 'Ishaq al-Kindi. Among the many medical works he translated were several by the great Greek phys-ician Galen, and others by the tenth-century Persian polymath Ibn Sina, known to the Latins as Avicenna. And he translated several books of logic and natural philosophy by the greatest of all ancient thinkers: Aristotle.

It is hard to recapture the revelation that Aristotle represented for medieval scholars. He had answers to all the questions they had been asking – and many others they had never thought of posing. It wasn't just the breadth of his studies – from the placement of a crocodile's tongue to the most fundamental structure of knowledge: it was the analytical clarity with which he laid out the thought processes that would lead to a satisfying answer.[21] Medieval scholars were so awe-struck at his achievements that they couldn't even name him. They simply called him the Philosopher.

To be sure, it was not easy to read Aristotle. Medieval natural philosophers were used to the *Timaeus*, by Aristotle's teacher, Plato.

The first half of this beautifully written description of the divinely ordered cosmos was translated into Latin in the fourth century, and the translator added a detailed commentary.[22] Although the *Timaeus* departs slightly from the entertaining dialogue form of Plato's other works, it is still an absorbing read, as the legend of the lost city of Atlantis leads into analysis of subjects like the four elements (fire, air, water and earth), the nature of time, the relation between the human body and the soul, and how our vision works. Aristotle's works were more voluminous, and more obscure. In part this was because, unlike Plato's dialogues, they were not written and refined for publication but were more like rough lecture notes. In part, too, it was a result of the translation process. The ancient Greek works had changed as they were translated through Arabic, and sometimes Spanish too. And many translators, including Gerard of Cremona, preferred to translate word by word, producing texts which, while faithful to the originals, were hardly written in flowing classical Latin.

If students – who could often join the foundational Arts Faculty aged as young as fourteen – struggled to know where to start with shelves of impenetrable Aristotelian prose, help was at hand. Like today's undergraduates, they had textbooks. Pithy summaries of the great philosophers made their ideas accessible, and the clearest of these became bestsellers. Even as the curriculum of the Arts Faculty was completely remodelled in the image of Aristotle's works – accommodating the 'three philosophies' of natural and moral philosophy and the more fundamental metaphysics alongside the old trivium and quadrivium – these summaries became the set texts for the masters' lecture series.[23]

So before the students dived into Aristotle's *Physics, On the Heavens, On Generation and Corruption, Meteorology, On the Soul, On Animals,* and his shorter works on psychology, respiration and ageing, all required as part of the natural philosophy course, they could warm up with the two introductory manuals of astronomy and cosmology: *De Sphera* (*On the Sphere*) and *Computus.* Each of these was allotted eight days by the Oxford timetablers. The *Computus* covered all the calendrical calculation we tried in the last chapter. It came in various versions, each adopted and updated as the science of the cycles developed. There were multiple texts called *The Sphere* too, but one was head and shoulders above the others for its accessibility and popularity. It was written in about 1230 by John of Sacrobosco.

Nothing much is known about this John. Forty years after his time, Robertus Anglicus, whom we last met chronicling the inventions of clockmakers, wrote an extensive commentary on Sacrobosco's *Sphere* for his students at the university of Montpellier. Near the beginning he asserted that Sacrobosco was, like him, English, but his claim has not been proved. The wandering antiquary John Leland (who, on his scholarly travels in the 1530s, was the last person to describe the St Albans clock before its destruction) tried to pin down the name 'Sacro-bosco' – holy wood – to a location. Failing to find settlements named Holywood on the map of England (though they existed in Northern Ireland and south-west Scotland), Leland opted to claim Sacrobosco for the Yorkshire town of Halifax – rather implausibly, since Halifax actually means 'holy hair'. While we cannot know where Sacrobosco was born, we do know where he was buried: in the de facto University church of Paris, in a tomb adorned with an astrolabe.[24] Evidently, by his death he was an established university master. He wrote a brief introduction to algorismus, drawing on the *Arithmetic* of Boethius, and his *Computus* was a commonly set text on that subject. But his *Sphere* was by far his best-known work, and still survives in hundreds of handwritten medieval copies. Monk-students, who arrived at university rather older and better educated than other undergraduates, were often exempted from the initial arts course.[25] But it is clear from the number of monastic manuscripts which include the *Sphere* – including the Merton Priory book we looked at in Chapter 2 – that canons and monks like John Westwyk were keen to work through this carefully arranged primer anyway.

In a beautifully simple text, whose four chapters together are about the length of one chapter of this book, Sacrobosco set out the basics of medieval knowledge of the universe.[26] He drew on a range of sources, especially Ptolemy and al-Farghani (Alfraganus) – another Abbasid astronomer whose work Gerard of Cremona had translated – but also quoted classical poets like Ovid and Virgil. He began with Euclid's geometry, defining what a sphere is, and then described the spheres of the heavens and Earth. A consummate teacher, he built up layer on layer of complexity, explaining the varied motions of the stars and planets, the ways that day-lengths and stellar visibility depended on your location and the season, and how eclipses work.

Students and masters read Sacrobosco's *Sphere* avidly for hundreds of years after his time. In Oxford's oak-panelled halls lecturers worked through it systematically, expounding Sacrobosco's succinct prose for the benefit of their fascinated pupils (image 3.2). Some wrote up their lectures as extended written commentaries on the core text, so we have a good idea of the ground they covered. Let us take a moment to imitate those scholars and have a closer look at one part of the treatise. We shall focus on Sacrobosco's explanation that the Earth is round.

Today it is widely assumed that medieval scholars thought the world was flat, but that is a myth largely invented in the nineteenth century. It was popularised in a work by Washington Irving that can be charitably called 'imaginative history', *The Life and Voyages of Christopher Columbus*, published in 1828. Irving pictured his hero, inspired by 'natural genius', arguing that it was possible to sail westward to the Indies, against fierce objections from ignorant churchmen at the Spanish court.[27] Irving's story was picked up by anti-religious writers and used as an emblem of a general conflict that they imagined was being waged between science and religion, in which a few brave individuals struggled against the suffocating power of the Church.[28] No such simplistic conflict existed. In fact, Columbus' geographical assumptions were based on the work of a contemporary of John Westwyk, the Paris master and later cardinal Pierre d'Ailly, who himself drew heavily on Sacrobosco's *Sphere*.[29]

Sacrobosco explains that the heavens are a huge sphere, with the planets set in smaller spheres nested one inside another (like a Russian doll). Beyond the seven planets – which, you will recall, included the Sun and the Moon – were two outer spheres: the fixed stars and the 'first moved', the engine of the daily rotation of the heavens. The innermost planetary sphere, with the shortest cycle, was the Moon. Citing Aristotle's *Meteorology*, Sacrobosco placed four more spheres within the sphere of the Moon, for the four elements: first fire, then air, then water, and finally, the heaviest element, earth, at the centre of everything (image 3.3).

As evidence for the earth's roundness, Sacrobosco pointed out that

3.2. Frontispiece to Charles Kyrfoth's *Computus manualis*, a basic textbook published in Oxford in 1519. Note the classroom equipped with a clock, hourglass, armillary sphere and astrolabe. Styles of dress had changed since John Westwyk's time, but the curriculum was much the same.

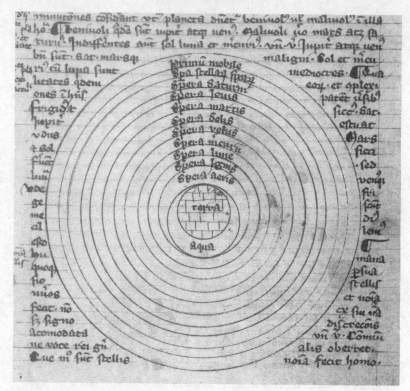

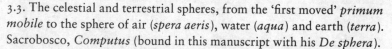

3.3. The celestial and terrestrial spheres, from the 'first moved' *primum mobile* to the sphere of air (*spera aeris*), water (*aqua*) and earth (*terra*). Sacrobosco, *Computus* (bound in this manuscript with his *De sphera*).

the stars rise and eclipses occur at different times as you travel east or west. And as you travel north or south, he added, you see different stars altogether. If the earth was flat, he explained, the same stars would rise at the same time for all observers. It only seems flat, he said, 'because of its great size'. Yet compared to the firmament, it must be infinitesimally small, since exactly half of the sky and stars are always above the horizon. The seas, like the earth, must also be round, since a lookout stationed at the top of a ship's mast can see further than someone standing on deck. Also, Sacrobosco suggested logically, just as water droplets form beads on leaves, so the seas 'naturally seek a round shape'. Aristotle had one more argument, which Sacrobosco did not use: whenever we watch a lunar eclipse, the Earth's shadow on the Moon is always round.[30]

The next question, for a consummate geometer like Sacrobosco, was obvious: if the Earth is a sphere, we can easily work out its size. The oldest estimate of the Earth's size appears in Aristotle's treatise *On the Heavens*, written in the fourth century BCE. There the Philosopher notes that 'those mathematicians who try to calculate the size of the Earth's circumference arrive at the figure 400,000 stades'.[31] Aristotle only mentions this to support his arguments that the Earth must be round – otherwise it would not have a circumference at all – and that it is small relative to the stars. A *stadion* was the length of a stadium – rather like journalists' habit of estimating areas in terms of today's football fields – but that length could vary, between about an eighth and a tenth of a mile. Aristotle's estimate thus came to 40,000 or 50,000 modern miles. The true circumference is about 25,000 miles, so Aristotle's figure was of the right order of magnitude, albeit not particularly close.

Aristotle was rarely interested in numbers – his specialism was explaining causes, answering 'how' and 'why' questions. So it is hardly surprising that he didn't explain the methods 'those mathematicians' had used. Towards the end of the following century, however, another Greek philosopher named Eratosthenes did explain how he had found the Earth's size. When Sacrobosco described, with his characteristic pithy clarity, how any student could carry out the calculation on a clear starry night, he cited Eratosthenes as an authority for his own estimate: 252,000 stades.

Those 252,000 stades are extremely close to the correct value. Where did they come from? Not from precise measurement, but by a chain of educated guesses – that was all the Greek astronomers wanted. Eratosthenes observed that at the ancient city of Syene, on the Nile in southern Egypt, the Sun was directly overhead at noon on the summer solstice. In other words, Syene lay on the tropic of Cancer. At the same time in Alexandria, the Sun was not quite overhead. If, looking up at the sky, he imagined a vertical circle which ran down from that zenith to the southern horizon, kept descending to a point directly below his feet, and came back up the other side to rise above the northern horizon and reach all the way to the zenith once more, the Sun was just a fiftieth of the way round that circle. So, since the Earth was a sphere, the distance from Syene to Alexandria must be a fiftieth of the way round the Earth. Here Eratosthenes assumed that Alexandria, where the winding River Nile fanned out into the Mediterranean Sea, was due north of Syene. He took the distance between the two cities to be 5,000 stades. If one fiftieth of the Earth's circumference was 5,000 stades, the full circumference must be about 250,000.[32] Later astronomers adjusted it to 252,000 stades, simply because that number is easily divisible by 60 and 360. With that convenient rounding, Sacrobosco could say that, for every degree of the Earth's circumference, the distance is 252,000 ÷ 360 = 700 stades. (A rather smaller – and less accurate – estimate of 500 stades per degree, reported by Ptolemy and Alfraganus, was seized on by Columbus to boost the feasibility of his proposed voyage west to the Indies.)[33]

Neither Eratosthenes nor Sacrobosco saw any need to measure these distances. Eratosthenes' figures of a fiftieth of a circle for the Sun's zenith distance at Alexandria, and 5,000 stades to Syene, are clearly just round numbers. He certainly did not hire someone to pace out the distance step by step, as some fanciful retellings have claimed.[34] His point – echoed by Sacrobosco – is simply that the Earth *could* be measured, with knowledge of its sphericity and the basic techniques of geometry.

However, the same sphericity that made it possible to have a good guess at the Earth's size presented scholars with a new problem. If both earth and water form spheres, and if earth is the heaviest element, with its natural place within the sphere of water (image 3.3), why is the land not covered with water? Why, asked some students safe in their Oxford classrooms sixty miles from the sea, haven't we all drowned? In fact, that question was based on a misunderstanding of Aristotle's cosmology, and the Philosopher had several answers to it. The easiest was this: just because the natural place of earth was beneath the water, that did not mean that it was all there, all the time. A mountain stream will carry pebbles downhill with it, and so over time will erode a gorge, but it will take much longer to carry the whole mountain down to the sea. In any case, it was thought, the mountains themselves contained water – without it, they would crumble to dust, as mud does if you dry it out. Even if – as Aristotle supposed – the universe is eternal, infinite time would not be enough to separate the elements into distinct spheres, since they are always changing into one another.[35]

Sacrobosco had an even easier answer: he didn't care. Here we must understand a vital disciplinary distinction – one that may feel as odd to us as our distinction between mathematics and music would seem to a medieval scholar. For them, astronomy and cosmology were almost entirely separate. As part of the quadrivium, astronomy was a quantitative mathematical science: it measured the movements and positions of the heavenly bodies. Cosmology was part of natural philosophy, and asked qualitative questions: 'what' and 'how' rather than 'when' or 'where'. So, for astronomers, the question of the elemental spheres simply did not arise. Their science was largely instrumental: if a spherical Earth correctly placed the Pole Star at 40 degrees above the horizon in Toledo, and 52 degrees in Oxford, there was no need to worry about the water. Criticising Sacrobosco for not properly addressing that question would be like criticising the illustrator of a biology textbook for not giving his bodies hats or nail polish.

For Sacrobosco, indeed, astronomy was instrumental in a very literal sense. His *Sphere* was not just about the sphere that makes up the heavens. It was also – or even mostly – about the brass sphere commonly used to model them: a sort of skeleton globe of the heavens known as an *armillary sphere*.[36] If Sacrobosco was thinking more about showing students how a moving model can display the universe's workings than about the philosophical implications of Aristotelian cosmology, it is not surprising if he neglected to worry about the sphere of water.

He did not ignore the issue completely, but did restrict himself to the slightly unsatisfactory comment that 'dry land blocks the flow of water, to protect the life of living things'. Some of his later readers took this to mean that the sphere of water had been displaced so that it no longer surrounded the earth completely, leaving a little poking out the top. When these readers drew the spheres, they drew the water off-centre, so the earth came into contact with the sphere of air (as in image 3.3, where *terra* just touches *spera aeris*). Pushing the Earth out of the water on just one side had obvious implications for the habitability of other parts of the earth, especially the southern hemisphere. Other readers acknowledged alternative possibilities. Our old friend Robert the Englishman touched on the issue in the carefully structured series of fifteen detailed lectures on the *Sphere* that he delivered to his attentive students at Montpellier. He suggested that divine will, or perhaps the astrological influence of some star, might have dried out part of the earth. But he also pointed out that the earth could absorb part of the water. Later philosophers developed this idea to suggest that the Earth's weight was unevenly distributed. So its centre of gravity could remain at the centre of the universe, while its geometrical centre could be off to one side.[37]

A quite different suggestion was that, as Aristotle had explained in his treatise *On Generation and Corruption*, the elements are constantly transforming into one another. As a result, the spheres of earth and water are entirely mixed.[38] Aristotle did not explain how elemental water might transform into earth, though, as we shall see, some influential medieval thinkers did. Even so, the advantage of this explanation was that it would work even if earth and water had been created as separate spheres, for the process of constant transformation would have mixed them up. And it was perfectly compatible with the biblical account of creation, in which God had gathered the seas apart from the land on the third day.[39]

One argument of Aristotle's, however, was certainly not compatible with biblical creation: his belief that the universe was eternal. For centuries before they discovered Aristotle, Christian philosophers had been well used to dealing with theories that contradicted the faith. The Early Church Fathers likened pagan philosophy to the Egyptian gold and silver that the biblical Israelites took with them on their exodus from slavery: it might be tarnished by association with the pharaohs but was still potentially precious. Even problematic or irrelevant doctrines, they said, could contain nuggets of useful learning. St Augustine took up the same analogy in the early fifth century. He warned against pursuing knowledge in a spirit of arrogance, but recognised that the methods and insights of the natural sciences could provide support to theology. He also pointed out that if Christians mishandled Scripture to support non-sensical beliefs about nature, this would bring their holy faith into disrepute.[40]

Using a metaphor much like his north African contemporary Martianus Capella's personification of the liberal arts as bridesmaids, Augustine suggested that the pagan sciences could be in the position of handmaiden to religion. They were inferior, but indispensable – and trusted with considerable autonomy. On Dante's journey from Inferno to Paradise in the *Divine Comedy*, the Italian poet recognised that non-Christian philosophers could hardly go to heaven, but he placed them in the next-best place – the lush meadow of Limbo:

Colà diritto, sovra'l verde smalto,	There before me, on the enamelled green,
mi fuor mostrati li spiriti magni,	great souls were shown to me;
che del vedere in me stesso m'essalto.	I glory still in having seen them.
. . .	. . .
Poi ch'innalzai un poco più le ciglia,	When I raised my eyes a little higher,
vidi'l maestro di color che sanno	I saw the master of those who know
seder tra filosofica famiglia.	seated in philosophic family.
Tutti lo miran, tutti onor li fanno:	All look up to him, all do him honour:
quivi vid' ïo Socrate e Platone,	I saw there Socrates and Plato,
che'nnanzi a li altri più presso li stanno;	nearest to him, ahead of all the rest;
. . .	. . .
Euclide geomètra e Tolomeo,	Euclid the geometer, and Ptolemy,
Ipocràte, Avicenna e Galïeno,	Hippocrates, Avicenna and Galen
Averoìs, che'l gran comento feo.	Averroës, who wrote the great commentary.

Io non posso ritrar di tutti a pieno,	I cannot describe them all in full,
però che sì mi caccia il lungo tema,	my great theme drives me ever onward,
che molte volte al fatto il dir vien meno.[41]	so words often fail to match the deeds.

At the highest point of all, closest to Paradise, Dante did not name 'the master of those who know'. He did not need to. References to Aristotle in the *Divine Comedy* exceed everything except the Bible itself. His pagan science was certainly not a threat to theology.

Despite this, in the thirteenth century the Church authorities faced something they were not used to. It was not just that the breadth and influence of Aristotle's newly translated writings overwhelmed earlier scholarship – so that the entire structure of the curriculum had to be remodelled to accommodate them. They also arrived at an awkward moment. The bishops were increasingly concerned that the universities were too independent. They feared that the autonomy for which students and masters had fought so hard would limit the Church's power to control how matters of faith might be interpreted in the classroom. People in authority have always worried about what gets taught, but in an era of unprecedented institutional change, the church leaders had good reason to feel uncomfortable. The universities were new and potentially dangerous places.

In addition, these were troubling times for the wider Church. Ever since the reforms of the 1050s to 1080s, which had secured the Church's independence from secular rulers and given the clergy greater moral authority, the popes had encouraged ordinary people to lead a more active religious life. It had worked. Pilgrimage had increased, corrupt priests had suffered effective boycotts, and thousands of Christians had even been willing to die for their faith in the Crusades. But mass movements were hard to control. The Pope might feel for people dissatisfied with rapidly changing social conditions, especially in the growing cities, and he could certainly approve of the desire to lead a simple life. But he did not always like the results. Spiritual movements sprouted across Europe in the twelfth century. They were often loosely organised, and it was sometimes hard to separate saints from demagogues. Rejection of worldliness and embrace of the apostolic life, preached by Francis of Assisi, was provocative but acceptable. Denial of the Church's authority and its monopoly on the interpretation of Scripture, which the Waldensian movement espoused, was dangerous. The idea that the world and Christianity had both been created by Satan, as the Cathars claimed, was plain heresy.

In such a tense climate, it was reasonable to be suspicious of new books that presented unfamiliar ways to understand the world. In 1209, the same year that the Pope gave his personal approval to the first small band of Franciscans, he launched the Albigensian Crusade against the heretical Cathars of southern France. The following year, the Archbishop of Sens, whose province included Paris, banned the masters there from reading Aristotle's books of natural philosophy.[42] The ban was confirmed five years later by a papal legate.

It had little effect. In the first place, the 1215 prohibition applied only in the Arts Faculty. In the higher Theology Faculty, students could roam unrestricted through Aristotle's natural philosophy – as could anyone at all who studied in private. Many arts masters quietly ignored the ban anyway. And at other universities, including Oxford, it carried no weight at all.[43]

The scholars knew a good thing when they saw it, and Aristotle's works were simply too useful to ignore. And like the Franciscans, who had initially aroused suspicion within the Church, if the texts were not heretical they could in fact be deployed as a weapon against heresy. Their coherent, convincing account of nature might help win over wavering believers. In 1216, the year after the second prohibition, the Pope licensed a new Order of Preachers, explicitly tasked to combat heresy. Nicknamed the Dominicans after their Spanish founder, they were to be among the most enthusiastic readers and developers of Aristotelian scientific theory in the medieval universities.

In 1228 the newly elected Pope Gregory IX renewed the prohibition on the Paris masters. But the study of Aristotelian science was actively encouraged at the university Gregory founded in Toulouse the following year. It was founded as part of the Treaty of Paris that finally ended the Albigensian Crusade, and its mission to combat heresy would use all the Egyptian gold available.

Yet even as Gregory signed that treaty in Paris on Maundy

Thursday of 1229, the university there was on strike. A conflict between the students and civic authorities was ablaze. Drunken students had smashed up a tavern. Being clerics, they were under the jurisdiction of the Church courts, but the townsfolk wanted the queen regent, Blanche of Castile, to punish them. The heavy-handed city guards killed several students, the Church refused to get involved, and the masters walked out in protest. It was international news: brother Matthew Paris in faraway St Albans chronicled the events in detail, blaming Blanche for 'a woman's impulsiveness and the haste of an agitated mind'. King Henry III of England issued an open letter inviting the masters and students to continue their studies under his protection. Many did indeed leave Paris to study elsewhere, including at Oxford, Cambridge and the new school at Toulouse.[44] Not everyone in England was happy with the influx of international students: the existing students protested against the rent rises that Oxford landlords imposed to take advantage of the Parisians. Nor did they appreciate King Henry's tightening of university discipline in response to rising disorder blamed on the new arrivals. But Oxford did benefit from the fresh injection of students, masters, their books and ideas.[45]

The Paris strike lasted two years, until Pope Gregory finally gave in to the masters' demands. He placed the university under his personal protection, guaranteeing their right to strike in future and allowing them to continue lecturing during the summer vacation (which was limited to one month). Crucially, although Gregory maintained that the books of natural philosophy would remain prohibited until they had been purged of theological error, he promised that the prohibition would not be enforced. It was probably in this year, 1231, that Sacrobosco wrote his *Sphere*, in which he was able to cite Aristotle without fear of consequences.[46]

By mid-century, the issue had almost been forgotten. In 1255 the Arts Faculty issued a new curriculum that included all Aristotle's known works. Students read them avidly, often alongside the lucid commentaries of Muhammad Ibn Rushd (Averroës), who had worked in Almohad Spain in the late twelfth century. Yet there were still troubling inconsistencies between these works of natural philosophy and the core tenets of the theologians. And as the masters of Paris became more confident in their discussion of Aristotle and Averroës, and bolder in their claims for the power of philosophy, trouble was brewing.

As we have seen, the Arts Faculty was supposed to be only a school of basic training. Masters would teach there for a couple of years before progressing to study in a higher faculty such as theology. But as the Arts Faculty transformed into a philosophy faculty in the image of Aristotle, some masters were now content to spend their entire careers there. They wrote treatises exploring tricky questions that arose in Aristotle's writings. These so-called *quaestio* treatises took their structure from the debates that had first been held in theology before spreading to other faculties as common classroom exercises. The texts were originally just a report of the discussion, which followed a set format of argument, counterargument and a final decision by the master. Increasingly, though, masters would refine the reports for publication. This allowed them to present radical theories with the excuse that they were simply exploring contrasting ways of interpreting set texts and resolving contradictions between them. It also meant that science was shaped by the conventions of scholastic logic.

The theologians observed all this with concern. They saw the most ambitious arts masters overstepping the boundaries of what they should be teaching and straying into theological issues. For example, there was the question of how we can obtain universal knowledge – knowledge of things in general – based on our limited experiences of particular things. That may sound like irrelevant philosophical abstraction but, if you ignored it, how could you justify coming up with scientific theories about the generalised workings of nature? How, for instance, can you say anything conclusive about trees in general, when you have seen only a limited, possibly unrepresentative, sample of trees? In his commentary on Aristotle's treatise *On the Soul*, Averroës had argued that all humans must share a single united intellect that allows us to have such universal knowledge. The implications of this unity of the intellect for individual human souls were as problematic as the idea that the world is eternal.

Leading Paris theologians, including the Dominican friar Thomas Aquinas, wrote texts attacking these 'Averroist' notions. In 1270 the Bishop of Paris, Étienne Tempier, stepped in, denouncing thirteen erroneous doctrines and excommunicating anyone who taught them.[47] Unlike the earlier prohibitions of whole books, these were condemnations of particular ideas. The bishop's brief announcement did not name any masters accused of supporting such false theories, but everyone knew who he had in mind. The most notorious of the so-called Averroists was an arts master from Brabant (in modern-day Belgium) named Siger. Siger had, it was true, defended the unity of the intellect and the eternity of the world in his writings. But that defence was carefully qualified. He made clear that his goal was only to elucidate the meaning of previous philosophers, and he stressed that when faith conflicted with reason, faith was more reliable. After all, faith was God-given, while reason relied on humans' imperfect senses.[48] And it was not a crime to discuss false ideas – not yet, anyway. No less a scholastic luminary than Aquinas had spent years reading the works of Aristotle and Averroës in order to sift out the Egyptian gold from the heretical dross.

But Aquinas was a regent master of the theology faculty. Siger was trespassing on the theologians' turf. Under pressure from above, in 1272 the arts masters introduced new faculty rules restricting their own freedom to discuss theological questions. But it was not enough. In 1277 Bishop Tempier went much further than before, issuing a list of 219 condemned propositions. The ideas he condemned included the familiar heresy that the Earth was eternal, but many more of them were physical theories that threatened to limit the power of God. It was an error to believe, said Tempier, that God could not create more than one universe, nor make a vacuum by moving the heavens in a straight line. Tempier was not trying to impose any particular beliefs – that there were really two or three universes, for example. His point was only that God could have made the world any way He wanted. Other condemned propositions were straightforward rebukes of the arts masters' arrogance. They had, Tempier implied, been making the shocking claim 'that there is no more excellent state than to devote oneself to philosophy' and teaching subversively 'that a theologian's preaching is based on fables'.[49]

The 219 articles were assembled by a panel of sixteen theology masters, apparently in some hurry. Tempier was probably obeying a papal directive, or else reacting to the recent summons of Siger to appear before the Inquisitor of France. But the condemned propositions went way beyond what Siger had written – and in any case, he was apparently acquitted of heresy. On the one hand, Tempier condemned ideas that no sane philosopher believed, such as the infamous doctrine of 'double truth', which said that two contradictory statements – for example, that the world was created *and* eternal – could *both* be true. On the other, the condemned articles included positions held by many perfectly respectable theologians, including Aquinas himself. At this point Aquinas was safe, having died in 1274, but the fact that the condemnation was issued on the third anniversary of his death may reflect the desire of some rival theologians on the panel to hammer an extra nail into his coffin.[50]

Unlike previous rulings that had prevented only the teaching of false theories, in 1277 the mere discussion of such questions was banned outright. Yet some historians have seen this Church condemnation as a surprisingly positive moment for medieval science. They argue that the arts masters were too tied to Aristotle. Being forced to loosen their loyalty to his principles may have freed the philosophers to consider other ways of viewing the world. Being more open-minded about the power of God encouraged them to consider things previously thought impossible, like the existence of a vacuum. That said, for most young astronomers reading their Sacrobosco or musing on the material of mountains, it made no difference at all. They wanted to explain or predict what they could observe, and cared little for such hypotheticals.[51]

Eleven days after Tempier's condemnation in Paris, the Archbishop of Canterbury condemned thirty doctrines that were allegedly being taught by masters at Oxford. The underlying cause was the same difficult negotiation of disagreements between Greco-Arabic philosophy and biblical authority, but the doctrines condemned in Oxford were mostly about logic and grammar, as well as the ever-complex question of the relationship between the body and the soul.[52] In any case, the prohibitions were largely ignored. As in Paris – where the condemnations were also partially rescinded after Aquinas was made a saint in 1323 – the students who simply sought to get their heads around the classic texts of the quadrivium and did not stray into theology until properly qualified were left to get on with it.

*

How strongly did these shockwaves of condemnation ripple out to the great Benedictine houses, in Glastonbury and Durham, Westminster and St Albans? Questions of speculative cosmology may have been distant from their concerns, but nonetheless they felt they were missing out on something. They had been slower than the friars to see the potential of university education. Now they were losing benefactors; they were losing the brightest recruits; they were losing opportunities to educate those monks they had managed to recruit.[53] In September 1277, six months after the Archbishop of Canterbury had stepped into the Oxford curriculum debate, the Benedictines of southern England agreed to set up a college at Oxford. 'Let learning bloom again,' they vowed, making it clear they realised they had some catching-up to do.[54]

Things took a while to get going, in part because some independent-minded abbots were fiercely opposed to having this new expense imposed on them by the centralised Benedictine chapter. The new Oxford college opened only in 1283, initially as a pilot project for fifteen monks from the abbey of Gloucester (image 3.4). But the pilot went well enough that in 1291 the Benedictine abbots agreed to throw Gloucester College open to all the monks of Canterbury province, which included most of England. A counterpart to house brothers who travelled from the north of the country, named Durham College, was founded in the same year.

Sending monks to Oxford was an expensive business. In 1277 the abbeys had agreed to support the new venture with tuppence in every mark (1¼ per cent) of their income for the year, and half that in subsequent years. It was not enough, especially after the Pope's 1336 order to send one in every twenty monks to the university. Apart from the cost of tuition, the students needed clothing and travel, candles and books, food and, of course, drink. The cost of graduation parties and gifts was a particular concern.[55] In the 1360s the provincial chapter, then under Thomas de la Mare's presidency, ruled that the minimum funding for a student was £15, plus travel expenses. This was equivalent to the annual wages of a master craftsman or lawyer. All but the richest monasteries struggled to finance their students, usually scraping together support from a patchwork of sources, including levies on the senior monks, supplying students directly with things like

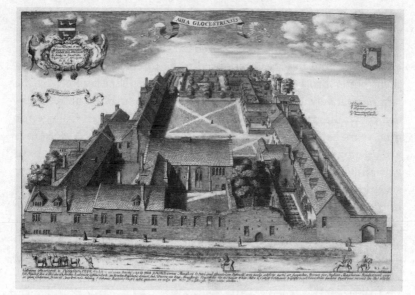

3.4. Gloucester Hall, Oxford, in 1675, including many buildings surviving from Gloucester College (with the St Albans shield carved above the right-hand gateway, now facing on to Walton Street).

candles and books, and seeking sponsorship from benefactors of the monastery. St Albans was able to fund their monks much more generously than other houses, in part by demanding donations from the dependent priories, which graduates would often go on to manage. Even so, the abbots were constantly looking out for new sources of student funding. When Richard Missenden closed the doors of Beadlow priory in 1428, the St Albans abbot seized the chance to divert a chunk of that dependent house's income towards the upkeep of the Oxford students.[56]

Time at Gloucester College was, then, exceedingly precious. The main reason so many students left without incepting as a master was that it was simply not cost-effective to stay at the university long enough to take the degree. A full course of studies leading to a doctorate in theology would take seventeen years. The half-decade Richard Missenden spent at Gloucester College was a more common stint, and many monks had to be content with much less than that. They were often allowed to stay in Oxford only for short stretches of the year, either because they were needed in the monastery or because there was so much demand for rooms at the college.[57]

Their limited time was supposed to be directed towards making them better brothers. That meant studying either theology or the administratively useful subject of law. The clockmaking abbot Richard of Wallingford, for example, attended Gloucester College for nine years before qualifying to lecture on the *Sentences*, the standard theology textbook first used in the old cathedral schools. Richard was an unusual case: he had already studied in the Arts Faculty for six years before he took monastic vows at the age of twenty-two. His initial study had been sponsored by his local priory, a daughter house of St Albans, but the prior of Wallingford was apparently unable to support him any longer without a firmer commitment on his part.[58] Three years at St Albans culminated in Richard's ordination as priest, and he was then permitted to pack his bags once again for Gloucester College.

The college stood in lush meadows donated by a Gloucestershire baron. Outside the north gate of Oxford, it was safely removed from the worst temptations and dangers of student life but close to the clean water of the winding River Thames. In its routine and rules, it was a hybrid of a university college and a Benedictine priory.[59] The students were mostly reliant on teaching and resources within the college and were normally barred from studying with non-monks. Alongside attending lectures and structured discussions, they were still expected to participate in divine offices, and to practise preaching in both Latin and English. Despite this, they clearly contrived to study widely and have active social lives. Richard of Wallingford wrote five treatises in astronomy while working towards his theology degree. He was an exceptional case, no doubt, but other monks were certainly willing to enrich their preaching with the Egyptian gold of science. We find, for example, notes on the theological implications of Aristotle's views on the creation of matter jotted in the margins of a commentary on the Philosopher's treatise *On Generation and Corruption* used by the St Albans students.[60]

As for their social lives, some brothers clearly took full advantage of the opportunities on offer in a university town. A few years before John Westwyk's time, a Franciscan friar from Cornwall named Richard Trevytlam wrote a poem, 'In Praise of the University of Oxford', that was really more complaint than praise. He compared the university favourably to Athens and Paris but bemoaned declining standards there. Above all, he criticised the behaviour of the monk-students. Leaving the holy observance of their monastery, he said, they take up feasting and hunting, read banned books and provoke the poor friars with divisive preaching. He singled out three particular culprits. One, a monk from Glastonbury Abbey, regularly drinks until he can no longer speak or stand, but after sleeping off his stupor slightly he preaches against the faults of others:

Nutant vestigia, caligant oculi,	His gait is unsteady, his eyes are a blur,
Lingua collabitur, pes deest gressui . . .	His tongue starts to tremble, his foot will not stir . . .
Tamen in crastino cum sol caluerit,	Yet on the next morning – the sun's in the sky,
Digesto paululum vino quo maduit,	The wine that he soaked in is almost quite dry –
Hic plebi predicat et fratres inficit.	His sermon the friars will hotly decry.[61]

The poem was written at a time of intense rivalry between the friars and monks within the university. But there is evidence from Glastonbury Abbey itself to show that it was not just anti-monastic slander. Around the same time Trevytlam was writing in the 1360s, a rather efficient abbot sent – and kept copies of – a succession of letters to his monks at Gloucester College.[62] First he urged one senior monk to keep an eye on two recent arrivals, John Lucombe and Robert Sambourne. Then, when the two were caught hunting, fishing and trespassing, they were summoned back to Glastonbury to explain themselves. (It seems that hunting, which you may remember Chaucer's greasy-faced prelate also practised, was a particular weakness of monks.)[63] Four years later the abbot wrote to Sambourne, who had spent all his money, encouraging him to live more frugally and to do some teaching work to support himself. The advice apparently helped, as Sambourne was put in charge of a new student the following year. But his colleague John Lucombe could not stay out of trouble. In 1366 the Glastonbury abbot received a letter of complaint from abbot-president Thomas de la Mare himself, acting on a report from the prior of the students at Gloucester College. The details of Lucombe's misbehaviour were not specified – though the word Thomas used, *incontinentia*, hints at something of a sexual nature – but it is clear that even a conscientious abbot sometimes struggled to maintain order at a distance.[64]

A common crime was losing the precious books the monks had been given by their abbey. This happened to that St Albans commentary on *On Generation and Corruption*, which an Augustinian canon was able to buy cheaply not long after John Westwyk's time.[65] Even at bargain prices, though, books were precious commodities. Made from the carefully stretched and scraped skins of sheep or calves, inscribed in ink prepared using the acidic galls found on oak twigs where wasps had laid their eggs, the plainest textbook is still an impressive feat of medieval craftsmanship. The St Albans monks mostly brought books with them, borrowed from the cloister collection for a year at a time. Likewise, the Merton Priory book that features Sacrobosco's *Sphere* alongside the cylinder dial and calendar mnemonics was also lent to an Oxford student. Students could acquire additional books at the university. Those were mostly second-hand, but sometimes a monk-student had to make his own handwritten copy of an essential text.[66] Stationers in Oxford hired out exemplars for students to copy, though the system was not as efficient as in Paris, where books were lent in sections (called *peciae*) at fixed rental prices, so multiple students could copy the same text simultaneously.[67] The copies might be bound, but that was not inevitable or immediate. The dirty first folios of many surviving books testify to long periods without covers. When the loose booklets were bound they might be sewn only into a thickish sheet of parchment, rather than hard boards, and they could be accompanied by texts that were much older, with unrelated contents.

Every authority from the Pope downwards had set out strict regulations about the care of books. The very frequency of such rulings, and the curses written into many St Albans books promising anathema to anyone who might steal or deface them, suggest that books were not always well looked after.[68] In a popular work named *Philobiblon* (*The Love of Books*), the Bishop of Durham, Richard of Bury, described what he used to see in his time at Oxford:

> some headstrong youth lazily lounging over his studies, and when the winter's frost is sharp, his nose running from the nipping cold drips down; nor does he think of wiping it with his pocket-handkerchief until he has bedewed the book before him with the ugly moisture . . . He does not fear to eat fruit or cheese over an open book, or carelessly to carry a cup to and from his mouth; and because he has no wallet at hand he drops into books the fragments that are left.[69]

As common as this kind of neglect was the pledging of books as security for loans. Books were hard currency. Students who, like Robert Sambourne, got themselves into financial difficulties, could pawn their books in exchange for cash at one of the university's loan chests. If the loan was not paid off within a year, the book would be sold, and that is probably how the St Albans Aristotle came into the hands of that opportunistic Augustinian.

Since students were responsible for their own books and the first colleges were loose communities, libraries were not an immediate or automatic feature of university life. But as scholars donated books to their colleagues, and books passed freely between student monks, so communal collections soon sprang up.[70] At first, the books – and some scientific instruments – were kept in locked chests, which might only be opened once a year for scholars to return the resources they had borrowed and make a new selection. Later, the colleges built library rooms, where reference books could be placed on open shelves. They were often chained in place, as much to symbolise the scholars' communal ownership as to prevent theft. In the 1420s the abbot of St Albans paid for and stocked a library building for Gloucester College, as well as a new chapel. The extent of the abbey's domination of the college in John Westwyk's time and the following few decades is shown by the fact that the St Albans coat of arms was carved above the college's main gateway. Although Gloucester College was disbanded at the Dissolution of the Monasteries, and its buildings later incorporated into what is now Worcester College, the decorated medieval gateway still survives in Oxford's Walton Street (image 3.4).

It is clear, then, that the abbots of St Albans were not deterred by the cost and disciplinary risks of sending students to university. As they saw it, their investment not only yielded the indirect benefit of bringing the prestige of learning to the monastery, but directly enhanced the education of the monks. A steady flow of brothers along the fifty-mile road through the Chiltern Hills plugged the monastery into an enormous network of scholarship, and the returning students brought the world's ideas – and books – back with them. For the universities were truly international. The first pan-European language of scholarship – Latin – allowed masters to work freely from Paris to Padua, Cambridge to Cologne. And the cross-border networks of religious orders, and common curricula, meant a friar could transfer as easily between theology faculties as the employee of a multinational company might relocate from New York to Shanghai today.

The most illustrious scholars, such as the Italian Dominican Thomas Aquinas and his German teacher Albertus Magnus (Albert the Great), moved several times. They picked up new ideas and texts on their travels and brokered a lively trade in knowledge. Albert, for example, drew on his personal observation of mining, metallurgy and alchemy, as well as his eclectic reading, to shed new light on previously unanswered questions.[71] We saw one earlier, left by Aristotle: if the sphere of earth is not covered by the sphere of water, because the two elements are constantly transforming into one another, how does that elemental change occur? Aristotle had argued that what is solid is dry – stones do not flow – but also pointed out that when moisture is eliminated from a clod of earth it crumbles to dust. So why do mountains not crumble? Avicenna (Ibn Sina) answered this by arguing that what holds clay together, even when it is baked into bricks, must be a moisture more like oil than water. As the clay hardens into rocks, Avicenna suggested, all the watery moisture evaporates – but the oily moisture is left behind. The newly hardened rocks with their locked-in moisture are then either thrown up into mountains by earthquakes or left standing proud as winds and floods erode the land around them. Albert, writing in the 1250s, endorsed Avicenna's account and added the evidence of the abundant fossil beds he had visited in the area around Paris (though he suspected that the shellfish he saw there had sprung into life as the rock hardened).[72]

Albert's immensely wide range of interests won him the nickname *Doctor Universalis* – the Universal Teacher. He wrote on everything from geometry to medicine, logic to falconry. He accurately described the symptoms of lead poisoning and the morphology of countless plants. It is easy to see why popes in the 1930s, standing against an oppressive atmosphere of fascist anti-intellectualism, first canonised Albert and then made him the patron saint of scientists.[73]

Still, although Albert was called 'the Great' during his own lifetime, not all his contemporaries agreed with that assessment or approved of his eclectic dabbling. His most vehement critic was his Franciscan rival, the Englishman Roger Bacon. They probably met at Paris in the 1240s. For Bacon, writing his extended essays about the reform of learning for the Pope, Albert represented all that was wrong with the scholarship of his day. Albert was, Bacon admitted, hard-working, and had observed much. But this did not make up for his utterly inadequate philosophical training, for his lack of experience teaching the arts, for his ignorance of languages. Above all, he apparently had no knowledge of the two sciences Bacon thought were most important. *Perspectiva* and *experientiul science*. 'His writings are useless,' Bacon spat. 'He has corrupted the study of philosophy.' And yet the students of Paris, to Bacon's baffled indignation, saw him as an authority alongside Aristotle, Avicenna and Averroës.[74]

Leaving aside the bitterness of Bacon's intensely personal attack, it is worth pausing briefly to examine some of these criticisms. First, his accusation that Albert was ignorant of languages. It was true up to a point. Albert had, for example, got his understanding of Avicenna's geology not by reading the Persian polymath's *Kitab al-Shifa'* in its Arabic original but from the partial Latin translation by the English scholar Alfred of Shareshill (who may have been working at Oxford).[75] Yet that was merely testament to the immense importance of the translation movement of the twelfth and thirteenth centuries.

Bacon was a passionate proponent of intensive language-learning as a keystone of education and research. His criticisms of his colleagues' production and use of translations are forceful, repeated – and deeply unfair. He did not provide any real evidence for the flaws he claimed to have found in translations, nor for his own boasts of linguistic expertise. He went as far as to claim that the translations of Aristotle into Latin were so obscure and wrong, such a source of error and ignorance, that scholars would be better off not having them at all. 'If I had control over the [translated] books of Aristotle', he raged, 'I would have them all burned.'[76] Yet while Bacon underestimated both the quality and quantity of available translations, he was airing a common concern. Ever since St Jerome had translated the Bible, carefully explaining his choice to translate the sense of each phrase rather than word by word, scholars had been aware that translation was an uncertain business. Translating freely as Jerome did, and Cicero before him, risked making the work too much your own, subverting the intentions of the original author. Working word by word, as Boethius had done, preserved the original text but resulted in barely comprehensible Latin. Still, most translators – including the prolific Gerard of Cremona – chose the second option. They squared the circle by filling the margins of manuscripts with glosses and commentaries to make the meaning as clear as possible. Bacon, though, preferred Jerome's way. The real problem, he sneered, was that translators were not only linguistically incompetent but also lacked scientific training.[77]

Bacon admitted one exception to his wholesale criticism: his idol Robert Grosseteste, who had taught himself Greek in order to read and translate new works of theology and cosmology.[78] Grosseteste was lecturer to the Franciscans in Oxford in the early 1230s and was elected Bishop of Lincoln in 1235. That was before Bacon became a Franciscan, but they probably overlapped at Oxford, and Grosseteste certainly made an impression on the younger scholar. Grosseteste had written his own *Sphere* textbook, in which he supported the idea that the spheres of earth and water were as one. He also composed several short, mind-bending treatises on more speculative areas of natural philosophy.[79] These blended the cosmological ideas of Plato, filtered through the writings of St Augustine, with the newly arrived philosophical methods of Aristotle and Avicenna.

Plato's *Timaeus* had introduced a mathematical theory of light and vision, and Grosseteste put light at the very centre of his science.

He used its action to explain human understanding, the relationship of the body and soul, and even the structure of the universe. He had already proved that the universe could not be eternal, refuting the arguments of Aristotle's *Physics*. Now he showed how it was created using light. His explanation drew heavily on the geometrical optics of al-Kindi, yet another Abbasid polymath whose work Gerard of Cremona had excitedly translated. Unlike Aristotle, who did not think light moved at all, al-Kindi had pictured light radiating outwards in straight lines. Grosseteste's brief but visionary treatise *On Light* described an explosion of light at the beginning of time, oddly reminiscent of the Big Bang. The light spread outwards in all directions, pulling matter with it. Grosseteste argued mathematically that the light must be infinitely multiplied but the matter could only be finite. It could not extend infinitely without creating a vacuum somewhere. So when the matter had stretched as far as it could, the outermost celestial sphere of the firmament was formed. Then, as light was dispersed inwards, the other celestial and terrestrial spheres were created.[80]

It is easy to see, then, why Bacon thought Perspective – the science of light and sight – was such an important field. Studying it, he was greatly influenced by one ground-breaking thinker whose work had not been available to the older Grosseteste. This was a Basra born scholar named Ibn al-Haytham, known to Latins as Alhacen. Working mostly in Egypt around the year 1000, Ibn al-Haytham successfully merged mathematical analyses of light with the causal and medical approaches that other theorists had taken. Crucially, he overturned the *extramission* theory of Plato, Euclid and al-Kindi. According to that theory, the eye emitted visual rays, sent out to collect information from the objects they encounter. The problem with its reverse, the *intromission* theory, was that if light came radiating in all directions from all points on an object, when those rays hit the eyeball they would be hopelessly jumbled – how could the eye possibly make sense of them? Ibn al-Haytham resolved this question with some clever geometry. He suggested that only the rays hitting the eye at right angles would enter the eye unimpeded, to be picked up by the optic nerve. Other rays would be refracted, weakened and ignored.

An understanding of optics had real practical potential. Grosseteste and Bacon both wrote enthusiastically about how useful magnification could be. It will allow us to read tiny letters, they boasted, and even to count grains of sand. Bacon was also very excited by the idea of employing strategically placed mirrors in warfare. By the end of the thirteenth century, enterprising scholars in northern Italy, where there was a flourishing glass industry, were using carefully shaped and polished lenses to aid their reading and writing. Friars proudly announced from their pulpits that this ingenious new art allowed users to continue their work well into old age. The first spectacles were born.[81]

Judging by the scientific interests of the day, these truly were the Light Ages. Franciscans were unusually interested in light, which they took as the means of God's work in the material world.[82] But monks, too, naturally followed scientific developments in this field. Perspective could help unlock the cryptic beginning of John's gospel, which said that in God was life, 'and the life was the light of mankind; the light shines in the darkness and the darkness has not overcome it'. When the monks read there that John the Baptist was not himself the light (*lux* in Latin) but came as a witness to the light (*lumen*), they might well wonder about the difference between the two kinds of light, *lux* and *lumen*. Most philosophers, following Avicenna, used *lux* for light and its luminous properties in general, and *lumen* for instances of propagated light and its effects. Bacon tended to use the terms interchangeably. But in Grosseteste's theory the two kinds of light were integral to the different phases of creation. The immaterial divine light of the first Big Bang, multiplying infinitely outwards, was *lux*, while the bodily light that made the inner spheres of heavens and Earth was *lumen*. They had different origins and characteristics, not unlike how physicists today may think of light as having the properties of a wave or a particle.[83]

More fundamentally, though, most readers of John's gospel took *lux hominum*, the light of mankind, to refer to human understanding. Throughout the Scriptures the ideas of sight and thought, light and understanding, were intertwined, as when St Paul wrote that 'now we see through a glass, darkly', or in the Psalms, where the unfolding of God's words 'illuminates', gives light.[84] To some philosophers, that relationship between sight and light offered the key to understanding how the human mind can achieve knowledge of things that cannot be seen – including God. It is not surprising, then, that monks made copies of the most popular textbook on Perspective – even though that textbook had been written by a Franciscan who, as Archbishop of Canterbury in the 1280s, had come into fierce conflict with the Benedictine abbots.[85] John Peckham's *Essentials of Perspective* introduced the mathematics of light and vision, saying little about wider philosophical questions; but its main attraction was that it summarised Ibn al-Haytham's *Perspective* in one-tenth the length of the original work. This was the sort of knowledge that monks had the opportunity to access only through their participation in university life.

We should not ignore the other science that Bacon valued highly, though monks seem to have been less interested in it. This is what he called *scientia experimentalis* – the science of trial and experience. Here, too, Bacon was influenced by Robert Grosseteste. The ultimate source of knowledge, Aristotle had said, was the human senses (though Grosseteste was well aware that the senses were fallible – no match for the knowledge provided by divine illumination). In pure 'higher' sciences like arithmetic and geometry, logical methods could provide demonstrative proof of causes. But the 'lower' applied sciences like optics and astronomy could show only that correlations existed, not prove cause and effect. So an astronomer could observe that the Moon waxed and waned, but it was the geometer who explained why – because it was a sphere. Even then, the higher science could provide proof only for the lower – the geometer could not explain why the Moon was a sphere. Ultimately, philosophers had to fall back on their senses, both to establish the first principles of science and to test the conclusions of scientific arguments.[86]

So far, so loyal to Aristotle. But Grosseteste went further, explaining how philosophers could use repeated observation of particular events to establish universal principles.[87] Similarly, Ibn al-Haytham wanted scientific reasoning to be confirmed by the experience of rigorous controlled testing (the Arabic word he used, *i'tibar*, meant 'careful consideration'). Bacon embraced – and widened – this. His manifesto was to take the experimental practices that were common in occult arts like alchemy and natural magic, sift out the 'fraud' and 'illusion', then apply them in the established Aristotelian sciences. Philosophers, he said, should make more use of apparatus such as burning mirrors that focused rays of light to produce concentrated heat. He argued that *experimenta* – meaning anything from everyday experiences to designed experiments, as well as thought experiments – could demonstrate new truths that were outside the remit of existing sciences. Aristotelian philosophy could not explain the attractive force of magnets, or the natural magical powers of certain stones and plants, but speculative experiential science could catalogue such phenomena. It could also uncover new technologies for what Bacon saw as the essential task: defending Christendom against the imminent Antichrist. Such technologies might even include flying machines, 'in which a man may sit turning some kind of engine, so that artificial wings beat the air like a bird'. 'I have never seen one,' he admitted, 'but I know a wise man who has designed one.' He may have been referring to the daring venture of Eilmer of Malmesbury 250 years earlier.[88]

As an example of the power of experiential science to contribute to the all-important science of Perspective, Bacon made a study of the rainbow in the 1260s. He encouraged experimenters to examine its colours in sparkling crystals from Ireland or India, in dewdrops, or in water falling from the blades of a mill. Observing the rainbow itself, he pointed out for the first time that its maximum altitude was 42 degrees above the horizon. He rejected Grosseteste's theory that the rainbow was caused by a triple refraction through layers of cloud, and proposed a new theory of reflection. Bacon's new theory had its own problems, but its focus on individual drops of water was a step forward in understanding. Fifty years later another friar who had studied at Paris, a German Dominican named Theodoric (Dietrich) of Freiburg, would see how the rainbow could best be explained by a combination of refraction and reflection within those individual drops.[89]

As we read the books that monks brought back to their monasteries, we find few signs of interest in such speculative science. Monks were well aware that their time at university was limited. They had more urgent practical questions to answer, particularly in astronomy and the calendar. Still, some monks did make their mark on the fashionable problems of the age. One such was Roger Swineshead, a student from Glastonbury Abbey whom the angry friar Richard Trevytlam named as a rare model of how monks should behave at university. When Swineshead was at Oxford, in the 1330s, the cutting edge of natural philosophy was the question of how to quantify things that were usually thought of as qualitative, like hotness or speed. This was an area where Roger Bacon's exhortation to pay attention to alchemy was good advice.

Alchemy was the study of minerals, especially metals, and the processes through which they could be changed and purified. Its practitioners dissolved and distilled, heated and mixed, crystallised and filtered. As they uncovered the hidden properties of chemicals, they hoped to be able to prolong human life and produce precious substances. In the process, they learned a great deal about the elements from which all matter was made. Aristotle had defined each of the four elements as a combination of two qualities: either hot or cold, plus either wet or dry. Earth, for example, was cold and dry, while air was hot and wet. (You can probably guess the qualities of fire and water.) But while Aristotle had acknowledged that you could have more or less of such elemental qualities, measuring them was harder. If you put two differently sized lumps of Avicenna's clay into the same fire, why will they reach the same temperature in different times? And if – assuming they have not yet hardened into bricks – you combine together two lumps at the same temperature, why does the temperature not increase? A monk named Walter of Odington had an answer to this. He explained the difference between the intensity or degree of heat in an object (temperature) and its extension (the total quantity of heat, which we measure in calories). On this basis, since all metals were compounds of the elements, he was able to quantify their heat, wetness, and so on. If you combined any amount of substances with different degrees of contrasting qualities, he could predict the properties of the resulting compound.[90]

The idea of assigning numerical values to such qualities was rooted in medical practices, since – as we shall see in Chapter 6 – physicians mixing medicines needed to predict their heating and cooling effects.[91] But it also appealed to students of logic in Oxford, especially at the college that was then the richest and most independent: Merton College. They wanted new ways to answer fundamental philosophical questions, such as how far you could change or move a thing without making it something else. If my large black hound turned into a small brown terrier, would he still be my dog? Such questions had high-level theological implications, as they might help explain how the Eucharist could change its substance into the body of Christ while retaining the outward appearance of bread.

At a more restricted level, a group of Oxford scholars drew on the alchemists' combinations of contrasting qualities to explain mathematically how competing forces permit or prevent motion. If the hinges are loose on your garden gate, so the bottom grinds against the path, how fast will it move when you try to open it? Aristotle had implied that the velocity of an object like that gate was proportional to the force pushing it divided by the resistance to it. The problem with that formula, as a fellow of Merton College named Thomas Bradwardine pointed out, was that no amount of resistance, no matter how great, would stop the gate altogether. (When you divide by any finite number, no matter how big it is, the answer will always be more than zero: the gate will keep moving at least a little.)[92] He suggested that the ratio of force to resistance could be preserved if the velocity was *geometrically* proportional to that ratio. In other words, to double the velocity, you must square the ratio of force to resistance. This was a definite improvement, but it failed to explain what would happen if there was no resistance at all. So Roger Swineshead, the praiseworthy Glastonbury monk, suggested a model where the velocity was proportional to the force *minus* the resistance.[93]

That was just the simplest part of Roger's treatise *On Natural Motions*. In it, he asked – but could not answer – a new question: how could qualities that remained constant be compared with, or added to, qualities that were steadily changing? A group of scholars at Merton soon after Bradwardine, around 1340, applied themselves to these challenges with such success that historians have called them the Oxford Calculators. The first answer was given by William Heytesbury, in a textbook he wrote for first-year students of logic.[94]

He claimed that if you moved at a steadily changing speed you would cover the same distance as someone moving at a constant speed – if their constant speed were the average of your starting and finishing speeds. This 'mean-speed theorem' was a big advance. Nowadays we unthinkingly describe instantaneous speed in terms of a hypothetical distance covered in a set time – a glance at my speedometer giving miles per hour makes it feel inevitable. But the formulation of the mean-speed theorem and its concept of instantaneous velocity were anything but inevitable.

One of Heytesbury's Merton colleagues, named Richard Swineshead – possibly from the same Lincolnshire village as Roger – soon proved the mean-speed theorem. Richard's sixteen-part *Book of Calculations* was copied across Europe. It was so advanced that later writers dubbed him the Calculator. But his techniques were rather hard to follow, and few readers got through all sixteen tractates. Indeed, by 1500, Italian scholars were using his name as a scathing metaphor for the pointless abstraction they associated with English philosophy. Nonetheless, he continued to have his admirers in later centuries, including the great German mathematician Gottfried Leibniz.[95]

No one in Oxford could continue in the footsteps of the Calculators. In part this was because of the Black Death.[96] This devastating plague pandemic killed only one of the Merton mathematicians, Bradwardine, who died just a month after becoming Archbishop of Canterbury in 1349. But it is notable that far fewer scientific works were written at Oxford in the following decades, probably because so many of the next generation never made it to the Arts Faculty at all. Instead, the baton passed to Paris, which may have suffered less badly from the plague because it drew its students from a wider international pool. There, in the decades either side of 1350, two philosophers made impressive advances in mathematical physics. Jean Buridan developed a theory of impetus to explain how a ball keeps moving after you have thrown it (a problem Aristotle had been unable to solve), while his student Nicole Oresme proved Heytesbury's mean-speed theory with a beautifully clear graph – a method that could calculate the distance and mean speed even when the acceleration was not constant. Both the ideas of impetus and the mean-speed theorem were to have a significant influence on Galileo almost three hundred years later.

If only a few monks were, like Roger Swineshead and Walter of Odington, taken with such abstract natural philosophy, many more were enthused by the astronomy they encountered at Oxford.[97] It is astronomy that fills the books they brought back to the cloisters. Not only did they study textbooks on the established subjects we have encountered already, like the computus and the *Sphere*; they also embraced the new astronomical tools that were being developed in Spain and Paris and that came quickly to Oxford. We see this in the astronomical album of Adam Easton, who as prior of students at Gloucester College reported that misbehaving Glastonbury monk to his provincial president. An academic high-flier, Adam went straight from Oxford to become a cardinal at the papal court, but late in life he donated his books in two batches to his old priory at Norwich Cathedral. One, which later found its way to Cambridge University Library, is a wide-ranging collection of cheaply but carefully copied treatises (and scrawled verses abusing the 'fallacious friars').[98] It contains manuals on astronomical and mathematical instruments, guidance on surveying techniques and trigonometry, and updated tables to compute the positions of the planets or find the latitudes and longitudes of English towns. Essays explaining the astrological effects of the Moon in different constellations and how to predict the sex of a foetus – or if it will be twins – are followed by Robert Grosseteste's *Sphere* textbook, and instructions for using an astronomical calculator, which still survives in the medieval library of Merton College. The great manuscript cataloguer M. R. James, better known to non-historians for his atmospheric ghost stories, described Easton's collection as 'a labyrinthine book'.[99] James had seen more than a few of those, but Easton's manuscript simply reflects the eclectic astronomical interests of scholar-monks.

How far did those monks manage to keep up their interests when their brief sojourn in Oxford came to an end? That depended, of course – not only on how motivated each individual was to maintain his expertise, but also on the tolerance of his abbot for such studies, as well as the availability of books, which the next cohort of students might need. There are certainly some books that, despite being carefully stored in monastic libraries, show no signs that anyone read them there.[100] Yet at least some monks did keep in touch with their old colleagues and skills. In the 1370s a Merton College fellow named William Rede bound together a collection of astronomical writings, some of which he had bought from the executors of the plague-struck logician Thomas Bradwardine. Among the tracts and tables were two letters from a certain Reginald Lambourne. Lambourne described himself as 'a simple monk of Eynsham'. The first letter, addressed to 'my dear and reverend master', discusses the astrological – and especially meteorological – significance of the positions of Jupiter and Venus at the time of two lunar eclipses in 1363. The second, which Lambourne wrote to Rede himself, uses astronomical data to give a more general long-range weather forecast for the years 1368 to 1374. It, too, was addressed deferentially, to his 'most reverend lord'. Lambourne had been a fellow at Merton College in the 1350s before making his profession at Eynsham Abbey. But he was not willing to let his useful astronomical learning lapse, and took advantage of the abbey's location just ten miles upstream of Oxford to keep in touch with his old associates.[101]

Other monks had no need to take such personal initiative. At St Albans, where a succession of scholarly abbots had fostered an atmosphere of wide-ranging scientific study, the monks amassed a store of scientific books, which they eagerly read and re-read. Those who had studied at Oxford had special privileges, including access to the abbot's private library. John Westwyk was probably one such monk. As we shall see in the next chapter, he carefully read and annotated two works brought back from Oxford. Perhaps, then, he was using his access privileges to maintain and build on what he had learned at Gloucester College. Although we cannot know for sure that Westwyk himself attended Oxford, we can say for certain that he and his contemporaries were profoundly influenced by university developments, which saw the works of an international scientific fraternity of Jews and Muslims, Italians and Germans, given a proud place in English monasteries. In any case, Westwyk was still learning, as we shall see by looking closely at his annotations. It is time, then, to return with him to St Albans and see how the monks followed their fascination for astronomical instruments.

4

Astrolabe and Albion

Some came back with a doctorate, others after only a short summer. All the monk-students had to return to the cloister eventually. It must have been quite a wrench. As abbot of St Albans in the 1330s, Richard of Wallingford would voice his regret that he had left for university at such a young age, and had been distracted from theology by mathematical pursuits. But he recognised that it was such learning that had, by God's grace, let him leave behind his humble origins as the son of a blacksmith, 'lifted from the dung to sit among princes'.[1]

Doubtless, many monks felt the same ambivalence. Far from the pleasures of the university town, they had to re-accustom themselves to St Benedict's austere rule, where humility and obedience were paramount. They returned to the same place in the order of seniority that had been set on the day they made their monastic profession. Nonetheless, at St Albans the graduates were granted special privileges, including exemption from the midday Mass, opportunities to continue their studies, and the possibility of improved accommodation. These arrangements risked creating resentment in the monastery, and even undermining the authority of the abbot, which may be why graduates were often sent far away to manage one of the dependent houses. Those that stayed were expected to take up greater responsibility, such as preaching and contributing to the education of the next generation of monks.[2] They were also expected to harness their scholastic skills in the production of new books.

The abbey employed professional scribes for their most important devotional and record books, but the bulk of copying was done by the brothers themselves. Abbot Thomas de la Mare saw this as a way for scholarly monks to avoid sinful idleness. 'Let them be occupied

with set duties, each according to his abilities,' he exhorted: 'studying, reading and writing books; annotating, correcting, illustrating
and binding too.'[3] It is in such activity, about the year 1379, that John
Westwyk left his first definite mark on the historical record. While
a new scriptorium was gradually rising from its foundations, John
remained in the cramped old writing room. In this space, where for
much of the year cold hands would struggle to write by the light of
precious beeswax candles, he worked on two books.[4] The choice he
made tells us a great deal, not only about his interests but about scientific innovation in that era. The books were both about scientific
instruments.

Both books contain the same two instrument treatises by Abbot
Richard of Wallingford. Richard composed both treatises during his
final year at Oxford, 1326–7. Several of the surviving copies of these
two scientific works were made at St Albans abbey – and for good
reasons. The painstaking work of copying Richard's works allowed
the knowledge they contained to be shared within St Albans's network of daughter houses. It was also an important way to honour the
memory of an illustrious former abbot. And the act of copying was
itself a priceless learning opportunity.[5]

It will be clear from the manuscripts we have already examined
that no two books are the same. It is not just that a collection of scientific tracts often represents a unique selection – the copyist's personal
medley. Even within a single treatise, which could be as short as two
pages, each copy is different. The preparation of the parchment and
page layout, the size and formality of handwriting, the style – or
absence – of decoration, the frequency and accuracy of diagrams,
the completeness of the text, and even its title: they all vary enormously. On top of this, books were not static objects. Later users
could add colour or commentary in the margins, amend mistakes,
fill in diagrams where they were lacking, give the work a new title or
add the (supposed) author's name, or simply sketch in a doodle. So
reproducing manuscripts was always a continuous process, and the
boundaries between reading, copying and editing were blurred. It
is in this role of an active reader that we encounter John Westwyk,
adding some diagrams to a treatise by Richard of Wallingford called
the *Rectangulus*.

The rectangulus was a perfectly pared-down celestial calculator. In Chapter 3 we saw Sacrobosco modelling the heavens as a spherical cage – a ball made of brass rings called an armillary sphere. Such spheres, still hanging in Oxford lecture halls in the sixteenth century (image 3.2), had two practical functions. You could look across them to the sky, measuring the motions of the stars on their scales as the great Greek Ptolemy had instructed.[6] Or you could use them for teaching, especially to demonstrate the three main planes of astronomy: the horizon, the equator and the ecliptic (image 4.1a). Each plane is a flat circle across the middle of the heavenly sphere, like the seam on a cricket ball, and each has its poles – if you imagine a line drawn straight up and down from the middle of the circle, cutting through it at right angles, the poles are where that line touches the heavens. We have encountered all three planes already. We have observed the rising and setting of the celestial equator, carrying the stars in circles around the Pole Star and making a sundial work. We have also seen that the altitude of the Pole Star – the angle between the Pole Star and the plane of the horizon – tells you your

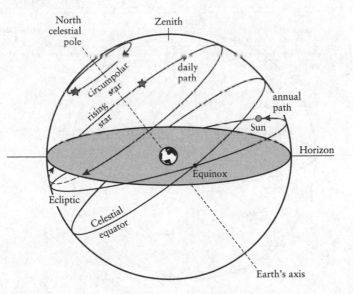

4.1a. The three heavenly planes: horizon, equator and ecliptic.

latitude. The horizon plane has its own 'pole' – the zenith, directly above our heads. And the third plane, the ecliptic, carries the Sun on its annual journey through the stars on a path angled at about 23½ degrees to the equator. The planets follow a similar path through the stars, though they stray a little either side of the ecliptic and sometimes seem to change direction.

Each plane had its own pair of coordinates for measurement (image 4.1b). You can measure a star's *altitude* above – or below – the horizon, as well as its *azimuth* (direction around the horizon, starting from north, like a compass). Or, in terms of the ecliptic plane, you can measure the star's celestial *latitude* as an angle north or south of the ecliptic, and its *longitude* around the ecliptic, starting from the 'equinox', where the ecliptic intersects the equator. Or the star can be measured from the celestial equator, in terms of *declination* north or south of the equator, and *right ascension* around the equator, also starting from the equinox. All the systems are useful for different purposes. We have already made ample use of altitude, have observed

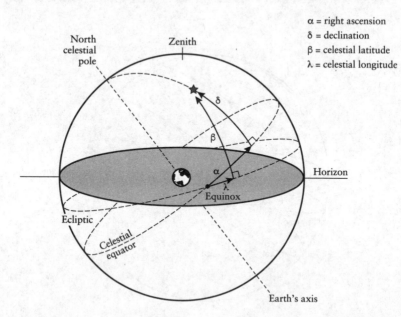

4.1b. Ecliptic and equatorial coordinate systems (see also image 2.10).

the Sun's increasing longitude as it moves through the zodiac signs and have considered its declination back and forth across the equator with the seasons.

If you found the last two paragraphs confusing, you're in good company. People have always had difficulty thinking three-dimensionally. That is why armillary spheres were so useful. The problem was, they were also extremely difficult and expensive to make. Only the most skilled craftsmen could forge their rings and engrave their scales with enough accuracy for high-quality measurement, and for precise conversion of measurements between sets of celestial coordinates. One solution for such practical coordinate-conversion was to dispense with the sphere and represent each of the three planes as a disc (image 4.2). It did not matter if those discs were set slightly apart, since they were measuring the angles of objects almost infinitely far away. The only essential principle was that the discs were angled correctly to each other. Placing them one above another in a pile produced an instrument called the torquetum (or turketum). This idea was already known in Muslim Seville in the 1100s, and by the end of the following century two astronomers – one from north-eastern France, the other Polish – had produced Latin manuals based on similar principles. One of them was copied into that textbook that a canon of Merton Priory took with him to Oxford.[7]

If you could simplify a three-dimensional sphere to a set of two-dimensional discs, you could also go a step further and reduce the discs to hinged pairs of arms. This is what Abbot Richard of Wallingford realised. 'I designed the rectangulus', he wrote, 'as an antidote to the tedious and difficult work of making an armillary sphere . . . as a means of determining the paths and places of the planets and fixed stars . . . and all the problems that can be solved by an armillary, an astrolabe or a turketum.'[8] Its arms pivoted in three dimensions, with a sight at the top for observation. Plumb lines hung down from the upper arms, allowing their angles to be measured against scales on the lower arms. In one sense, this was a simple device: relatively easy and cheap to make (especially if you substituted wood for some of the brass arms), and child's play to use for finding the positions of stars in any of the coordinate systems. But it was conceptually quite difficult, since it no longer looked anything

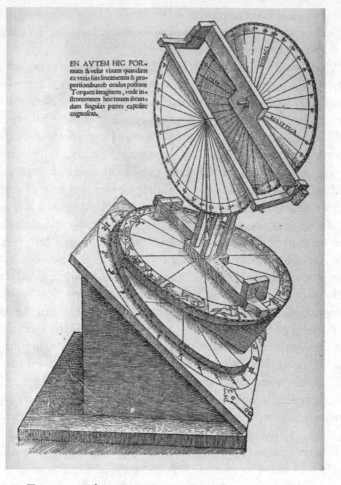

4.2. Torquetum, from Peter Apian, *Introductio geographica*.

like a sphere. To understand how each pair of arms contributed to a simulation of the heavens, diagrams were essential. But copyists did not always take the trouble to include them. So it was a charitable deed for fellow students of astronomy when John Westwyk added some diagrams to a copy of the *Rectangulus* that he found at St Albans in the late 1370s.

The scribe who had written out Richard's text had left room for the diagrams. He started the page with an expanded right-hand margin; then, after copying out eight lines, he began to indent the text on the left too (image 4.3). This created space on both sides for the diagrams. Scribes often did that. The idea was that they – or perhaps someone more expert in technical drawing – could return with an appropriate quill and ink to finish the job. It did not always get done. But John Westwyk did fill in the gaps in this manuscript – better a few decades late than never. This is the first surviving work in Westwyk's handwriting: a perfect match for the manuscript where he signed his name. Its confidence is striking. His diagrams, probably adapted from some in another manuscript, are neatly drawn, giving a sense not just of the geometry of the instrument but its physical presence. He included the wavy threads of the dangling plumb lines, details of the joints connecting the sturdy brass arms, and even some simple decoration at the top and bottom of the base column. He also labelled the diagram in his own words. In the top-right-hand corner of the page, he noted which of the arms were fixed in place and explained how their scales should be marked out according to the ruler he had sketched below.

It is entirely fitting that Westwyk's first foray into astronomical commentary concerns an instrument. Instruments were at the heart of medieval science. One French inventor, in a prologue preserved in Adam Easton's astronomical album, wrote that 'the noble science of astronomy cannot be properly understood without appropriate instruments'. Medieval libraries were full of them, stored and lent alongside the books.[9] Their significance went well beyond the narrow practical functions of taking observations or simplifying calculations. We have already seen that Sacrobosco's *Sphere* presented the universe for students to understand in the form of an instrument. Geoffrey Chaucer went further. Setting out to write a thorough five-part introduction to astronomy, apparently for his ten-year-old son, he built it around a guide to a single instrument: the astrolabe.

John Westwyk would later read and learn from Chaucer's guide. But even before Chaucer wrote it around 1390, Westwyk had already studied the astrolabe. He drew on his knowledge of the astrolabe when, in the second of those two books he worked

4.3. John Westwyk's annotated diagrams for the *Rectangulus* (1326) of Richard of Wallingford.

on at St Albans, he created new copies of the *Rectangulus* and *Albion* treatises. The *Albion* was Richard of Wallingford's most advanced invention – an instrument that made the rectangulus look like a child's toy, and even rivalled the great St Albans clock in its complexity. We will take a look at this planetary supercomputer shortly. First, though, just as John Westwyk did, we must get to grips with the astrolabe.

Portable, multifunctional and elegant, technically advanced and a status symbol, the astrolabe was the classic medieval scientific instrument. It represented the cutting edge of scientific knowledge, with all the exciting – and troubling – implications that might have. Medieval writers and artists placed astrolabes in the hands of sorcerers, of students, of wise Solomon himself. Its functions ranged from complex astronomy to simply telling the time. So let us first, like Chaucer's little son, learn how to tell the time with an astrolabe.[10]

The first and most important thing to remember is that this is a mobile device. That may be easy to ignore when we are most likely to see them fixed in a glass case, with their parts frozen in position (image 4.4). But these objects were frequently carried from place to place, and all their parts were made to move too. Just as the celestial spheres are in constant motion, so were the components of the astrolabe. It modelled the two most important motions of the heavens: the daily cycle of day and night and the annual passage of the Sun along the ecliptic.

The astrolabe pictured below has not moved very far during the seven hundred years since it was crafted. It is now at the Whipple Museum of the History of Science in Cambridge but was most likely made for the skies above Norwich, the prosperous trading city sixty miles north-east across the marshes and woodland of East Anglia. Engraved on the solid background is a grid dividing the heavens (images 4.5, 4.6). Lines of azimuth and altitude criss-cross the sky, laying down points of reference like the coordinates or contours on a map. Over those imaginary gridlines move the stars, always fixed in relation to each other (image 4.7). The stars often fall below our horizon, and the astrolabe shows that part of the sky too. There, where we cannot see, we have no need for circles of altitude – known

4.4. An English astrolabe (*c*.1340), fixed in place on a Perspex stand (diameter: 295 mm).

as *almucantars*, revealing the instrument's Arabic heritage – or the curved lines of azimuth. So, below the horizon, the plate is emptier. Many makers, including this one, engraved the space with curves to convert between equal and seasonal hours.

The circular lattice in image 4.7 is called the *rete*, named for its similarity to a net. (English retains the word 'reticulated' for the net-like colour patterns of some animals, such as giraffes or pythons.) Each of its sinuously curving pointers represents a notable star. Now, take this slim, delicately crafted rete and place it over the solid plate. Notice how the holes in the centre line up perfectly.

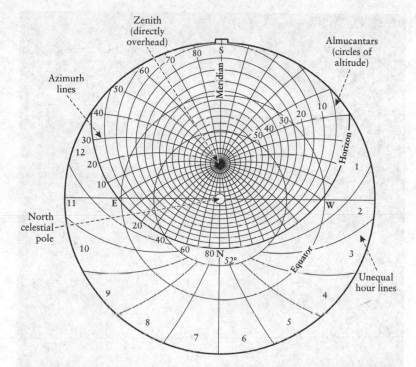

Zenith (directly overhead)

Almucantars (circles of altitude)

Azimuth lines

North celestial pole

Unequal hour lines

4.5. An astrolabe plate for 52 degrees latitude. Perhaps surprisingly for modern users, south is at the top, east on the left. Many astrolabes have a few separate plates which can be slotted in as needed (the *tang* sticking out at the top helps it fit snugly in the astrolabe). The Whipple astrolabe does not have such separate plates: these lines are engraved directly on to the main part of the instrument.

A snugly fitting pin through those holes will ensure that the rete rotates smoothly, just as the stars rotate around the north pole, and some lead inlaid at the back of the bottom makes it evenly weighted. Now turn the rete clockwise. Don't be afraid: these instruments were made to be handled – 'thou maist turnen [it] up and doun as thyself liketh,' said Chaucer.[11] This one is a few decades older than John Westwyk, and if it has survived since the early fourteenth century with only minor repairs, it must be reasonably sturdy. Pick a star – perhaps the pointer that curves almost into a circle just behind that watchful bird on the right-hand side (image 4.8). Its name,

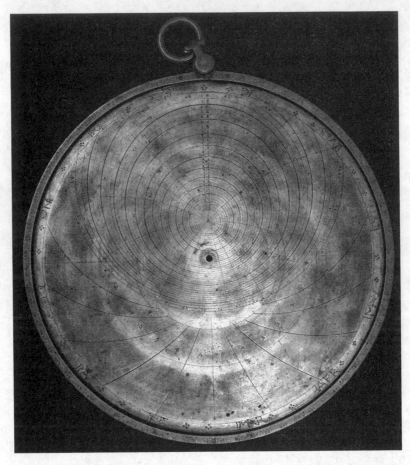

4.6. A plate with scales of azimuth and altitude for latitude 52 degrees (suitable for Norwich) engraved directly into the 'womb' of the Whipple Museum astrolabe. The hole in the middle is the north celestial pole. Slightly above it, curved lines of azimuth meet at the zenith, directly above the observer's head. Rippling outwards from the zenith are (not quite concentric) circles of altitude (*almucantars*). On this astrolabe they are marked every two degrees, with a thicker line every six degrees. The outermost arc is zero degrees, i.e. the horizon. In the much emptier space below the horizon are curves for reading time in unequal hours. They spread outwards from the tropic of Cancer and cross the celestial equator.

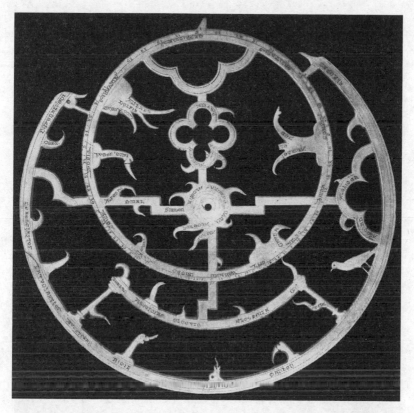

4.7. Rete of the Whipple astrolabe. Each curving thorn indicates a notable star. The hole in the middle is the north celestial pole. The largish, off-centre ring at the top represents the ecliptic, and the 90-degree arc below it is a segment of the celestial equator. Only one star pointer has broken in the last seven hundred years: the middle one of the triple pointer sprouting from the right-hand side of the ecliptic ring.

engraved in neat Lombardic letters, is Algorab. The Arabic word *al ghurab* means 'crow', for here we are in the constellation Corvus, even though that crow-like bird is facing away from it. In image 4.4 Algorab is already a little below the horizon, where the closely spaced almucantars give way to emptier space, and as you turn the rete, it will descend further. Keep turning clockwise, though, and it will start to rise, until it breasts the horizon on the left-hand side

4.8. The sharply curved pointer for Algorab, behind an attractive bird which points somewhat away from the constellation Corvus. At the top of the image is a short arc of the ecliptic, with Virgo and Libra visible.

of the instrument. That left-hand side is east, for all stars, like the Sun, rise towards the east. The top of the instrument is therefore the south, which confuses some people when they first handle an astrolabe today; but medieval users were apparently quite comfortable changing their perspective.

Turning the rete as you just have, more than twelve hours have passed before that first glimpse of Algorab. A full rotation of the rete is twenty-four hours, double that of today's clocks: a full revolution of the heavenly spheres or – if you must be modern – a full spin of the Earth on its axis. In practical terms, it makes no difference which way you see it. The astrolabe works equally well whether the

Sun goes round the Earth or the Earth around the Sun, since it only measures angles.

Either way, unlike the monks, who would wait in the pre-dawn cold for a key constellation to crest over the cloister, we are not accustomed to watching the stars for so long. In the time we have spent turning Algorab, we have seen many rise and set. The thorn on the end of the outer circle, for example, is labelled 'Alacrab'. In John Westwyk's day, it was also known as Cor Scorpionis, Scorpio's heart. But the maker of the Whipple astrolabe preferred to abbreviate the Arabic translation, Qalb al-'Aqrab. Now called Antares, it is the fifteenth-brightest star in the sky. We have watched it fall from view in the west. A little later, we have seen the brightest star of all rise. That is Sirius, in the mouth of the constellation Canis Major. Its pointer is right at the bottom of the rete, carved into the shape of a dog's head (image 4.9, plate section). Its long, outstretched tongue marks the Dog Star's precise position. Once again, the English maker of the Whipple astrolabe used the Arabic name, Alhabor.

The forty-one stars on this astrolabe would have been familiar to any astronomer. John Westwyk himself wrote out an almost identical list of stars fifty years after the Whipple astrolabe was made. He gave both the Arabic and Latin names: *algorah* and *coruus*; *calbalacrah* and *cor scorpionis*. Although Ptolemy listed more than a thousand in his *Almagest*, the same few dozen, with minor variations, appear again and again on short lists and instruments like these.[12]

As we rotate the rete over the grid of altitude and azimuth lines, we can predict where on the horizon a star will rise and the highest point in the sky it will reach. At that moment of culmination, the star crosses the meridian line running vertically down the top half of the astrolabe, before descending through the almucantars on the right-hand – western – side.

Ascending from east towards south, descending on the right-hand side . . . this all relied on an idea we have already encountered: that a sphere can be projected, or squashed down, on to a flat surface. Indeed, one Syrian scholar, writing around 1270, tells how such a squashing led to the instrument's invention. Ptolemy, out riding a donkey one day in the mid-second century, had dropped the armillary sphere he was carrying. The donkey trod on it and squashed it flat. And thus was born the astrolabe.[13]

That fanciful tale contains a golden grain of truth: Ptolemy did develop new techniques of projection. In his *Geography* he evaluated and extended earlier ways of showing the spherical globe – or at least that portion of it that was thought to be inhabited – on a flat map. In his *Planisphere* he did the same for the heavens, explaining the principles of *stereographic projection* devised by earlier astronomers like Hipparchus.[14] The stereographic method had two great advantages for instrument-makers and astronomers. First, for the instrument-makers: when you squashed a sphere down to an astrolabe plate, circles on the sphere remained circles, easily engraved on the astrolabe. And second, for the astronomers, angles observed in the heavens were the same on the astrolabe.

How did stereographic projection work? Imagine an impossible observer out at the south pole of the heavens, looking 'upwards' towards Polaris at the north celestial pole (image 4.10). She can see everything in the heavens. She is not concerned by how near or far each item is, only their angles – how close they are, from her perspective, to the vertical line up to the north pole. As she looks up, she traces each important line on a horizontal sheet of glass that stretches right across her field of view. This flat glass ceiling is the plane of projection. In this method, it is the celestial equator. Every celestial circle closer to her than the equator will seem larger than the equator, so she will trace them outside it. Everything north of the celestial equator will seem smaller and be traced inside it. As a boundary for her chart, she chooses the southern tropic of Capricorn. That will be the largest, outermost circle. The two tropic circles and the equator all lie flat across her field of view, so will be centred on the north pole. The ecliptic, at an angle to the equator, will be off-centre on her chart. She draws it as a circle that just touches the two tropics. The two places where the ecliptic line crosses the celestial equator are the two equinoxes. As with the ecliptic, we can draw the almucantars

4.10. Stereographic projection. Circles on the sphere remain circles on the astrolabe. The ecliptic is a circle passing through the solstices and equinoxes.

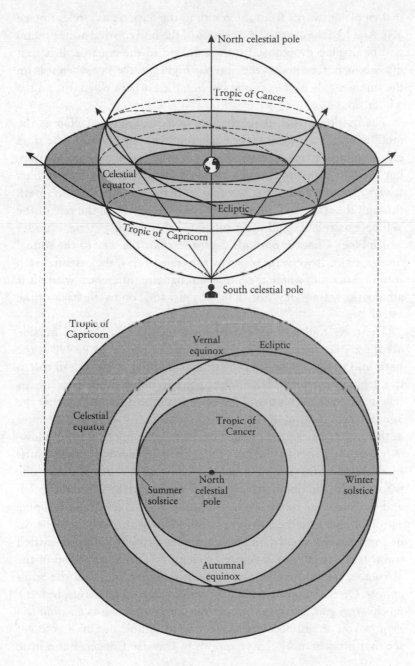

that ripple outwards from the zenith to the horizon as circles (image 4.5). And like the ecliptic, the centres of the horizon and almucantars will be displaced because they are angled to the equator. It is that displacement that makes the Sun rise higher in the sky when it is on the summer side of the ecliptic but hug the horizon when it is on the winter side.

The motion of the astrolabe rete over the engraved plate is the motion of the stars against the background of our horizon. The stars stay resolutely fixed in relation to each other, marching in lockstep, but the Sun wanders through them on its way around the zodiac. It was essential, then, for the rete to mark the Sun's annual path through the stars. Indeed, the only complete circle on the rete of the Whipple astrolabe (and most others) is that ecliptic circle – 'under which lyne', Chaucer reminds us, 'is evermor the wey of the sunne' (image 4.7).[15] If we want to know the time of day, the essential step is to find out just where the Sun is on that circumference: where it is among the stars today. For that, we must turn on to the back of the astrolabe.

There we find a double calendar (image 4.11). On the outside are the zodiac signs, with Sagittarius and Capricornus visible at the bottom. On the inside are the familiar months of the Julian calendar, from Ianuarius to December. The gap between the two reveals how the maker of this astrolabe dealt with the difficult fact that the Sun moves at a variable rate through the stars. As we saw in Chapter 2, this is today largely explained by the Earth's elliptical orbit. However, astronomers since before Ptolemy had achieved excellent results by thinking of the Sun's annual motion as an *eccentric* circle – one whose centre did not match the centre of the Earth. When the Earth and the Sun were furthest apart, in June, the maker of the Whipple astrolabe could see the Sun moving more slowly through the signs. So he engraved the months on an eccentric circle, with its centre shifted towards the twenty-fifth degree of Gemini. There, at the top of the astrolabe, the two circles were closest together, marking the Sun's *apogee*. On the opposite side, at its *perigee* above Sagittarius, he left a much larger gap. Thus, as you turn the pointer, known as an *alidade* – by now you do not need me to mention its Arabic origins – you will see that it takes more than a month to traverse Cancer, but a little

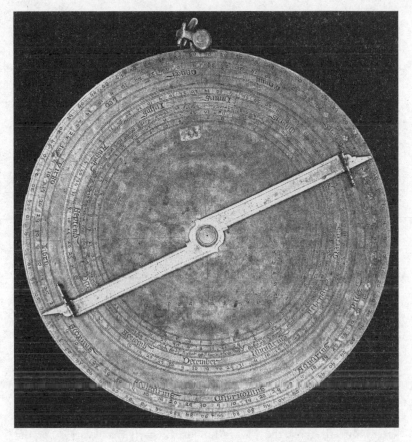

4.11. Back of the astrolabe. On the outside are the zodiac signs. Inside are the calendar months, with the centre displaced towards the end of Gemini (at the top) to account for the eccentricity of the Sun's motion; note the larger gap between the two calendars at the bottom). Inside the inner circle is a lightly traced shadow square, useful for surveying. In this image, the alidade is turned to an altitude of 22 degrees above the horizon.

less for Sagittarius. That is why the northern hemisphere summer – or at least the period between the spring and autumn equinoxes – is longer than the winter.

The names of Julius Caesar's months are very familiar to us. But they were much less relevant to John Westwyk. As we saw in that tiny pocket-book of watchman's instructions, monks preferred

4.12. Detail of the astrolabe calendar. Above the name of the month (March) there are three feast days: *cedde* (Chad), on 2 March, with ferial letter E directly above the 'M' of Marcius; Gregor[y the Great] (A) on the 12th; and the Annunciation of Mary (G), on 25 March, using the standard medieval form of '5'. Beneath is the start of Aries. It lines up with the end of 12 March, the date of the vernal equinox when this astrolabe was made.

to mark the passing days in feasts. No surprise, then, to find that within each month on the astrolabe are marked three to five holy days (image 4.12). This is where the craftsman had a chance to personalise his astrolabe. The stars on the rete, as we saw, had come from a standard list. Likewise, some of the days in the calendar, such as Epiphany, the feast of the Annunciation of Gabriel to Mary and – exactly nine months later – Christmas Day, were obligatory subjects for the engraver's burin. But that still left space for some personal choice, either for the customer who commissioned the instrument or the maker himself. Margaret of Antioch and Clement of Rome, for example, are hardly household names. In East Anglia, though, where this astrolabe was made, these saints had a devoted following. Some notable local churches were dedicated to them, and so are two spaces on this scientific instrument. The inclusion of St George, promoted as a patron saint of England by Edward III around this time, may be a mark of the maker's patriotism. He also included the martyred Archbishop of Canterbury Thomas Becket and a sprinkling of other English saints, such as Chad, the seventh-century bishop of the Mercians, on 2 March. Above the name of each feast we find its date, and the ferial letter A to G that allows us to identify which day of the week it will be.

Using this double calendar, we can easily find the position of the Sun on its slow journey around the ecliptic. All we have to do is turn the alidade so its pointer lines up with today's date, and read the Sun's celestial longitude off the outer zodiac calendar. For instance, on the feast of the Annunciation (25 March), the Sun was just coming up to the twelfth degree of Aries.

Now we know where to find the Sun on the ecliptic ring of the rete. John Wostwyk would have marked it in some way – perhaps with a little inked line. Or he could have taken a blob of melted beeswax from one of the candles in the writing room, rubbed it between his stiff fingers to make a soft little ball, and stuck it in the Sun's place on the ecliptic ring. The Sun moves only one degree each day, so that one mark will do for today. We can rub or scrape it off and make a new mark tomorrow.[16]

Before we turn to the front of the astrolabe, though, we need to do one more thing. Telling the time means setting the astrolabe to match the sky at this moment. We can do that by finding the altitude of anything identifiable. Any star on the rete will do, but we will use the Sun itself. Now, finally, is the moment to pick up the astrolabe. Astrolabes were meant to be easily portable, but this one is somewhat larger than average – about twelve inches in diameter, roughly the width of an open hardback book – so you might want to use both hands. (Or, like some medieval astronomers, you could hang it from a tripod.) Grasp the astrolabe firmly by the suspension ring at the top and let it hang vertically from your stronger hand. Use your other hand to steady it. Feel its reassuring solidity and take in its distinctive metallic smell. Turn it so it is edge on to the Sun and start to move the alidade slowly upwards to point at that day-star. Attached to the alidade is a set of sights. Adjust the alidade, millimetre by millimetre. Move it 'up and doun', says Chaucer, 'til that the streams of the Sunne shine through bothe holes' – or, looking at it another way, when the notched shadow of the upper sight falls precisely on the lower sight.[17] (If you are observing a star, you can hold the astrolabe high and look upwards through the sights, but trying to take a fine measurement while staring at the Sun is both difficult and unhealthy.) When you are happy that you have lined up the alidade perfectly, you can read its angle from the scale on the rim. In image 4.11 we can read that the alidade is 22 degrees above the horizontal.

Now all we have to do is turn back to the front and set the Sun's position for this moment. If today is the feast of the Annunciation, that means turning the rete until the twelfth degree of Aries, marked with our blob of wax, sits atop the 22-degree almucantar (altitude circle). It can do so in two places: one on the left, the east side, as the Sun rises towards the meridian, and one on the right as the Sun begins to fall in the west. To choose which is correct, we need only know whether it is morning or afternoon. For this, Chaucer's instructions remain as clear as the day John Westwyk read them: 'sette the degre of the sunne, in [the] case that it be *beforn* the middel of the day, amonge thine almykanteras on the *est* syde of thine astrelabie. And if it be *after* the middel of the day, sette the degree of thy sunne upon the *west* syde'.[18]

This, then, is the configuration of the heavens at this moment. All the stars are in place, with the sign of Virgo in the astrologically significant *ascendant* position. You could point out that Denebola has just risen, or that Betelgeuse (named 'Elgeuze' on this astrolabe) is about to cross the meridian. We can use any star to tell the time – we simply have to measure its altitude and position the rete accordingly. Whichever we use, our final step is to align the rule on the front with the Sun's position in the ecliptic. The time can then be read where the rule points like the hand of a clock to the raised edge of the astrolabe. This edge is known as the *limb*. Many astrolabe-makers made it easier to tell the time by engraving the limb with numbers for the hours, like numbers around a clock face today. The limb of our astrolabe is divided into 360. Helpfully, the maker punched little circular marks every 15 degrees, to remind us that the Sun travels 15 degrees every hour. So when we see that the rule points to 60 degrees, the fourth hour-marker round from the top, we know it is four o'clock in the afternoon.

How accurate is that time? The accuracy of astrolabes was limited by three factors. First, there were the simplifications in the basic design, like the fact that they rarely take account of leap years and gradually become less accurate as the precession of the equinoxes shifts the stars. Second, there were limits to their precision. Precision was not just limited by the craftsman's ability to engrave the scales perfectly; it would also be impaired if the astrolabe was not used at the latitude for which it was engraved. The third factor, of course, was user error. The first of these made little difference – that's why astrolabe-makers so rarely worried about it. As for the instrument's own precision, the scales on a well-made astrolabe like ours can easily be read to the nearest degree. Assuming the maker had divided the degrees reasonably equally, that gives the time to the nearest five minutes or so. Using the astrolabe at a slightly incorrect latitude might throw out your result by five or ten minutes more. User error, though, could make a much bigger difference. Misreading the altitude of the Sun or a star by just a few degrees could throw out your time by a quarter of an hour – and more if the star was near the meridian, when its altitude changed most slowly. 'I warne thee for evere', Chaucer admonishes, 'make thou nevere bold to take a just ascendant by thine Astrelabie, or else to set a clokke, whan eny celestial body [which you are measuring] be nigh the southe lyne. For sothly thou shalt err.'[19]

In that process of ascertaining the time with our astrolabe, we have encountered all its core components – and hinted at a huge number of potential functions. The rete of fixed stars with its ecliptic circle for the Sun, the background horizon grid, the rule for use in telling the time, the suspension ring, the double calendar of months and zodiac signs, the alidade with its sights for measuring altitudes – together, they offered dozens of uses. As Chaucer mentioned, you could set a clock – which at this time was not nearly as accurate as an astrolabe – or take the astrological ascendant. You could find the direction of north, the height of a building, the length of daylight, and much else. All of these components, and uses, would have been familiar to John Westwyk. And all were contained within a package that could comfortably hang from the belt of his black habit (monastic robes, alas, did not have pockets – though astrolabes were often stored in leather cases or fabric bags). Mobile data, it is clear, is nothing new.

Beyond the basic concept of a planisphere – the celestial sphere made conveniently flat – astrolabes could vary enormously in design. The flexibility of this instrument was certainly part of its appeal. The changes that craftsmen made tell us about their varied aims and the changing roles of instruments in astronomy and timekeeping in different parts of the world over more than a thousand years.

In second-century Alexandria, Ptolemy had described the concept of a planisphere, but his astrolabe was only a model of the moving heavens, not a measuring device. Its functions were primarily educational and symbolic, not observational. Within two centuries, though, craftsmen added the alidade for measuring altitudes and the astrolabe was complete. It was described by another Alexandrian astronomer, Theon, in the late fourth century. Only his table of contents survives, but soon instruction manuals for the construction and use of the instrument began to proliferate. From Greek, they were translated and adapted into Syriac and Arabic, and later Latin and Sanskrit as the instrument spread in all directions.

Writing in Sanskrit in the 1420s, the Hindu scholar Ramacandra Vajapeyin praised the astrolabe above all other scientific instruments. 'When you know the astrolabe well,' he enthused, 'you will know the universe like a fruit on the palm of your hand.'[20] Forty years earlier on the other side of the world, Geoffrey Chaucer expressed similar sentiments in English. His *Treatise on the Astrolabe* was an introduction to the mathematics of the heavens, as comprehensive as Sacrobosco's *Sphere*. It was written, he claimed, for a ten-year-old boy desperate to get his hands on – and head around – this gadget. 'Little Lewis my son,' he began, 'I realise your earnest desire to learn the theory of the astrolabe.'[21] But 'Little Lewis' may well have been a marketing tool. The title *Treatise on the Astrolabe* is a bland modern convention: the work's first title, given either by Chaucer himself or an early copyist, was 'Bread and Milk for Children' – equivalent to emblazoning 'For Dummies' on the cover. The multifunctionality of the astrolabe made it ideal as a starting point for science, and Chaucer's treatise was widely copied, including in monasteries.[22]

The instrument's symbolic value was almost as important as its practical utility. Rumi, the thirteenth-century Persian poet, wrote that 'love is the astrolabe of God's mysteries'. The astrolabe was a key to understanding – understanding both God and yourself. If Nature was a book which, like Scripture, contained clues to the divine plan, and if the world sphere, as Sacrobosco said, was a machine, then in the intricate movements of a man-made celestial machine you could find clues to the craftsmanship in Creation – a window into the mind of God. Moreover, studying the astrolabe could help you to find your place in the world, not merely geographically, but existentially too. Chaucer concluded the prologue to his *Treatise* by alluding to the moral necessity of finding one's correct position in the social order. 'God save the king,' he wrote, 'and all who obey him, each in his *degree*.' This political statement is no passing metaphor. Medieval science, we must remember, was not artificially separated from subjects that dealt with moral questions. Degrees of altitude and social status shaded subtly into one another. So Chaucer's treatise was far from just a specific instrument manual. It was not even just a general astronomy textbook. It was part of an all-round education.[23]

As the components of an ideal education varied from time to time and place to place, so did the components of astrolabes. Some astrologically inclined craftsmen, for example, found it convenient to engrave specialised astrological markings to make their calculations easier.[24] Other makers, meanwhile, were in no particular rush to tell the time, and removed the rule from their instruments. Without the rule, timekeeping was a little more laborious, but still perfectly possible, and the rete was left unobscured for better mapping the stars. In Islamic cultures, some craftsmen added sets of curves, making it easier for Muslims to pray at the Quranically prescribed times. Others might mark the direction of Mecca from a few different cities. Meanwhile, instrument-makers in the Latin West commonly filled some of the space inside their calendars with a shadow square, which made it simple for surveyors to compute the height of a building when they knew their distance from it, or vice versa (image 4.11).

The most common variation of all, though, was to include extra horizon grids to represent the sky at different latitudes. As we saw in Chapter 1, the altitude of the Pole Star above the horizon is equal to your latitude. Since the pole – around which all stars rotate – must always be in the centre of the astrolabe, if we want to take our astrolabe to a new latitude we must equip it with a new horizon. The Whipple astrolabe, which we have been working with, is rare in having only one grid of almucantars, engraved directly on to the astrolabe. Perhaps its first owner did not plan on taking it travelling, especially since it was substantially larger than average. But the vast majority of surviving astrolabes have the space within the raised limb filled with at least a couple of brass plates – and more likely three or four. The main body of the astrolabe then became known by a maternal metaphor as the *mater*, or mother, a container 'that receiveth in her wombe the thinne plates', as Chaucer told 'Little Lewis'.[25] Each plate would be engraved on both sides with grids for different latitudes. Thus the mobile medieval scholars could use their portable computers wherever they went. All they had to do was remove the central pin, take off the rule, swap over the plates and reassemble the device.

Besides being simply useful, all these variations were a way for craftsmen to display their virtuosity. One instrument-maker a generation after John Westwyk, a Parisian canon named Jean Fusoris, boasted that he could not only make astrolabes which accommodated the distortions of the four-year leap cycle but also knew how to compensate for the excesses of the Julian calendar. His marketing genius won him the business of King Henry V, but such dealings with France's enemy in the midst of the Hundred Years War got him into serious trouble, as we shall see in Chapter 6. Still, even if makers – none of whom were as ambitious, nor as prolific, as Fusoris – did not get themselves into such trouble, they still had to face the serious difficulties of manufacturing their instruments with satisfactory precision. To take only the most basic problem as an example: how could they ensure that the calendars were accurately divided into 360 degrees, and 365 days, with each division exactly equal? That was a question that continued to vex instrument-makers right up to the Industrial Revolution.[26] The most popular medieval manual instructed makers to mark off fifteen days between the start of December and the winter solstice. This left 350 days, to be split into seven arcs of fifty days, each then subdivided into five segments of ten days, and each segment halved. That was not easy. No wonder craftsmen sometimes took shortcuts. It was not unheard of, for example, to divide the calendar into twelve equal months, squeezing thirty-one days into thirty where necessary and stretching February to fill the same space. In any case, as Chaucer pointedly reminded 'Little Lewis', an astrolabe could never give predictions as precise as the ones provided by carefully computed tables, which all astronomers had to learn to use.[27] It was never intended as a tool for new scientific discoveries or testing theories but was rather a device for modelling and simplification, labour-saving and convenience.

It was also a design classic. Astrolabes featured the latest trends in architecture and the decorative arts, from Gothic tracery to tulips. The four-leaf design in the middle of our astrolabe, for example, was a ubiquitous symbolic motif. Representing the Christian cross or good luck, it was often painted in manuscripts or built into church window frames. John Westwyk's astrolabe may not have been so beautiful: we must beware of assuming that all medieval objects featured craftsmanship as impressive as the ones proudly displayed in today's museums. These may have survived only because they were precious, kept carefully and not in daily use. Although hundreds are extant (the first person to attempt a complete list of them was Derek Price in 1955), they were surely far outnumbered by functional devices made of parchment or wood.[28] Still, the number of ornate brass astrolabes that survive shows that very many of these instruments were elaborately decorated. They signalled their owners' education, good taste and high status.

Nevertheless, even the best gadgets have their limitations. For example, although an astrolabe could measure the altitude of the

Moon or planets, it could not reliably find their celestial longitudes or latitudes, or predict eclipses. The rete, remember, represented the fixed stars; it would be too unwieldy to show the paths of the planets, which wandered either side of the Sun's ecliptic line. There were instruments, though, that could do all of that. We find one in the hands of John Westwyk as he sat in the St Albans writing room, feeling its weight, measuring its width and marvelling at the mind of its inventor.

A generation after John Westwyk, the abbot of St Albans wrote a short history of invention. John of Wheathampstead was a proud patron of scholarship. He endowed the library at Gloucester College and commissioned the stained-glass cloister windows with their portraits of illustrious men of learning. His history of invention was part of a four-volume, fully alphabetised encyclopaedia covering the achievements of great men in all fields. He called it *The Granary*, apparently as a pun on his own name.[29]

It is sometimes said that the Middle Ages abhorred novelty, but there is no sense of that in *The Granary*. Under the heading 'Invention', Wheathampstead tells the story of everything from fire to trousers. The latter, he says, were invented by the legendary Assyrian queen Semiramis so that no one could identify her sex.[30] His account drew on the standard encyclopaedic authorities read by all scholars in his day, like Isidore of Seville, but he also incorporated his own research. This is particularly apparent in the section on scientific instruments. There Wheathampstead poured scorn on the common myth that the astrolabe was invented by the Egyptian king Ptolemy, pointing out that the astronomer who wrote the *Almagest* was not the same man at all. He also noted more recent inventions by the Muslim astronomers al-Battani and al-Zarqali – whom he knew as Albategni and Arzachel – and the Provençal Jew Profatius (Jacob ben Machir ibn Tibbon), as well as other scholars in France and Italy. Closer to home, he highlighted the achievements of a Glastonbury monk who, he said, had devised a ship-shaped sundial. And, he said, a wall-mounted dial which could tell the time in equal hours was invented by a certain monk of St Albans (and contemporary of John Westwyk) named Robert Stickford. Wheathampstead may have heard about

Stickford's astronomical exploits through his uncle, who had been at the abbey at the same time.[31]

Among all these illustrious names, one stood out: 'a man so well educated in the art of astronomy', adjudged Wheathampstead, 'that from his era right up to the present day no Englishman has arisen to match him'. This was Richard of Wallingford. And the pinnacle of Wallingford's achievements? It was not his astronomical clock, though Wheathampstead did acknowledge that in passing, together with the rectangulus. It was the Albion, he said, the planetary super-computer 'which, one reads, contains in itself the functions of all the other instruments'. Wheathampstead's source, which he was quoting almost verbatim, was the description written a century earlier by Richard of Wallingford himself.[32]

The name Albion was a masterstroke of branding. If the monks understood nothing else of Wallingford's instrument – if, indeed, they had not even read the complex treatise – they were convinced of its multifunctional power by its name alone. The St Albans chronicler spelled it out: it was short for 'all-by-one'.[33] It also alluded, of course, to the name of the protomartyr Alban. And it carried more than a hint of patriotism. Every educated Englishman of the day knew the old national myth: a band of Trojan exiles had followed a prophetic vision to a beautiful and fertile island called Albion. After defeating the giants who ruled it, the exiles had renamed the island Britain, after their leader, Brutus. The St Albans library possessed a well-thumbed copy of the tale in its quintessential telling by Geoffrey of Monmouth.[34] Geoffrey never explained why the island was originally called Albion – but a new prequel to the story, which became popular just at the time Richard of Wallingford was working, did exactly that. In this prequel, a Greek – or possibly Syrian – princess named Albina plots with her twenty-nine younger sisters to kill the husbands they have been forced to marry. When the youngest sister – who, unlike the others, truly loves her husband – exposes the plot, the other sisters are rounded up and set adrift on a rudderless ship. At the mercy of the waves, the ship eventually runs aground on an apparently uninhabited island, which they name Albion after the oldest sister. They live happily without men, but copulate with spirits in their sleep, spawning a race of giants who rule Albion until the arrival of Brutus.[35]

The St Albans monks avidly read these stories, woven through with biblical allusions and allegories. They would have been familiar, for example, with the story in Genesis that fallen angels had mated with human women, begetting a race of giants, which God had wiped out in the Flood. They had seen the blue-skinned giants sprawled up the margins of some manuscripts in their library. And they must have appreciated the detail in Geoffrey's version that the final giant of Albion was called Gogmagog, for the nations forming Satan's armies in the Book of Revelation were named Gog and Magog.[36]

Richard of Wallingford, too, may well have had giants on his mind as he composed his *Albion* treatise. Its principal source texts had also drifted to England from Greece and Syria: Euclid, Ptolemy and al-Battani. Soon after Richard completed it in 1327, abbot Hugh of St Albans died. Richard was visiting St Albans at the time, and some monks whispered that he had used astrology to prophesy the abbot's death. On 29 October, the day when a new abbot was to be elected, a Mass was held in honour of St Alban. Richard was invited to preach. He chose as the theme of his sermon the challenge issued by the giant Goliath to the Israelites: 'choose a man from among you'. Richard clearly saw himself as the biblical hero David, who accepts the giant's challenge. He had already taken on gigantic scientific challenges in improving the instruments of his towering predecessors. Now he felt ready to tackle the even greater task of returning prosperity to an indebted and embattled abbey.[37]

The new abbot was elected, as normal, by a delegation of senior monks. It was conventional for the chosen candidate to hesitate for a short while – genuinely or not – before accepting the enormous responsibility laid down by the Benedictine Rule. When Richard was duly appointed, he consented rather quickly. Although he claimed to be perplexed by the election result, some monks suggested scepti-cally that even his brief hesitation was 'more feigned than from the heart'. Richard was keen to offer an excuse for his haste, as the abbey chronicler recalls:

> Brother Richard of Wallingford himself would frequently say that when he celebrated – not without tears – the Mass that day, in honour of Saints Alban and Amphibalus, he had a feeling of such faith in a

peaceful election, to the honour of God and His Church. He kept that firmly in his heart, and resolved that whoever might be chosen that day – chosen by God for His Church – he would accept (so far as was in his power) without any argument whatsoever. And because he felt this faith, when he himself was later elected he consented more readily, for fear of disrespecting the Holy Spirit.[38]

Richard's first duty was to visit the papal court at Avignon, to have his appointment confirmed. But as soon as he had completed that arduous – and expensive – process, he began the task of reforming the morals and finances of the monastery. His predecessor had, over the previous eighteen years, neglected to collect rents, allowed the monastery buildings to fall into disrepair, and turned a blind eye to shocking behaviour by the brothers. On one occasion, when Queen Isabella came to visit, the townswomen of St Albans had protested with babies at their breasts, telling the queen they had been raped by monks. Only the queen's inability to understand English had allowed the brothers to cover up the scandal. Richard imposed penance on those found guilty of 'sins of the flesh'. However, his judgement on the senior monks who had not paid the tithes they owed the abbey treasury was far harsher. They were sacked from their senior roles within the monastery, removed from their privileged places in the church and dining hall, sentenced to perpetual silence, excommunicated in writing, and subjected to twice-weekly corporal punishment. In practice, Richard probably never intended to carry out that severe sentence, and a group of older monks quickly persuaded him to amend it. He ordered the offending obedientaries to do secret penance instead. But the damage to relations within the community was done. 'From that day', the chronicler reported breathlessly, 'some false brothers began to collude – in fact, it would be more accurately called a conspiracy – against the Abbot.' The conspirators plotted to depose him from the abbacy, claiming concern for his health. For his leprosy was already apparent.[39]

It might seem surprising that a diagnosis of leprosy did not end Richard's career. Lepers were indeed often isolated from medieval society, for fear of contagion, but they were treated with genuine compassion. Christ, after all, had associated with lepers. Following

His example, St Albans abbey ran two houses for men and women – mainly local monks and nuns – afflicted with the disease (or other conditions with similar symptoms). And if leprosy was often seen as a divine punishment for sinners, it could also be viewed the opposite way: as an earthly purgatory permitting the most devout to pass straight to heaven. Either way, there was never any suggestion that Richard himself would be confined to an institution. 'Such was his sanctity and *science*', declared the chronicler Thomas Walsingham, 'that no-one, whether resident or visitor, shunned his table or his company.'[40] Walsingham's Latin word *scientia* means a broad range of knowledge, but there is no question that the greatest respect was reserved for Richard of Wallingford's expertise in the mathematical arts.

John Westwyk would have heard all these stories, swirling among the monks from half a century earlier. Walsingham wrote up his chronicle in the newly constructed scriptorium after Westwyk left St Albans, but he was likely already compiling materials when Westwyk was sitting near him in the cramped old writing room. Westwyk would have heard how Wallingford sold a cache of precious library books to that bibliophile bishop Richard of Bury, perhaps to buy influence at court. He would have heard how the visiting King Edward III criticised Wallingford's lavish spending on his clock, when cloister walls still lay in rubble – and how Wallingford had an immediate riposte to the king (spoken 'with all due respect', the chronicler reassures us): that his successors as abbot could repair the fabric of the monastery, but none of his successors could complete this project he had begun. Above all, Westwyk would have heard of Richard's scholarly works: his commentary on the Rule of St Benedict; his compilation of the provincial Benedictine constitutions; the many new books and instruments of arts and sciences that 'he wrote, compiled, taught and invented'. Westwyk would have heard how Richard frequently expressed pious regret that his scientific pursuits had been a distraction from theology – yet Westwyk must also have noted that Richard's rapid advancement to abbot, and his later reputation, both rested on his science. In his portrait in the abbey chronicle the leprosy-scarred blacksmith's son Richard appears in action, his abbot's mitre tucked under a table, engraving an instrument that looks much like the Albion. And when John turned the pages of the *Albion* treatise itself, he would have read Richard's devout hope that 'its design could direct the minds of many people to higher things'.[41]

The chronicler boasted that Richard's inventions were 'unheard of before his time' but, as Westwyk was well aware, an important function of the Albion instrument was to bring together and refine the attributes of earlier instruments. That, after all, was what the name All-by-one referred to. To improve instruments, to refine their functions and make astronomical calculations more straightforward, was a primary goal of scientific thinkers like Richard. They wanted to enhance the computing power of their inventions, but they also wanted to make the best possible model of God's creation. If the world could be made predictable, understandable, then to improve the instruments that replicated its workings was to imitate God. The big difference, of course, was that God had created the world machine from nothing, whereas inventors like Richard stood on the shoulders of giants (to use a metaphor coined in the twelfth-century cathedral schools) and gave their predecessors due credit.[42]

So when John came to make a new copy of the *Albion* treatise, he added two pages of his own commentary about the relationship between Richard's compendious invention and some of the older instruments it incorporated. The first of these was the *saphea* of Arzachel – or, to give them both their full names, the *safiha al-shakkaziyya* of Abu Ishaq Ibrahim al-Zarqali. Al-Zarqali ('the blue-eyed one') worked in Muslim Al-Andalus in the late eleventh century: first in Toledo and later, when the city was threatened by the forces of Christian Castile, in Córdoba. He was a prolific astronomer, compiling user-friendly tables and developing new theories to account for long-term changes in the motions of the Sun and stars. Abbot John Wheathampstead's *Granary* mistakenly gave him credit for the cylinder dial, but he did invent several new instruments. His saphea came in two versions, both based around a universal astrolabe plate. Why the simpler of the two was called *shakkaziyya* is a mystery, but it was probably a corruption of the word for 'herbalist', after the Toledo apothecary believed to have invented its universal projection.[43]

The saphea was universal because – unlike normal astrolabes – it could be used at any latitude. The plate was engraved with both equatorial and ecliptic coordinates, so it could be used to convert stellar positions between them. Despite its versatility, it was never as popular as the standard planispheric astrolabe, since it was simply too complex for most makers and users. It was not too complex for Richard of Wallingford, who acknowledged his debt to al-Zarqali's design when he added it into his Albion. But having the highest expectations of his readers, he felt no need to explain that part of his multifunctional instrument. 'Anyone who wants to learn the art of constructing the saphea can easily do so,' he remarked dismissively, 'so there is no point spending any longer describing these things.' As for how to use it, he added off-hand, 'the saphea has its own treatise'.[44]

John Westwyk did not think its functions were so self-explanatory. Showing the same spirit of charity that had led him to complete the *Rectangulus* diagrams, he attempted to fill in the gap Richard had left. On a blank page at the end of the *Albion* treatise he wrote an explanation in simple but clear Latin. '*Quantum ad sapheam*' ('Concerning the saphea'), he began, with the extra-large initial Q in red ink, its tail curling down six lines of the margin. He described the main markings on the Albion's saphea plate, noting that its outer circle was the meridian – unlike on a normal astrolabe, where the meridian is the straight line running down the middle – and pointing out the two equinoxes, Aries and Libra, sharing pride of place at the centre of Arzachel's projection. Then, with another curling red capital Q, he did the same for the astrolabe. He remarked on the differences between the Albion's astrolabe plate and the design he was used to. Some were obvious, such as the inclusion of a zodiac band on the plate itself – of course, it is on the rete of a normal astrolabe. Other upgrades were more subtle. 'Notice,' he remarked in admiration, 'how the names of the [zodiac] signs are written so that each sign begins at the end of its name, and how this makes them easier to read.'[45]

But the Albion was much more than an upgraded astrolabe or saphea. It was intended, Richard of Wallingford explained, as a sort of geometrical almanac, a planetary computer. Its movable parts were pre-programmed, engraved with algorithms which, moved into

the correct positions, could solve hundreds of astronomical problems. So while it was shaped like an astrolabe and, as Westwyk
found, could even include an astrolabe among its brass discs, the rest
of it was entirely different. Unlike the astrolabe, which replicated the
motions of the heavens on a flat disc in a visually intuitive way, the
Albion was an extremely sophisticated mechanical calculator. Richard did not cut discs to represent each planet's path through the
heavens; instead, he engraved scales with the theoretical components
of their motions, which Ptolemy and his successors had calculated
and refined. Turning the plates into position and drawing a thread
across them was like looking up data in the tables of an almanac and
instantaneously performing a calculation with it. Once you had mas-
tered its seventy-plus scales, you could compute planetary positions
and speeds, times of conjunctions, and eclipses – really anything an
astronomer might need.

With its non-circular, non-uniform, even spiral scales, the Albion
was not an easy instrument to understand. And Richard's description was not easy to follow. We can see John Westwyk struggling to
understand it on every page of the copy he made of Richard's treatise. It was not a simple copy, but a compilation of Richard's own
version of the treatise with an updated version made by a Franciscan friar in Oxford named Simon Tunsted. Westwyk's reading was
an intense activity. It involved constant cross-referencing of the two
copies and comparing both with a brass Albion instrument he held
in his hands. On the very first page of his copy, he spelled out his
scholastic aim:

> Master Richard, abbot of the monastery of St Albans, first composed
> this book; and through it he devised and made that marvellous instru-
> ment called 'Albion'. But later a certain Simon Tunsted, professor of
> sacred theology, changed certain things not only in the book but also
> in the instrument, as will be clear to scholars in this book. Also, he
> added certain things.[46]

Making it 'clear to scholars', in Westwyk's conscientious world, meant
highlighting every difference between the two versions. Sometimes
all he needed to do was to underline an implanted passage. He did
so, for example, where Richard had noted that annual computations

began on 1 March – a common custom of astronomers who found it convenient to place the February leap day right at the end of the year. Tunsted disagreed, insisting that the base data should be calculated for January. This was probably because he was using the latest 'Alfonsine' tables, computed by astronomers working for King Alfonso of Castile, which arrived via Paris soon after the *Albion* was written.

In that instance Westwyk had only to use a neat line of crimson ink to mark Tunsted's insertion. At other times, though, he had to work much harder. In the worst case, having painstakingly – and near-perfectly – copied out the seventeen tables of planetary data that came with the treatise, he was stunned to realise that one of them was invalid as a result of Tunsted's change. It provided a key component of the Moon's position, measured from the equinox. Tunsted, though, had decided that he could make calculations easier by measuring that parameter from the Sun's own position instead. This left Westwyk with little choice but to copy out a whole new table. He drew a special little symbol of dots and diamonds to show where the table should be inserted eleven pages earlier. Next to it he noted, with just a hint of dutiful exasperation:

> The Lord Abbot put the mean longitude of the Moon on his spiral, but Master Simon put the elongation of the Moon from the Sun on his spiral . . . so I wrote this table so that anyone can do it this way if he pleases.[47]

John was doubly unfortunate, because the table he added is full of copying errors. Copying tables full of apparently random numbers was an important but taxing task. It was easier – provided that you had an exemplar to copy – than computing a new one from scratch, but it required a plentiful supply of the key monastic skills of concentration and precision. Even the best copyists might misread a number, or accidentally skip or repeat a line. But Westwyk was unusually good at it. He copied the original *Albion* tables from the manuscript where he had filled in the missing *Rectangulus* diagrams (we can tell that that manuscript was his exemplar because he repeated one rather unusual spelling), and did so with very few errors. But the extra table that Tunsted's revision required did not appear in that manuscript. Westwyk had to find it in another set of astronomical tables, and to judge by the number of mistakes, he picked one that had been badly copied from its own original source.[48]

Without his source, we cannot be absolutely certain who made the mistake. It is possible, of course, that despite Westwyk's earlier near-perfect copying, his own concentration levels had dipped. By now, on the eighty-eighth page of the treatise, he was certainly feeling frustrated. He let his dutiful demeanour slip an inch, with a comment that had nothing to do with what he had just copied. Pointing out that Richard had used a particular scale to calculate the equation of time – the difference between true solar time and mean clock time – he observed wearily that 'Simon [no 'Master' this time] works in another way, as was shown under Use 18 – and also in other places, many of which seem inconsistent.'[49] As veiled as Westwyk's neat Latin is, this is a moment of refreshing frankness, as his editorial task began to weigh more heavily.

He did take the opportunity to cut a few corners. His drawings of the Albion's scales are, in places, markedly more slapdash than the manuscript he copied. Richard had given step-by-step instructions for how to lay them out with compass and ruler – twenty-four equal divisions here, an intersecting diameter there – but Westwyk's diagrams look a little rushed. In one place he missed out a diagram entirely. In its place, he asked his reader to 'note that the figure of the scales of the first limb of the first face should be in this space, but it is very plainly inscribed on the instrument, so it is omitted here'.[50] No need to draw a picture of the instrument if you are lucky enough to have one in your hands.

That instrument was quite possibly Richard's own – maybe even made by the blacksmith's son himself. Westwyk at one point amended a statement of the instrument's diameter; and in his commentary on its astrolabe face he remarked on a few very specific details, such as that the azimuth lines were labelled at the horizon. He was evidently interested in both the instrument and Richard's theoretical treatise, and his final pages of commentary wandered from one to the other in a stream of connected but disorganised thoughts. He pointed out how neatly Richard's design separated the eclipse markers from the instrument's main scales to avoid confusing the user, and then digressed to discuss al-Battani's theory of eclipses. In the last of his notes he abruptly stopped describing the layout of the spiral scale, apparently remembering that Simon Tunsted had changed it. He left the middle third of that page blank. Perhaps he ran out of time to complete his notes before delivering the book to its intended recipients.[51] But who were those intended recipients? And why did Westwyk go to so much trouble to produce this careful compilation?

Hostility to Richard of Wallingford's rule as abbot, back in the 1320s and 1330s, had not just come from within the monastery. He also had serious trouble in the town. Affairs of the cloister were, of course, not isolated from secular life. The monks had daily dealings with lay-people, from sourcing food to welcoming pilgrims. Even the brass for scientific instruments came from craftsmen whose usual business was making high-end copper kitchenware. Most importantly, the monastery raised much of its revenue from rents, local taxes and fees for the use of its water-mills. Richard's predecessors had failed to protect the abbey's financial interests, leaving legal disputes to drag on for decades. Although Richard's first concern was to reform the morals of the monks, he soon realised that to get a grip on the abbey's spiralling debt he had to assert the abbey's authority over the town.

A few months before Richard's election, the St Albans townsmen had persuaded the newly crowned Edward III to grant them a charter of privileges. This, believed the burgesses, gave them the right to operate their own hand-driven mills for grinding corn and fulling cloth. Richard, however, argued that the abbey had a monopoly on these profitable processes. The dispute rapidly escalated to summons and counter-summons. When the abbot's constable came to arrest a leading burgess, the burgess resisted and both men were killed in the ensuing riot. Legal proceedings initially went against Richard, but through a combination of lobbying local lords, challenging the composition of juries if they included St Albans men, and offering generous hospitality to the judges, the tide of decisions began to turn in his favour. When a settlement was reached, it was quickly breached by the townspeople, but Richard soon forced them into complete submission. They gave up the new charter of rights and surrendered all their hand-mills.[52]

To underscore his victory, Richard used their millstones to pave the floor outside his parlour. The townsmen might not often see them there within the monastery, but this was a statement to the monks as well. The abbot was well aware that several of the brothers, either through family ties or regular contact, had formed close bonds with the citizens of the town and were dangerously close to disloyalty. For the most treacherous monks, the only solution was to do what St Albans abbots had done for centuries: exile them to the furthest outposts of the abbey's network.[53]

The remotest St Albans cell was at Tynemouth, in the far north of England. That was where abbots most commonly sent rebellious monks, and the mere threat of that clifftop priory was, according to the chronicler Matthew Paris, enough to make a stubborn monk weep and beg for mercy. On the other hand, it was also a posting that provided a formidable challenge for outstandingly ambitious monks. The Benedictine president Thomas de la Mare had proved himself over nine years as prior there before succeeding to the abbacy of St Albans in the midst of the Black Death. And when relations between the abbey and townspeople soured again fifty years after Richard of Wallingford's time, Abbot Thomas must have remembered its inhospitable potential.[54]

The 1370s were a decade of increasing taxation, as England's war with France became ever more costly. The poll tax of 1377 was particularly unpopular, levied at a flat rate of fourpence per head on all adults over the age of fourteen. The finances and military fortunes of the boy king Richard II did not improve, and his councillors imposed another poll tax in 1379. A third attempt to collect the tax in 1381 sparked the Peasants' Revolt. The revolt burned fiercely at St Albans, where a group of angry townsmen stormed the abbey. Apart from breaking open the gatehouse jail and forcing the abbot to issue a new charter of rights, they smashed the old millstones taken by Richard of Wallingford. They distributed the shards as symbols of their victory, reported the horrified chronicler Thomas Walsingham, like pieces of sacramental bread. A few monks, forewarned of the marching rebels, had already fled to the comparative safety of Tynemouth. John Westwyk, though, was not among them, for he was already in the far north.[55]

The Tynemouth monks paid their poll tax. It was, by national standards, a wealthy priory, so on the sliding scale levied in 1379 they incurred a higher tax rate than their brethren in smaller houses. The register of the seventeen monks who each handed over forty pence is the only surviving list of brothers from the whole history of the priory. As so often for medieval historians, financial records are our best source of information about much more than matters of money. But Westwyk's name is not in this list. He was not yet at Tynemouth in the summer of 1379, when the taxed monks were listed. But he certainly left St Albans by sometime in 1380, when Thomas Walsingham compiled a list of the brothers for a sumptuous book of the abbey's benefactors. And by 1383 he had been at Tynemouth for long enough to want to leave the priory in the most dramatic fashion, as we shall soon see.[56]

We can catch only a tantalising glimpse of Westwyk's shadow on the road to Northumbria in 1379 or 1380. The timing of his move, at a moment of increasingly febrile relations between abbey and town, may suggest he was one more in the long history of exiled brothers. But whether he was sent as a punishment, to serve as a teacher, or to prove himself as a potential prior – as with so much in the eventful life of this monk – we cannot be certain.

Still, we do know one thing he brought with him. It was that book. He copied the *Albion* and *Rectangulus* treatises on to creamy-coloured parchment of good quality. The 160 pages were uniformly sized, and fifty were left blank. They must surely have come from the abundant stocks at St Albans. On the first page, after stating his aim of comparing the two versions of the *Albion*, Westwyk identified himself and expressed his hope of salvation. This was the gift tag that allowed Kari Anne Rand to solve the sixty-year-old mystery of the 'Chaucer' *Equatorie* manuscript, after hunting through hundreds of manuscripts to match the handwriting that Derek Price had found in 1951:

> Master John of Westwyke gave this book to [the priory of] God and
> the blessed Mary and St Oswin, king and martyr, at Tynemouth; and
> to the monks serving God there. May the soul of the said John and
> the souls of all the faithful, through the mercy of God, rest in peace.
> Amen.[57]

This book, then, was John Westwyk's gift to Tynemouth. Within it he added a new astronomical table adapted to the priory's higher latitude. Perhaps he had enough warning of his departure to draw it up at St Albans – or perhaps he produced it after his arrival at the clifftop priory. It is even possible that he copied the entire *Albion* treatise under northern skies, filling some of the blank leaves of parchment he had brought with him. His exemplar, the manuscript where he had supplied the missing *Rectangulus* diagrams, also found its way to Tynemouth sometime before 1450.

The precise movements of Westwyk and his pen are impossible to retrace. What we can know, though, is that it was at St Albans that he learned the importance of instruments: where he studied them, used them, and understood how they could be improved. Such learning, it was clear, was an act of respect to his monastic forebears, and to God.

5

Saturn in the First House

In the autumn of 1095 the Earl of Northumberland made his last stand. Scion of a powerful Norman family, Robert de Mowbray had gained the earldom because his uncle was a trusted advisor of William the Conqueror. He was 'a man of huge size', according to one Norman chronicler: 'strong, dark and hairy, bold and cunning, with a grim and severe demeanour; he was more often taciturn than talkative, and when he did speak he rarely smiled'.[1] After the Conqueror's death in 1087, Robert repeatedly rebelled against the new king, William Rufus. He broke with his former regional ally, the Bishop of Durham, and began to cause havoc, attacking peaceful merchants and even plotting to kill the king. William Rufus summoned the earl to his court, but Robert flatly refused. So William mustered an army and overwhelmed Northumberland, driving the earl back to his furthest outpost at Bamburgh. On a dark night, Robert slipped through the siege. He sailed south in a single ship with thirty men, hoping to surprise the garrison at Newcastle. But they were forewarned and he was forced to flee to his final fortress, ten miles downriver on an outcrop high above the mouth of the Tyne. There he held out through six days of fierce fighting. Finally, with all his men killed or in chains, and himself wounded in the leg, he was captured inside the clifftop church.[2]

The windswept crag overlooking the North Sea at Tynemouth has seen many conflicts. Hadrian's Wall, the northernmost limit of the Roman empire, ended just a few miles inland at the fort of Segedunum. Medieval rulers, too, recognised the value of the headland at the south-east corner of Northumberland, connected only by a slim neck of rock, with cliffs or steep banks on all remaining sides.

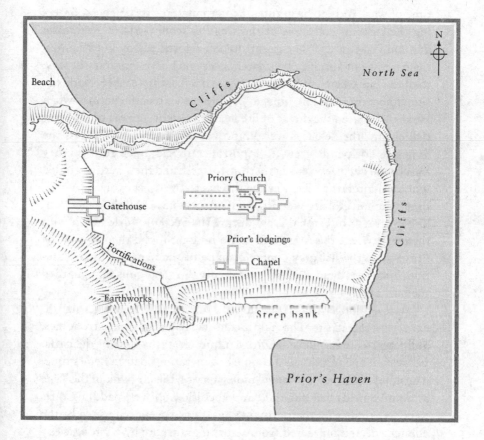

5.1. Plan of the surviving buildings of Tynemouth Priory. The cliffs to the north and east, and steep bank to the south, meant that the only viable access was through the gatehouse to the west. The space between the church and prior's lodgings was occupied by buildings including the cloister, chapter house, refectory and dormitory. Farm buildings occupied the northern part of the headland.

In the 1290s King Edward I made it a pivotal fortress in his wars against the Scots, and his successors expanded the fortifications he funded. Its defensive significance peaked in John Westwyk's lifetime. Richard II called it 'a castle and refuge for the whole country in time of war'. In 1390 he granted £500 to upgrade its defences, ordering the customs collectors at the wealthy wool ports of Newcastle and Hull to chip in.[3] The grant funded a fortified gatehouse across the narrow land bridge. This gatehouse, much more military in style than the one erected at St Albans a quarter-century earlier, made the site almost impregnable (image 5.1). By the sixteenth century, when Henry VIII faced the threat of French invasion in support of his Scottish enemy, the defences were improved with cannon. Then as the Tyneside dockyards grew in importance in the nineteenth century the coastal defences were again transformed, and they were upgraded with anti-aircraft artillery as the threats to the coast evolved. A children's author, Robert Westall, recalled growing up in the shadow of 'the great cliff of Pen-bal-crag' during the Second World War: 'with the castle where the Army still stood on guard, and the ruins of the Priory, and the flat grey concrete shape of the new coastguard station, and the tall radio-masts, and finally the great guns that guarded the harbour'.[4]

Yet even longer than Tynemouth's life as a fortress is its significance as a sacred site. The eighth-century historian (and astronomer, and finger-counter) Bede recorded a miracle at the base of the cliffs, performed by St Cuthbert a couple of generations before Bede's time. The monks of a newly founded monastery on the far bank of the Tyne at South Shields had mounted an expedition upriver, paddling rafts to collect timber for their growing buildings. Swept out to sea by the ebbing tide and prevented from returning to shore by a strong westerly wind, they seemed certain to drown. The wave-tossed monks were saved only by the young saint's prayers. To the amazement of the pagan crowd jeering from the clifftop, the wind miraculously abated and the rafts were set safely on land near the monastery.[5]

Bede's short account reveals two themes that are at the core of Tynemouth's history. First is the importance of the region as a centre of early English Christianity. Bede's own monastery of Jarrow was just a few miles away, and the pioneering Abbess Hild built her first nunnery nearby. It was in that seventh-century climate of conversion, Bede recalled, that Oswin, a pious local king, was betrayed to a rival and murdered.[6] Only four hundred years later was his body discovered under the church floor at Tynemouth, and the priory became a shrine in his honour. Countless hermits made their homes on the outlying rocks dotted along the fifty-mile stretch of coastline up to the Holy Island of Lindisfarne.

The precarious relationship the monastic communities had with the sea is the second theme that comes strongly through chronicles like Bede's. The water was not only a source of food and means of transport; it was also a reminder of the power – and occasional mercy – of God. The North Sea carried missionaries, fishermen and traders, but also swarmed with pirates and raiders. In 875 a Viking army occupied Tynemouth, burned its small church with the monks inside, and made the headland their base to conquer the whole of Northumbria. The monastic life was not firmly re-established on the clifftop until the 1070s, spurred by the discovery of St Oswin's relics. It was populated by brothers from St Albans after the rebellious Earl Robert de Mowbray took the priory from his rival the Bishop of Durham; the earl expelled the Durham monks, and made it a dependent cell of the influential southern abbey. The appreciative St Albans chronicler recorded how Robert himself, after a long imprisonment, ended his days as a pious monk at the mother house.[7]

That outpost was what awaited John Westwyk on his long journey north from St Albans abbey to Tynemouth Priory. The walk of more than 250 miles would have taken around two weeks, including a stop-off with the seven brothers of Belvoir, the St Albans daughter house halfway along the road.[8] Although monks made the journey fairly frequently, it was a serious undertaking. As John trudged north the landscape must have felt increasingly alien from the gently undulating Hertfordshire farmland that was all he had ever experienced. The Northumbria countryside was notoriously lawless, with border brigands, known as shavaldours, an increasing danger. The clifftop itself was home to a permanent military garrison and periodic populations of Scottish prisoners of war. And the Tynemouth priors had a reputation both for asserting their independence from St Albans's authority, and for bemoaning the priory's crumbling walls. At times, admittedly, their complaints were unjustified: successive English kings had considered it comfortable enough for extended royal visits in the early 1300s, and a prior in Richard of Wallingford's time had extended the already impressive church with a new Lady Chapel. Yet mere months before Westwyk's arrival, the prior wrote to King Richard II, complaining that

> flooding and erosion by the sea has cast down a great part of the walls; and the rents of the said priory are utterly insufficient to repair them . . . because a great part of the said rents lie near the Scottish border and are destroyed by the enemy. Therefore the said Prior and Convent beg our lord the King and his council to assign them some reasonable aid.[9]

Westwyk can hardly have been optimistic that he would have much opportunity to continue his scientific studies at Tynemouth. Yet he was to be surprised: despite the disruptions of border defence, there was certainly space for science at the northern house.

John realised that with every step he took towards the north pole, one of the astronomical tables he had so carefully copied would lose a little of its accuracy. That table gave the ascension-times of each degree of the ecliptic. Time, we know, is measured by the rising and

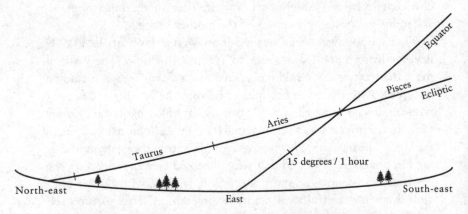

5.2. The horizon at Tynemouth. As the heavens turn, different arcs of the ecliptic and equator will rise in a given amount of time. Here, in the two hours it has taken for 30 degrees of the equator to rise, more than 60 degrees of the ecliptic (two signs) have risen.

setting of the equator. The celestial equator (or, if you prefer, the Earth) turns at a constant rate, and as each 15 degrees of the equator come up above the horizon, an hour passes. But the ecliptic is angled to the equator, so a given arc of the ecliptic will not usually ascend in the same time as an equally sized arc of the equator. As the equator keeps its position constant on the eastern horizon, the ecliptic moves (image 5.2). Fifteen degrees – one twenty-fourth – of the equator rise in an hour, but sometimes it will take more than an hour for a fifteen-degree chunk of the ecliptic to rise, and sometimes less. It was the ecliptic, let us not forget, that carried the Sun and signs of the zodiac, and shepherded the other planets on paths close to its own. So the table of ascensions works in much the same way as when we placed today's Sun on the ecliptic of our astrolabe with a blob of wax in Chapter 4. The day from sunrise to sunset always lasts as long as six signs – half a revolution – of the ecliptic. If you know what sign the Sun is in today, and you know how quickly those degrees of the ecliptic will rise, you can quickly calculate the length of the daylight hours.

The ascensions of the signs were hugely important to medieval astronomers. In particular, as we shall soon see, they were an essential component of the science of astrology. But for John Westwyk the problem was more immediate, more personal. It was part of his duty to Richard of Wallingford.

Richard of Wallingford had taken the 'All' part of his all-by-one Albion's name quite seriously: he had designed it to be self-contained, as far as possible. Unlike other planetary computers, whose users had to start by looking up reference data in tables, most of the reference data was already engraved on the instrument. This would have been an enormous relief to anyone who had struggled first to locate a particular table within a voluminous almanac, then to pull out just the right bit of data, and finally to set the dials and threads of an instrument to the precise configuration of the heavens. Richard's method meant no more looking for, and looking through, elusive tables – tables that, as we saw in the last chapter, would most likely contain at least some copying errors. Just as the back of an astrolabe incorporates a circular table of calendar months and zodiac signs, so Richard's Albion was engraved with the regularly repeating components of all the observable motions in the heavens. All the user had to do was look up a starting position (known as a *radix*) for a certain time, such as the beginning of the year, and then set the instrument for the intervening time, to predict an eclipse or find the position of any planet.

Yet Richard did include tables as the fourth and final part of his *Albion* treatise. These were to be used only once – when the instrument was first made. Richard, after all, had grown up amid the stifling heat of the Wallingford smithy: he was well aware that craftsmen might make his invention incorrectly, especially if they failed to hammer its brass sheets perfectly flat. He drew up tables for the final part of the treatise to help makers mark up its seventy-odd circles, or so that users could check that the divisions were accurate 'to a hair's breadth'.[10]

One of those Albion scales allowed users to track the signs as the ecliptic ascended above the horizon. The horizon itself changed, of course, as astronomers travelled north and south. The altitude of the pole rose and fell, the days lengthened and shortened – and the rising times of the zodiac signs altered too. If you wanted to use an astrolabe at a new latitude, you could simply switch in an appropriate plate for your new horizon. But the ascensions app on the Albion did not have that functionality. Instead, Richard instructed, makers should engrave it for use at 'a town or latitude where we intend to stay for a long time and make many observations'.[11] He explained how the irregularly spaced divisions of this scale could be laid out using a table of ascensions for the desired latitude. Then he supplied a table for his own latitude, at Oxford: 51° 50´ (fifty minutes being fifty sixtieths of a degree, just as minutes of time are sixtieths of an hour). Richard had surely not anticipated how far he would have to travel as abbot: south to Avignon for the Pope's blessing, and north to inspect the monks of St Albans's dependent cells. Even so, his table worked fine at St Albans, which was at almost exactly the same latitude as Oxford.

Tynemouth, however, was more than 3 degrees further north, at 55 degrees. If John Westwyk truly wished to validate Richard of Wallingford's legacy at the northern priory, he had to make Richard's tables usable in their new home. And this time he could not just copy one. Tables for the 55th parallel, where astronomers were as rare as

1.4, 1.5. Initial decorations for the months of October and November, in a St Albans calendar.

2.5. Abbot Richard of Wallingford, scarred by leprosy, with his monumental clock. From the St Albans Book of Benefactors.

2.9. Principal face of Richard of Wallingford's clock. The unequal hours are the inner circle of numbers from 6 to 6, read on the curves running inwards from them. Note also the sun showing the true solar time, and the golden dragon with its red tongue and tail marking the lunar nodes (1:4 scale reconstruction).

4.9 The head and curving tongue of the dog star, Alhabor (Sirius). The rete here is in place at the bottom of the astrolabe; note the 180-degree marking and twelve o'clock midnight mark on the limb.

5.10. The Coldingham Breviary. A Benedictine monk kneels before the Virgin and Child. Text, probably in the hand of John Westwyk, explaining how to calculate the times of the new Moon.

6.2. Matthew Paris's map of Britain (*c*.1255). Scotland is shown as almost a separate island, joined to England only by the bridge at Stirling. A box alongside Lincolnshire (second from bottom on the right-hand side) reads 'hec pars respicit flandriam ab oriente' ('this part faces Flanders to the east').

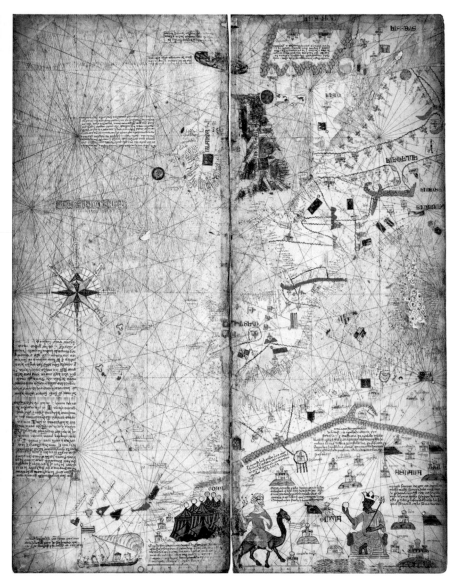

6.3. Western Europe, part of the *Catalan Atlas* (probably by Elisha ben Abraham Cresques, 1375). This luxurious production is heavily influenced by portolan charts, with criss-crossing rhumb lines indicating direction and coastlines traced by abundant harbour names, though it also includes significant inland detail. It features the earliest surviving compass rose, towards the left of the map. The island of Mallorca is striped in Catalan-Aragonese red and gold.

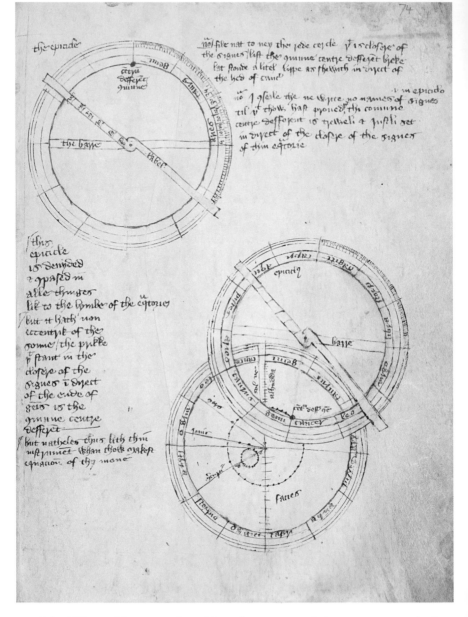

7.9. John Westywk's equatorium. Note the common deferent centre marked with a red blob in the top image. Next to it is his multilingual advice to the craftsman: '*Nota* I conseile [counsel] the[e] ne write no names of signes (*i.e. in epiciclo*) til that thow hast proved that thi comune centre defferent is treweli [truly] and justli set.'

a peaceful pirate on the North Sea, could not be pulled out of any old almanac. There was nothing for it but to compute a new table to track the rising signs of the ecliptic.

Fortunately for John, Richard gave him a clear steer – albeit a misleading one – of where to begin. It is right there in the table heading. 'This table,' Richard asserted, 'was calculated and drawn up according to the instructions in the second book of the *Almagest*.'[12]

'Almagest' was actually a nickname. Its original Greek title was the rather bland 'Mathematical Compilation', but it so impressed medieval Arabic scholars that they called it 'The Greatest': *al-megiste*. Their opinion was shared by astronomers for the next millennium, right down to Copernicus, who based his own epoch-making masterwork, *On the Revolutions of the Heavenly Spheres* (1543), closely on the methods and structure of the *Almagest*. Its author, the great Ptolemy, composed it in Alexandria around 150 CE. At this late stage of ancient Greek history Alexandria was ruled by the Roman empire, which is why Claudius Ptolemaeus had a distinctly Roman first name. He wrote in Greek, but we cannot know whether his ancestors were Greek or Egyptian. The name Ptolemaeus, though, may refer to a suburb of Alexandria, indicating that his family had been in Alexandria for several generations at least.

In that second book of the *Almagest* Ptolemy had drawn up tables of ascensions for a range of latitudes from the equator to the great Russian River Don.[13] At one extreme (the equator) the longest day – and every day – was twelve hours long; at the other it lasted seventeen hours. But even seventeen hours, corresponding to a latitude of 54° 1′, was not long enough for Tynemouth. So John had to look to the Alexandrian astronomer's explanation of his complex calculations, in order to try and replicate them.

Ptolemy realised that his readers might find this difficult. He built up the mathematics in stages, drawing on his predecessors' theories of the geometry of spherical triangles. Now, you may have studied some trigonometry at school. But unless you were at school in the 1950s, when spherical trigonometry still filled chalkboards around the world, all those triangles you carefully sketched with pencil and protractor were on flat surfaces. Ptolemy's geometry, too, was built up from flat triangles – but the angles and lengths that really mattered

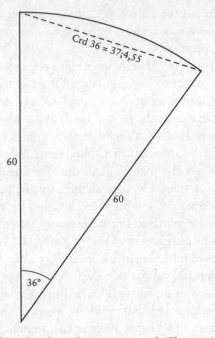

5.3. One of the first chords Ptolemy computed. The arc is part of a circle with radius 60. The chord is the dashed line joining the two ends of the arc. Its length is given in the standard sexagesimal notation we encountered with John Westwyk in Chapter 1. The units are followed by a semicolon, and further sexagesimal fractions are separated by commas. (Since this particular arc subtends an angle of 36 degrees, the length of the arc itself is one-tenth of the full circumference of the circle, or 12π = 37;41,57°.)

to him were traced on the curved surface of the celestial sphere. There the angles in a triangle no longer add up to 180 degrees. And that is just the start of it.

The foundation of the geometry in the *Almagest* was Ptolemy's table of chords. These chords are not musical – though the word reminds us how the medieval quadrivium of sciences brought together the ratios of geometry and the harmony of tonal intervals. A chord is the straight line joining the two ends of an arc (image 5.3). In the very first book of his masterwork, Ptolemy drew up a table relating the lengths of chords to their arcs, and to the corresponding angle at the centre of the circle. This relationship is very similar to the sine function you may remember from school, but of course Ptolemy could not press a button on a pocket calculator: he derived each value in the table from the geometrical principles Euclid had laid out.[14]

Ptolemy's table of chords is the earliest surviving trigonometrical table. This incredibly useful tool gave the chord for every half-degree from half a degree to 180 degrees. Ptolemy referred to it constantly throughout the rest of the *Almagest*. With some clever manipulation, he could use it to answer almost any question in mathematical astronomy. Predicting the length of the longest day in exotic regions he could hardly dream of visiting was just the start. There was no need to develop the sine, cosine and tangent functions we use today (though the shadow square we saw on the back of the astrolabe was a handy tangent calculator).

Ptolemy's ideas were taken up and substantially developed by geometers in India and the Islamic world. English astronomers were still coming to grips with their advances in John Westwyk's century. Richard of Wallingford, for example, compiled a four-part treatise on trigonometry early in his career, putting together passages he had read on chords, sines and other functions. Later, only shortly before leprosy claimed his life, he revised his treatise to incorporate the work of Jabir ibn Aflah, a Muslim working in twelfth-century Seville.[15] But for most of his astronomical purposes, Richard was able to get by with Ptolemy's less streamlined ideas.

The most useful of these was a staggeringly powerful two-part theorem. This theorem is usually credited to Menelaus of Alexandria, who lived in the century before Ptolemy – but although it is known as Menelaus' Theorem, it was probably not Menelaus' invention. It allowed mathematicians to calculate the lengths of arcs that intersected on the curving surface of a sphere. Ptolemy seems to have relied on an earlier version of the theorem, since he did not mention Menelaus when he explained it – even though elsewhere in the *Almagest* he gave Menelaus credit for precisely observing the Moon and stars in Rome. But whoever should be credited with its invention – and we shall probably never know for certain – it was an essential tool for astronomers wanting to predict and measure motions in the heavens.[16]

Ptolemy first proved Menelaus' Theorem, then used it to measure the simplest kind of heavenly rising – at the Earth's equator. At this unique latitude, where the north pole of the heavens is on the horizon, all the stars rise vertically (image 1.3). This rising at right angles to the horizon is the reason that distance measured along the celestial equator is called *right ascension*. And at that unique latitude, where the celestial equator is at right angles to the horizon, it is fairly easy to calculate the segment of the equator that rises in the same time as a certain segment of the ecliptic. To find just when a certain star sign will rise, or the precise length of the day, we need to know only two things. First is the distance between the equator and the point of the ecliptic that is just rising at the moment we desire. This is the *declination*. The Sun's declination changes through the seasons as it crosses the equator to north and south, moving back and forth along the ecliptic. Second is the angle between the equator and ecliptic – the *obliquity*.

Ptolemy covered all this in the first book of the *Almagest*. He provided a table of declinations and showed how to observe the obliquity using two large-scale instruments. Then, in the second book, he went a stage further. Using Menelaus' powerful theorem once again, he showed how to go from the rising-times of the signs at the Earth's equator – the right ascensions – to the rising-times anywhere else in the world. Since the signs now no longer cross the horizon vertically, those are not right ascensions but *oblique ascensions* (image 5.4). This required a further calculation, adjusting for the tilting horizon at different latitudes.

Sitting in the St Albans writing-room, John Westwyk must have visualised the Pole Star standing high in the Northumbrian sky as he carefully went through Ptolemy's steps. There was scope for some shortcuts, since tables of right ascensions were commonplace. Indeed, Richard of Wallingford had thoughtfully provided two right ascension tables in his *Albion*, with the same values accumulating from different starting points. John could certainly take that ingredient off the shelf. He used a table he had already copied with such care that he had been able to spot and correct the only mistake in his exemplar.[17]

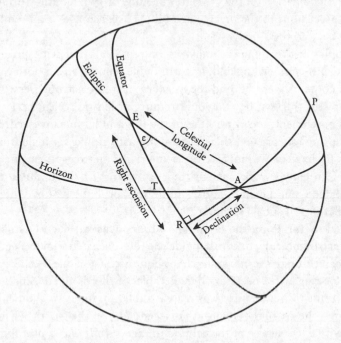

5.4. General theory of the ascensions on the sphere. The equator and ecliptic meet at E, the equinox, and the angle between them is the obliquity (ε), about 23½ degrees. The obliquity of the ecliptic is the same anywhere in the world, but the angle between the horizon and the celestial equator changes as you travel north or south. If you stand at the (terrestrial) equator the north celestial pole (P) is on the horizon: the triangle side RA will lie right on the horizon, and the celestial equator rises vertically (imagine everything except the horizon pivoted clockwise around A, until RAP is horizontal and ETR is vertical). And since R is then on the horizon, ER and ET will be the same. In that case, we can calculate the ascension ET = ER (the time taken for segment EA of the ecliptic to rise) from the right-angled spherical triangle EAR, using ε and the declination AR. But if you are not at the equator, so R is not at T, the oblique ascension ET must be found by subtracting TR (the *ascensional difference*) from the right ascension ER. The ascensional difference is a function of the observer's latitude (φ), since angle ATR is 90 degrees – φ (remember that the height of the Pole Star indicates your latitude).

But the calculation for the 55-degree latitude of Tynemouth – to adjust right ascensions to the particular oblique ascensions – required him to go back to the *Almagest*.

Copies of this monumental work were not all that widespread in fourteenth-century England. Its formidable reputation cut both ways, as its contents were beyond the mastery of many astronomers. And to copy all thirteen books would require at least 120 leaves of precious parchment, to say nothing of the ink and labour involved. It is perhaps not surprising, then, that many astronomers – including at times Richard of Wallingford himself – relied on excerpts or summaries of Ptolemy's work, such as the anonymous *Little Almagest* that circulated from the mid-1200s.[18] John, too, probably got by with the instructions in such a summary.

Luckily for John, the St Albans library was sufficiently stocked with astronomical works to provide the reference material he needed. He went through the 360 values of oblique ascension one by one, carefully looking up the necessary data in tables of chords and declinations. The result of his labours was a neat table, giving – to the nearest minute – the arc of the equator that ascended over the North Sea horizon with each degree of the ecliptic (image 5.5). It was a useful tool. The length of any day, for example, was easily found based on the position of the Sun on the ecliptic. So, on the longest day, when the Sun was at the cusp of Cancer, John simply subtracted that value in the table from the one 180 degrees further on, on the opposite side of the heavens. Converting that arc of the equator to hours, at a rate of 15°/hour, gave a longest day of seventeen hours and thirteen minutes. From there, too, it would be a moment's work to convert between the unequal seasonal hours and the equal clock hours for any day of the year. Yet despite its usefulness, this is the kind of manuscript page that most modern readers will skip over without a thought. Nothing – except perhaps the word 'tynemuth', inserted beneath the heading as an afterthought – betrays the work that went into it.

We can, however, tell just how much work went into it if we take apart the table. As we have seen, it was made by adjusting a table of right ascensions for the latitude of the Tynemouth horizon. Fortunately, the values in that prior table are arranged symmetrically. Just as the equator, ecliptic and horizon cut across the heavens and circle

5.5. *Tabula ascensionum signorum in circulo obliquo in latitudine 55 graduum*: Table of the ascensions of signs on the oblique circle at a latitude of 55 degrees: 'Tynemuth'.

back to their starting points, a graph of the right ascensions will be symmetrical, with the value at 1 degree matching those at 179, 181 and 359 degrees. In addition, a graph of the adjustments for latitude, the *ascensional differences*, is also symmetrical. But, crucially, it is not symmetrical in quite the same way. So with a little basic arithmetic, we can separate out the two components of each number in John's table.[19]

This deconstruction process uncovers two important details. First, we find that John Westwyk did indeed, as we suspected, use Richard of Wallingford's table of right ascensions as a ready-made ingredient in his own table. The numbers match up almost perfectly. But the second, and rather more revealing, detail comes when we isolate the values for the obliquity of the ecliptic. This angle between the equator and ecliptic, you may recall, was about 23½ degrees. Ptolemy had calculated it to the nearest second of arc. His value was closer to 24 degrees – 23;51,20°, to be precise, which is 23.86 in our familiar decimals. Later astronomers had competed to work out their own improved and updated values of this parameter, which, they realised, slowly changed over the centuries. Now, each component of John's table – the right ascensions and ascensional differences – incorporated a value for the obliquity. But the two components did not have to use the same value. When we delve into Richard of Wallingford's own table for the Oxford latitude, we find that Richard used two different values. Neither value matched Ptolemy's. And when we unpack John Westwyk's table we confirm that one component, his right ascensions, depending as they do on Richard's table, incorporate the same obliquity as Richard's (23;35°). The abbot got that value, like much else, from the Syrian astronomer al-Battani. But what of John's own adjustments, the ascensional differences for the latitude of Tynemouth? John had gone back to Ptolemy, and he proved it by dutifully using Ptolemy's value for the obliquity. In other words, when Richard wrote in his table heading that the oblique ascensions were computed according to the instructions in the *Almagest*, he was telling a half-truth. But Westwyk took Wallingford at his word, The abbot had not been entirely faithful to the great Alexandrian astronomer. But John Westwyk was able to honour them both.[20]

*

John Westwyk must have questioned whether all that computation was worth the effort. As he waited for a ferryboat to cross the Tyne on the final leg of his long walk north, he might have wondered whether his new brothers would make much use of the contribution he was offering to the priory library. The St Albans chronicle gave him little grounds for optimism. Matthew Paris had written specifically of the books that had been confiscated from one high-ranking St Albans monk banished to Tynemouth.[21] The list of rogues sent away to the priory was long. One such, according to Paris, had been 'a Lucifer among angels, a Judas among apostles, a worthless hypocrite among monks' named William Pigun. Pigun had been exposed as a double agent at St Albans after he forged a charter for a local baron involved in a legal dispute with the abbey. Sent to Tynemouth, he continued cursing the abbot. Justice was served on a dark night, reported Paris, when the gluttonous and inebriated Pigun took a trip to the dormitory toilet:

> Nodding his head, he began to sleep and, in sleeping, to snore with a hideous noise. And so he slowly slipped from drunkenness into sleep, from sleep into sudden death. Perhaps he was overcome by cold; but more likely, I believe, he was struck down by divine retribution. For when he had ceased his throaty grunting, these words were unmistakably heard in the privy where he sat dying: 'Take him, Satan.'[22]

There had undoubtedly been improvements in the century since Matthew Paris had written his chronicle. Thomas de la Mare's nine-year stint as prior, a generation before John Westwyk's arrival, had seen substantial building work, including a new brewery and dormitory for the monks. The dormitory even featured new toilets, flushed with piped water out to the sea below. These improvements were funded by increased income from rents, coal mines, and a fish market the priory had set up in nearby North Shields.

Yet no amount of redecoration could improve the weather. It was probably during de la Mare's priorate that one monk penned a complaint to a former cloister-mate back in St Albans. His description was long-winded, melodramatic and crammed with references to classical and more contemporary poetry, as well as to the Church Fathers and the Bible. But it was nonetheless heartfelt:

Since you were curious, dear brother, to learn about this place and its customs, and wanted me to tell you everything, good and bad, about the seashore and its inhabitants, I gladly obey . . .

Our house is confined on an exposed outcrop, surrounded by the waves on all sides, except one gateway cut from the cliff, almost too narrow for a cart . . . Night and day the waves rage, gnawing the hard rock with constant pressure so that the cliff now hangs heavy . . . Extremely dense and gloomy fogs come forth from the sea like dark smoke from the cave of Vulcan. These fogs dull the eyes, hoarsen the voice and constrict the throat, so that delicate air, imprisoned in the chest, cannot enter and exit as it freely should . . .

Spring with its flowers is outlawed there; summer warmth is banned, but the north wind and his allies stay permanently, as if King Aeolus [of the winds] claims our land as his capital city, chafing the country with deadly cold and snowy shackles. This unspeakable north wind rules the waves, which roar and rage . . . they bring forth bitter foam which, stirred up by the force of the winds, invades our homes and falls in clumps like pumice-stones on the castle.

For the local inhabitants, he felt a mixture of pity and disgust:

The greatest grief is to witness the peril of shipwrecked sailors, their rafts wrecked, masts swaying, keels between reefs and rocks, with no nail left to hold their timbers together. The sailors, their limbs numbed by cold, sink like lead in the violent waters, and no human power can prevent their deaths since, as a certain poet said, 'if my ship is dashed on the rocks, I have only the words "But Thou [O Lord, have mercy upon us]." ' Such misfortunes often fall on our tearful eyes.

No turtle-dove is heard in our land. The nightingale does not deign to visit, since bare branches offer no possible place to perform, nor is the breeze soft enough to channel harmoniously through its windpipes . . . but grey birds do nest on the rocks, and greedily feed on the cadavers of the drowned. Their harsh and horrible cries are an ominous warning of future storms . . .

The men living by the seashore are like Moors, the women are like Ethiopians, the maidens are filthy, the boys are as black as Hebrew boys . . . They eat sea-weed which is blacker than ink. This plant, which grows on rocks, lacks sweet flavour or good smell; it nauseates rather than nourishing the stomach . . . The women of this land use it as if it were an aromatic herb, whence their own colour takes on the colour of that plant.

The fruit trees are like shrubs, not daring to raise up their branches, for the sea strips and spoils their flowers and leaves . . . So fruit is hardly ever found. A sweet red apple is, as the poet might exclaim, 'like a black swan'. Indeed if, against all expectation, apples should spring forth, then they are dry and shrivelled, lacking juice and flavour. Their extreme bitterness will set your teeth on edge.

Beware, then, dearest brother, lest you fall to this place, deprived of all comfort, lacking all solace and joy.[23]

This dejected monk could find only two things to praise: the beauty of the newly enlarged church and the abundance of fish to eat. And even the fish gave cause for complaint, he remarked, since the monks grew tired of eating the same food day after day. (So it was perhaps unfortunate that when two successive queens stayed at Tyne-mouth Priory in the early fourteenth century, both their husbands, Edward I and II, sent them a thoughtful gift of fish. The luxurious pike, bream, eels and sturgeon may not have been as appreciated as their senders had hoped.)[24]

Some monks, of course, went to Tynemouth precisely for its trials. If the wind on the clifftop itself was not bitter enough for them, the priory had its own offshore hermitage where they could test themselves further. Coquet Island, twenty miles up the coast, was, for some, an ideal place to imitate the apostolic life. Conversely, it seems other monks appreciated that Tynemouth offered them greater flexibility and looser rules than might be enforced at St Albans. A generation after John Westwyk's time there, for example, the abbot of the mother house was alarmed to discover that the Tynemouth monks were using the church as a theatre. He forbade them from performing plays for the local laypeople to celebrate the feast of St Cuthbert.[25]

What we can piece together of Westwyk's time at Tynemouth sug-gests he enjoyed some of such freedom. The priory library had no more than a dozen or so books, compared to the hundreds at St Albans, but he certainly made use of them. In one, a copy of Bede's *Ecclesiastical History of the English People* bound with some shorter

5.6. Opening page of a manuscript from Tynemouth. A large initial T begins the phrase *Trine de[us] da ne dicar tua gr[ati]a vane Joh[ann]es de Westwyk* ('Threefold God, grant by your grace that I be not spoken about falsely. John Westwyk'), showing some imitation of the twelfth-century writing style above it. The missing chunk of parchment (top right) corresponds to a decorated initial letter that began the text on the other side of the leaf.

Church histories, he wrote his name. The first page of the manuscript begins with a list of the bishops of Lindisfarne in twelfth-century handwriting (image 5.6). Immediately underneath, halfway down the first page, John had a play at imitating that style of penmanship in vibrant blue ink. A bold, broad initial T begins the phrase (in Latin) 'Threefold God, grant by your grace that I be not spoken about falsely.' As an imitation of the 250-year-old handwriting, it began quite convincingly, but by the end of the line John had reverted to his normal writing, and his name 'Johannes de Westwyk' is quite recognisable. Both prayerful and playful, John here hints that he has been – or expects to be – the victim of some injustice.[26]

Despite the monks' eloquent complaints, the wintry North Sea climate was not an insurmountable obstacle to science. The calculations of astronomy, in any case, did not always require star-gazing. Still, if the Tynemouth monks were interested in forecasting a clear night, and did not find the screeching grey gulls sufficiently reliable auguries, or if they wanted to know when the bitter wind would once more stir up a hail of sea-foam, their celestial science could help them. Astronomy itself – and its sister science of astrology – could predict the weather. For such prediction, the tables John Westwyk had painstakingly copied and re-computed would come in very useful.

Weather forecasting is an ancient science. John Westwyk, as we saw in Chapter 1, would have encountered some rudimentary meteorology among the rural rhythms of his childhood. Farmers tailored their agricultural practices to the climatic cycles, timing seasonal activities like ploughing and harvesting according to the changing lengths of the day and the visibility of stars. Day by day, they also made use of weather lore. Take, for example, the theory that 'red sky at night' presages a fine following day. The monks of St Albans must have known that adage from the Gospel of Matthew, where Jesus cites it as an example of common knowledge. They would also have found it in the *Natural History* of Pliny, written around the same time in the late first century CE. The first nineteen books of this thirty-seven-part monument of Roman natural philosophy were in the monastery's library. In the eighteenth book, after describing the biology and cultivation of a host of different grains, Pliny gathered together indications of all kinds of weather based on observations of the Sun, Moon, stars, clouds, animals and plants.[27]

Medieval astronomers, however, could do better than such indications. Dark clouds in the east might, as Pliny suggested, be a sign of imminent rain, but the real causes of changes in the elements were further away. Aristotle, seeking to explain the continual changes that occur on Earth, had ascribed them to the heavens, for only the circular motions of the heavens were constant. At the start of his *Meteorology*, he took it for granted that all power of change in nature came from the heavens. The movements of the Sun, he pointed out, indisputably bring life to the Earth. Elsewhere he suggested that women's menstruation, while not reliably regular, tended to follow the phases of the Moon. Ptolemy, in turn, noted that the Moon governs the tides.[28] From such basic facts, it became universally accepted that events on Earth were a microcosm of what occurred in the heavens: the motions of the fixed and wandering stars. But it remained to be worked out precisely how the stars affected the Earth, and it was a matter of frequent controversy how far human minds could detect and predict those influences.

This was astrology. From the intuitive understanding of celestial influence, a complex predictive science developed. We now think of astrology as a pseudo-science, but highly intelligent scholars studied it logically and diligently throughout the Middle Ages and well into the early modern period. While its theories and parameters were continually challenged and refined, the basic principles remained as laid down by Ptolemy in his *Tetrabiblos* – 'Four Books' – on astronomical prediction, known to medieval astronomers by its Latin title *Quadripartitum*. Ptolemy presented this work as a companion piece to the *Almagest*, just as astrology was the little sister of astronomy. From the outset, Ptolemy acknowledged the uncertainties surrounding this 'second and less self-sufficient' science. He argued that scholars should 'never compare its perceptions with the sureness of the first, unvarying science', i.e., astronomy. The reason some people thought the second science useless, he claimed, was only because some arrogant practitioners made exaggerated claims for its power. Careful astronomers should not, he argued, be put off attempting 'such investigation as is within the bounds of possibility, when it is so evident that events of a general nature draw their causes from the enveloping heavens'.[29]

The basic principles of astrology were simple. Each planet (including, of course, the Sun and Moon), had its particular powers. Saturn, for example, was a cold, dry planet, said to govern old age and agriculture, among other things. Hot, moist Jupiter, in contrast, was the planet of nobility and reconciliation, governing judges and religious leaders. Changes on Earth were caused by the motions of the planets – since all change, as Aristotle defined it, is motion – but the way to track and predict such changes was to observe the *positions* of the planets at a given moment.[30] At certain positions they had particular power. At the meridian, obviously, since the Sun is highest and strongest at noon, but also at the horizon, when a planet first rises. The fixed stars, too, had their powers, but since they circled constantly their effects were most noteworthy when a planet moved among them. They could then either strengthen or diminish its effects. The planets could also enhance or cancel out each other's influence, when they passed close by one another or faced off across the sky.

If you could chart such complex influences with precision, you could predict a great deal about the future. When is the safest time to make this journey? Where are my missing valuables? How will I die? To address such questions, astrologers drew on intricate theories. Undoubtedly the most influential was that of the ninth-century Muslim Abu Ma'shar, known to Latin scholars as Albumasar. He blended Ptolemy's principles and Aristotelian physics with Indian and Persian ideas – such as studying vast multimillennial cycles to find recurring configurations of planetary positions – to form a complete and convincing synthesis. Coupled with the increasingly precise astronomy available to Latin scholars from the twelfth century, it was an irresistible combination. It was, in fact, Abu Ma'shar's astrology that gave many Latin readers their first introduction to Aristotle's natural philosophy.[31] And it was a short summary translation of Abu Ma'shar's extremely lengthy writing, known to Latins as the *Flowers of Astrology*, that influenced much of the astrology John Westwyk encountered.

One text John surely studied was Richard of Wallingford's *Exafrenon* ['six-part work'] *on Weather Forecasting*. Despite its title, the *Exafrenon* went beyond weather forecasting. (As with many medieval treatises, the title may well be an addition by a later copyist or cataloguer.) The worked examples in the sixth and final part were, to be sure, exclusively meteorological; and Richard finished with the well-worn story of Thales, the Greek who demonstrated the value of philosophy by predicting a bumper olive harvest, renting all the region's olive presses and making a fortune. The rest of the treatise, however, was a more theoretical introduction to the techniques required to draw up an astrological prediction. Here, unlike most astrological writers, Richard indulged his personal passion for precise calculation, giving clear mathematical instructions. He laid particular emphasis on Abu Ma'shar's theory of the Lord of the Year. This theory stated that one planet would have particular influence over the whole year, based on its position at the spring equinox, when the Sun moved from the sign of Pisces into Aries. But which planet would it be? That was decided according to the all-important division of the sky into *houses*. A planet in the first house was most likely to be the Lord. That is where those tables of ascensions, over which we saw John Westwyk slaving so diligently, came in very useful indeed.[32]

The houses addressed an important problem in astrology: when were planets particularly powerful, and how did their effects vary? One answer to this question, proposed by Ptolemy, was simply to theorise that the zodiac signs were responsible. Thus Leo, the 30-degree segment of the ecliptic where the Sun sat during the hottest days of summer, might be thought the natural home or 'domicile' of the Sun. Leo was therefore assigned the elemental qualities already associated with the Sun: hot and dry. Cancer, the neighbouring sign, belonged to the next-brightest heavenly body, the Moon, and was seen as cold and moist, like the Moon (which does, after all, govern the tides).[33] Each planet would be dominant when it was in its home sign – the Sun in Leo, or Mars in Aries – and weakened when it was in the zodiac sign directly opposite. In addition, there were subdivisions of a third of a sign, a fifth of a sign, and even individual degrees of the zodiac where a planet might gain or lose influence. Astrologers could then tally up the total influence, or *dignities*, of each planet.

Yet all this was insufficient, since it failed to take account of the planets' positions in the plane of the horizon. Were they high in the middle of the sky on the meridian, or just making their presence felt as they ascended? The solution was to divide the sky into another set of houses, according to those two key positions: *midheaven* and the *ascendant*. Three of the twelve houses lay in the angle between the ascendant and midheaven – between the degree of the ecliptic just rising and the bit crossing the meridian at the same moment. There were three more between midheaven and the ecliptic degree just setting, and the remaining six were similarly distributed on the other side of the ecliptic.

One advantage of this system was that the houses changed as the horizon changed. Moving significantly north or south shifted your horizon and could thus shift the boundaries of the houses. If a planet thereby slipped from one house into another, that might make all the difference to an astrological prediction. As Richard of Wallingford explained, a planet in the first house was prime candidate to be Lord of the Year. It would influence the whole year's weather forecast for a region. If it was Saturn, Richard wrote, 'he shall make so cold a winter in the north country that will kill well nigh all the beasts of the land, and grip the buds with dry cold'. But a small shift of the houses might make Mars the lord, and this would 'soften the northern winter' – good news for the frostbitten brothers of Tynemouth.[34]

The houses had another advantage over the old signs, at least for geometrically minded astronomers: their sizes could vary according to how quickly they rose. Thus each house had a fairly allocated period of influence. To calculate it, the tables of ascensions were invaluable. The most popular way to locate the start and finish of the three houses between the ascendant and mid-heaven was to say they would each take the same amount of time to rise above the horizon (image 5.7). That meant that those three would represent equal segments of the equator but unequal segments of the ecliptic. Astronomers used tables of ascensions to convert between the two. The basic geometry, like cutting a cake with cuts right across the centre, made each slice of the ecliptic the same size as the one opposite it – the first and seventh houses, for example. Then the remaining six houses would also represent equal rising-times, each with its opposite.

In practice, the procedure was simple. The hardest part was to establish the precise moment of the Sun's entry into Aries, since the year was about eleven minutes shorter than 365¼ days. Well aware of that, medieval astronomers compiled quick reference tables to track the shifting difference between the calendar year and the *tropical year* from spring equinox to spring equinox.[35] Once you knew the time for which the houses were to be divided, the rest was easy – provided you had some tables like the ones John Westwyk had copied and computed. John would have followed a nine-step process (image 5.8) to find the boundaries of the houses and draw up a horoscope, requiring only some reference to tables and a little arithmetic.

Astrologers often liked to lay out the houses as a diagram, in a geometrically pleasing pattern of squares and triangles. John

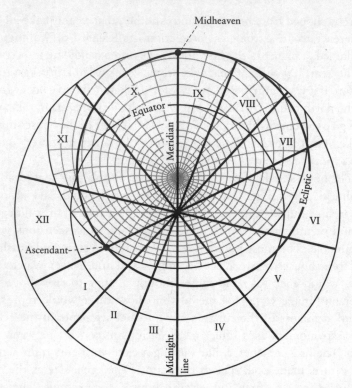

5.7. The houses, laid out with the ecliptic on an astrolabe plate for the latitude of Tynemouth (55 degrees). Houses I to III and VII to IX are all equal in right ascension (arc of the equator), as are the other six houses. Opposite houses (e.g. I and VII) are also equal in longitude (arc of the ecliptic).

was no exception. One horoscope diagram survives in his slightly scruffy handwriting. It shows the boundaries of each house and the positions of the planets within them (image 5.9). John must have drawn it up as an astrological exercise, for although he computed the house boundaries for a latitude around 51 degrees, he carefully copied the planetary data and accompanying text from a sample horoscope. The original specimen was computed by the eighth-century astrologer Masha'allah ibn Athari, a Jew working in Abbasid Baghdad.[36]

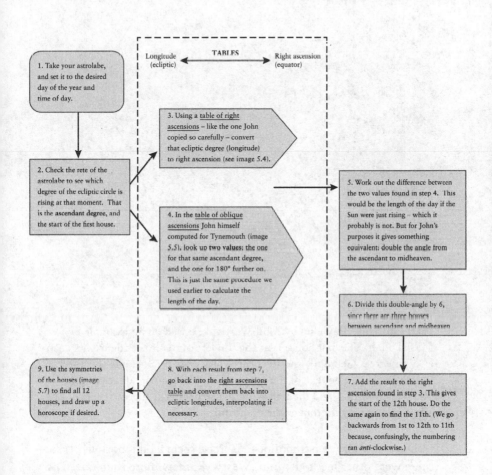

1. Take your astrolabe, and set it to the desired day of the year and time of day.

2. Check the rete of the astrolabe to see which degree of the ecliptic circle is rising at that moment. That is the **ascendant degree**, and the start of the first house.

TABLES

Longitude (ecliptic) ← → Right ascension (equator)

3. Using a table of right ascensions – like the one John copied so carefully – convert that ecliptic degree (longitude) to right ascension (see image 5.4).

4. In the table of oblique ascensions John himself computed for Tynemouth (image 5.5), look up **two values**: the one for that same ascendant degree, and the one for 180° further on. This is just the same procedure we used earlier to calculate the length of the day.

5. Work out the difference between the two values found in step 4. This would be the length of the day if the Sun were just rising – which it probably is not. But for John's purposes it gives something equivalent: double the angle from the ascendant to midheaven.

6. Divide this double-angle by 6, since there are three houses between ascendant and midheaven.

7. Add the result to the right ascension found in step 3. This gives the start of the 12th house. Do the same again to find the 11th. (We go backwards from 1st to 12th to 11th because, confusingly, the numbering ran *anti*-clockwise.)

8. With each result from step 7, go back into the right ascensions table and convert them back into ecliptic longitudes, interpolating if necessary.

9. Use the symmetries of the houses (image 5.7) to find all 12 houses, and draw up a horoscope if desired.

5.8. Steps to divide the astrological houses, using a table of right ascensions and a table of oblique ascensions for your latitude.

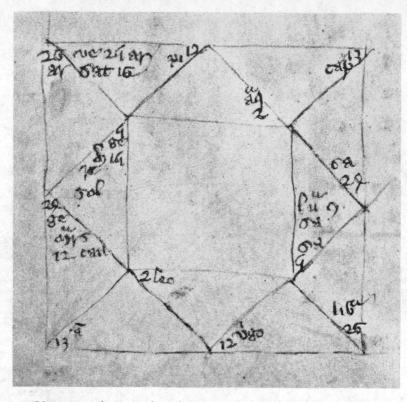

5.9. Horoscope drawn up by John Westwyk, based on an example by Masha'allah. The first house is at the top left of the inner square. Its cusp is at 5 degrees Ge[mini], and the others proceed anticlockwise (so the fourth house can be read at the bottom left of the inner square as the 2nd degree of Leo, and the following house begins at the 12th degree of Virgo, at the very bottom of the diagram).

Does that sort of astronomical exercise seem a bit laborious? If so, there were other methods. John Westwyk knew more than one. The most explicitly astrological chapter of the *Albion* treatise explains how to use the Albion to compute the boundaries of the twelve houses. When John came to copy it, he could not resist adding a new paragraph. The same houses, he explained, could be found using an astrolabe.[37]

For monks who were not so gadget-obsessed as John, there were still more alternative techniques. Tables might simplify the task. One monk at Tynemouth shortly after John's time certainly thought it needed simplifying. He took the book John had brought north to the priory and filled half a dozen of its precious spare pages with a ready-made table of houses. Using a table of houses required almost no calculation whatsoever. You only had to find the location of one house, generally the ascendant – as in the first two steps of John's process. All the other house boundaries then simply fell out of the table.[38]

Whilst the table of houses may have been easy to use, however, it was hard work to draw up. We can be sure that the Tynemouth brothers in their damp, draughty cloisters did not do that computation. For one thing, the table is very poorly copied. The Tynemouth copyist gave up halfway through the first page and had to start all over again. On another page he accidentally repeated a row; and in a column on the same page he apparently suffered a lapse of concentration, got lost in a sequence of numbers, and had to fill in random values to catch up to where he should be. A second hint that the tables were not drawn up at the clifftop priory is the fact that the houses are computed for a latitude of 51;50°, down in Oxford. There is no table heading to make this clear – you have to dive down into the underlying parameters to find out – so the monks were presumably unaware of this fact. They may not even have realised it mattered. Not all were as expert as John.

Even so, it is striking that at least one Tynemouth monk felt that such a table of houses was a fitting use for seven of the fifty spare pages of parchment that John had generously provided along with Richard of Wallingford's *Rectangulus* and *Albion* treatises. Tables of houses were widespread in varied user-friendly designs – the Oxford friar Nicholas of Lynn for example, included one in his astronomical calendar. They were so popular simply because the houses were the basis of almost all astrological prediction. Each house, according to Abu Ma'shar and most other astrologers, indicated particular things. The second house, just below the horizon, governed wealth, while the fifth was the place of children. Generous Jupiter, if placed in the second house, might have quite different financial implications from Mars, which was associated with fraud and theft.[39]

Thus astrology could do far more than predict the weather.

Richard of Wallingford, let us not forget, was rumoured to have predicted the demise of his predecessor as abbot of St Albans. If, as everyone agreed, the planets affected the elements, then everything made of elements was subject to their influence. Everything down here on Earth was celestially swayed, at least to some extent.

The idea that humans, made of elemental matter, were a microcosm of the universe was most evident in medical theory. Each part of the body from head to foot was ruled by a zodiac sign from Aries to Pisces. 'A doctor cannot cure,' wrote Robertus Anglicus in his commentary on Sacrobosco's *Sphere*, 'if he does not know the cause of the disease. And that cause cannot be known if the motion and position of the heavenly bodies are not understood. They are the cause of every condition of things down here below.' This theory did not just help explain the origins of illnesses; a wise physician could use it to plan any medical intervention, to treat the right part of the body at the right time.[40]

Some astrologers went further. If human life was governed by the heavens, the moment of birth or conception was surely crucial. A nativity chart, showing the configuration of the heavens at that moment, could reveal much about an individual. Sometimes called judicial astrologers, they helped guide clients' choices or answered questions according to the state of the sky.

The idea that our fortunes might be foretold in the stars was deeply problematic. How could heavenly influence constrain our God-given free will? John Westwyk's contemporary John Gower thought this dilemma a fitting subject for his Middle English poetry:

Benethe upon this Erthe hiere	Beneath upon this Earth here
Of alle thinges the matiere,	the matter of all things
As tellen ous thei that ben lerned,	(or so learned people tell us)
Of thing above it stant governed,	is governed by things above
That is to sein of the Planetes.	– that is to say: the planets.
The cheles bothe and ek the hetes,	Both cold and hot weather,
The chances of the world also,	and the events of the world
That we fortune clepen so,	which we call fortune,
Among the mennes nacion	among human nations:
Al is thurgh constellacion,	they are all caused by
	constellations.

Wherof that som man hath the wele,	So some men have wealth
And som man hath deseses fele	and some men have many hardships,
In love als wel as othre thinges.	in love as well as other things.
The stat of realmes and of kinges	The state of realms and of kings
In time of pes, in time of werre	in peacetime, in time of wars
It is conceived of the Sterre:	is born from the stars:
And thus seith the naturien	so says the natural philosopher
Which is an astronomien.	who is an astronomer.
Bot the divin seith otherwise,	But the theologian says otherwise:
That if men weren goode and wise	that if men were good and wise
And plesant unto the Godhede,	and pleasing to God
Thei scholden noght the sterres drede.	they have nothing to fear from the stars.[41]

Of course, finer minds than Gower's had given the matter much thought. St Augustine himself had admitted that the stars influenced many things on Earth, from the seasons to sea-urchins. But he denied that they could affect the human mind and its free will. After all, he pointed out, a twin boy and girl can be conceived at the same moment, in the same place, under the same celestial configuration. If the stars are not strong enough to set such a fundamental physical quality as the sex of those twins, he queried sceptically, how could they possibly constrain their life choices? On the other hand, Augustine's contemporaries were content to employ astrological symbolism in their Christian sermons. They plotted a sacred cycle from Aries, the lamb of God, to Pisces, whose two fishes represented the two races of Jews and Gentiles given new life in the waters of baptism.[42] Even without foretelling the future, astrology could help make sense of a divinely connected cosmos.

A later giant of theology, Thomas Aquinas, took a slightly different view. Like Augustine, he accepted that drought or rain could be predicted by observing the stars. Conversely, he also accepted that it would be superstitious, even demonic, to try to have certainty about future human actions. But there was a grey area. If, Aquinas pointed out, celestial bodies could influence human bodies, and if the human will is affected by bodily needs and appetites, then the heavens might just influence people's minds and behaviour. If I get grouchy when I am hungry, the astrally affected elements might just override my free will.[43]

Such reasoning did not impress the Paris bishop Étienne Tempier. Among the 219 philosophical propositions condemned in 1277 were several astrological ideas. These included the suggestion that health and sickness, life and death, were assigned according to the stars, or that our free will is subject to the power of celestial bodies. Perhaps most damaging to astrology was the condemnation of the idea that celestial configurations and their effects would repeat themselves – in 36,000 years, according to Tempier's list. Condemned or not, 36,000 years was clearly far longer than human records could chart, so no one could know what would happen next time. That point was developed to devastating effect by Nicole Oresme in 1360s Paris. In a treatise aimed at 'astrologers who think they are wise men, but make fools of themselves', he applied his mathematical skills to show that celestial motions were incommensurable. The relative positions of the planets, as he demonstrated, never repeated themselves in exactly the same way. So astrologers, lacking precise prior examples, had nothing reliable to base their predictions on.[44]

Oresme made other logical arguments, in a volley of polemical pamphlets fired at astrologers in the decades after 1350. Even if the celestial motions did repeat themselves closely enough for predictive purposes, he thought, the astrologers' actual calculations were hopelessly inaccurate. Other Paris scholars joined the assault. For example, one slightly younger colleague, an expert astronomer originally from Germany named Heinrich Selder, focused his attacks on the tables astrologers used for their data. He poured particular scorn on the doctrine of the Lord of the Year. The whole idea that the Sun's shift into a new star sign could completely change the power of another planet, he spluttered, was patently absurd. And he updated Augustine's point about the variable fortunes of twins, telling the story of a pair of conjoined twins. They had not only been conceived but also born at the same moment, yet they had different personalities and one slightly outlived the other. Selder concluded that the only thing the heavens really affected was the weather.[45]

Such violent debate belies the stereotype of the Middle Ages as an era of scholastic conformity. Yet even sceptics like Oresme and Selder did not go quite so far as to deny celestial influence entirely. The question then remained: how might that celestial influence actually work? It was not the only instance of unexplained action at a distance. So, perhaps, some scholars speculated, it was a similar mechanism to the mysterious power of magnets. On the other hand, the unarguable effects of the heavens on weather seemed to suggest that it had something to do with heat and light. Perhaps, then, the planets conveyed their influence in the same way as they could be seen?

One theory of sight suggested that they transmitted a tiny image of themselves to the eye. An alternative theory described rays travelling in straight lines, a geometrically attractive concept that was naturally suited to the neat diagrams of Oxford textbooks. The Oxford master and Bishop of Lincoln Robert Grosseteste picked up the theory of rays from the ninth-century optical theorist al-Kindi. Al-Kindi had also written a treatise on astrological weather forecasting which circulated widely among the mathematicians of European universities. But for Grosseteste, to seek any answers in the stars beyond an indication of imminent weather was utter 'futility and falsehood'. 'Judicial astrologers', he thundered, 'are both deceived and deceivers. Their doctrine ... is dictated by the Devil, and their books should be burned.' Even so, Grosseteste could still draw on the theory of rays, which he used to explain why there were two tides per day. Rays from the Moon, he argued, were reflected back from the heavens to the far side of the Earth, so the water was pulled in two directions simultaneously.[46]

His disciple Roger Bacon followed suit, but in more detail and with rather different conclusions. Geometry, as we saw in Chapter 3, could explain why a jumble of rays at different angles create a single image on the eye – because the most nearly perpendicular rays are the most powerful. The same theory, Bacon believed, also explained diverse astrological influences. He pictured rays converging in a sharply pointed cone shape. 'To each unique point on the earth,' Bacon explained, 'come the apexes of unique cones, and each point is the centre of its own horizon.' 'Thus,' he continued, 'two twins in the mother's womb are assigned different natures, and so later will have different characters, and will pursue different crafts, and distinct occupations throughout their lives.' Bacon's analogy between light and celestial influence, with its precise cones of rays, was a neat explanation, even for the problem of conjoined twins. Other questions, though, remained unanswered. Why, wondered some sceptical philosophers, did some planets have a heating effect, while others caused cold?[47]

A further vexed question was how the planets' influence on Earth varied with their distance from us. Ptolemy had suggested that the planets act mainly to modify the influence of the Sun and Moon. Thomas Aquinas, by contrast, argued that the closer a planet was to the perfection of heaven, the more powerful it was. The planets that took longer than the Sun to traverse the stars – those with orbits longer than a year – had spheres above that of the Sun. These 'superior' planets – Mars, Jupiter and Saturn – outside the Earth in today's solar system, were responsible for longer-term and perhaps further-reaching changes. They still had their effects on the weather, of course. Geoffrey Chaucer had rather disingenuously disavowed judicial astrology immediately after explaining it in some detail to 'Little Lewis'. But he employed a downpour of planetary proportions to enable a romantic encounter in the tragic tale of *Troilus and Criseyde*:

The bente moone with hire hornes pale,	The bent Moon, with her horns pale,
Saturne, and Jove, in Cancro joyned were,	Saturn, and Jupiter, were joined in Cancer,
That swych a reyn from heven gan avale	so that such a rain from heaven began to fall
That every maner womman that was there	that every kind of woman that was there
Hadde of that smoky reyn a verray feere.	Had of that smoky rain a real fear.[48]

Beyond their immediate impact on the weather, though, when two or even all three of the superior planets came close together very significant events might result. One such triple conjunction happened in the spring of 1345. Astrologers in both England and France had checked it out in advance. They noted that the three successive meetings of Mars with Jupiter, Mars with Saturn and the so-called 'great conjunction' of Jupiter with Saturn would all take place just before the spring equinox, when the character of the coming year was set. A lunar eclipse was also due in the same timeframe. The consequences might include famine and major political change, long-lasting and severe.[49]

Three years later the Black Death struck Europe. The astrologers naturally returned to their old prognostications, like self-critical economists after a stock-market crash, asking how successfully they had foreseen it and whether they could have done more. One Oxford expert discussed whether the pandemic was caused by the stars or, alternatively, was punishment for the sins of humanity. He concluded that 'there is strong evidence that the mortality was produced by God in the first way: that is, by the lunar eclipse and the great conjunctions as natural instruments. That mortality and the other effects,' he insisted, 'were predicted beforehand. And that prediction was fully founded on the books of astronomers.'[50]

It was the primary authority on great conjunctions – the Persian astrologer Abu Ma'shar – whom the monks of St Albans chose to commemorate in their cloister windows. There was certainly wide-ranging interest in astrology at the abbey. The monk-chronicler Matthew Paris produced a fortune-telling manual, which his successors at the abbey used and copied. The book was equipped with a *volvelle*, a numbered rotating disc that was spun like a roulette wheel to obtain a random number. That number could then be deployed in the rest of the book to provide answers to a host of everyday questions: money, travel, marriage, and so on. Matthew realised that such material was theologically dubious, but he was not unduly concerned. Christians must of course look to the integrity of their Catholic faith, he emphasised. But fortune-telling did not endanger free will, for 'nothing is inevitable if human precaution can prevent it'. Divine anger, he reminded his readers, could be averted. It was the job of monks to reconcile themselves to God through humble prayer.[51]

At some other houses the monks went rather further, potentially provoking divine anger by practising scholarly magic. The boundaries between astrology and magic could be blurred. Magic itself was not a simple category; already in the Middle Ages words like 'magic' and 'necromancy' were umbrella terms for a range of disparate practices. At one end of the spectrum was *natural magic*, which exploited the mysterious properties of nature. The grease from a lion, according to one manual, could be used to ward off wolves, while bathing in ass's milk would add lustre to your skin. At the other end of the scale were practices that used objects – talismans or images – to harness the power of the cosmos. This was known as *image magic*. One popular guide to astrology, whose author concealed his identity but may have been the Dominican polymath Albertus Magnus, divided image magic into three kinds. The first kind – which he calls 'abominable' – imbued power into images by invoking spirits or using smoke. The second kind – 'a little less disagreeable but still detestable' – achieved the same effect by writing powerful words. The anonymous author warned his readers that such necromancy was often covered with a respectable veneer of more mainstream astrology.[52]

However, a third kind of image magic, according to the same manual, 'eliminates the filth' of the other two. Its practitioners created images by channelling the power of the heavens directly. After all, if you could predict the influence of the stars, you could certainly change your behaviour to heighten or avert events. And it was widely accepted that the influence of certain stars was stronger on certain elements, since each thing down here below was a microcosm, imbued with qualities that matched some parts of the heavens above. So it is hardly surprising that magicians looked for ways to use earthly materials to attract or channel celestial powers. The poet John Gower, narrating in *The Lover's Confession* how Alexander the Great was partly educated by an evil magician, passed on a popular list of fifteen gemstones and plants that could do just that:

Nectanabus in special,	. . . especially Nectanabus,
Which was an astronomien	who was an astronomer
And ek a gret magicien,	and also a great magician
And undertake hath thilke emprise	and undertook this task:
To Alisandre in his aprise	to Alexander, in his education,

As of magique naturel	in order to know natural magic,
To know, enformeth him somdel	[he] informed him well
Of certein sterres what thei mene;	about certain stars: what they mean
Of whiche, he seith, ther ben fiftene,	of which, he said, there were fifteen
And sondrily to everich on	and individually to each one
A gras belongeth and a Ston,	belonged a herb and a stone
Wherof men worchen many a wonder	through which men work many a wonder
To sette thing bothe up and under.	to build things up or bring them down.
To telle riht as he began,	To tell it right as he began,
The ferste sterre Aldeboran,	the first star, Aldebaran
The cliereste and the moste of alle,	the brightest and largest of all
Be rihte name men it calle;	(men call it by that true name),
Which lich is of condicion	which is alike in nature
To Mars, and of complexion	to Mars, and in complexion
To Venus, and hath therupon	to Venus, and so has
Carbunculum his propre Ston:	ruby as its stone.
His herbe is Anabulla named,	Its herb is named spurge
Which is of gret vertu proclamed.	which is credited with great power.[33]

The fact that Gower calls this 'natural magic', rather than image magic, might be considered evidence that he was only dabbling in the occult sciences. But really it is a reminder that practices which aimed to study or to exploit the heavens and Earth could not be easily categorised. After all, their practitioners might have good personal or commercial reasons to keep them mysterious. Gower's source text for this magical lore was sometimes attributed to a mythical figure, Hermes Trismegistus ('thrice-greatest Hermes'), and sometimes to the biblical patriarch Enoch – whom several students of magic thought was the same person. Yet it was also copied alongside quite mainstream astrological texts by Richard of Wallingford and Robert Grosseteste. Gower may well have found it at the Augustinian priory in Southwark, where he lived in the latter part of his life. And a group of Benedictine monks at St Augustine's Abbey in Canterbury made a collection of more than thirty texts covering a broad range of scholarly magic, some blended with Christian rites. Magical study, they thought, could serve the needs of their monastic community. There is certainly no evidence that it got them into trouble with God or their superiors.[54]

However prevalent such practices were at other Benedictine houses, there is no evidence that the monks of Tynemouth slipped from astrology into magic. As for John Westwyk himself, his interests were clearly more mathematical. Word of his expertise in mathematical astronomy may, indeed, have spread beyond the formidable walls of Tynemouth priory. Seventy-five miles up the coast, in the Scottish borderlands, was the priory of Coldingham. The surviving jewel of the priory is a beautiful breviary – a service book with the texts of the psalms, a liturgical calendar tailored to the priory's practices, and other religious reference materials. It contains one full-page illumination showing a Benedictine monk praying to the Virgin Mary and Child (image 5.10, plate section). But if we can haul our eyes away from that stunning painting just for a moment, right underneath it we are stunned to discover handwriting uncannily like John Westwyk's. There John – if indeed it was he – found space to insert a short paragraph of instructions, a fragment of a longer guide. The fragment explains tersely how to find each new Moon using some new columns of numbers in the breviary calendar: '. . . if the dot is *after* the number, that hour *after* noon is the conjunction. If you are in the second cycle, use the number written at the lower left; and so on with the new Moons of the other cycles . . .' These dots and columns, based around the Golden Numbers we encountered in Chapter 2, give the times of every new Moon for four cycles of nineteen years. The instructions were in French.[55]

These brief instructions raise several intriguing issues. John could certainly have travelled north to Coldingham, imparting some of his astronomical expertise while he was there. Even with the risk of shipwreck, it would have been a manageable sail up the coast – the hermitage at Coquet Island was on the way. But the breviary was not necessarily there when John came to write in it. It could have come to him. Coldingham was a daughter house of Durham Cathedral priory, where the manuscript was most probably made, and the Borders priory was the subject of frequent disputes between the English cathedral and the Scottish king. In 1378, two years before John's arrival at Tynemouth, King Robert II had expelled the Durham monks from Coldingham, replacing them with brothers from the Scottish abbey at Dunfermline.[56] Some fled south to the mother house. They would surely not have wanted to leave behind a book as precious as this. The breviary was recorded in Durham 150 years later, so this could have been the moment it moved. If so, the migrating monks may have brought it to Tynemouth, or John may have annotated it at Durham when the two priories were on rather better terms than Earl Robert de Mowbray had left them.

This exquisite breviary is a reminder that monks and their books were more mobile, and more multilingual, than is often realised. Manuscripts frequently moved between monasteries – this one probably travelled to Oxford too – and generations of hands leafed through and annotated them. If monks found a useful text, whether sacred or scientific, they might well copy it into a spare space, no matter what language it was in.

For John Westwyk, the very language of these Norman-inflected French instructions might have been useful. Perhaps, as the Norfolk mystic Margery Kempe did a generation later, John was practising his French in preparation for a long journey.[57] We have followed him from the manor of Westwick, to St Albans and on to Tynemouth; now for a moment we are uncertain whether his science took him across the border into Scotland, or indeed down to the scriptorium at Durham Cathedral. Yet one thing we do know is that his travels were to continue, and that they would take him far across the sea. For in the sweltering summer of 1383 John Westwyk found himself on crusade.

6

The Bishop's Crusade

In 1383 John Westwyk marched behind the banner of the Holy Cross. Crusading, by his day, was an ancient institution. In 1095 the Pope had called Christians to arms, uniting great lords, devout pilgrims and ambitious adventurers in a penitential quest to conquer the holy places of their faith and resist the expansion of Muslim rule in the near East. As a mass movement, it was a stunning success. The foremost preachers of the era recruited tens of thousands of zealous Christians to a new kind of holy war. But 1095 was further in the past for John Westwyk than the French and American Revolutions are for us. The Crusades had lost much of the momentum of their early victories. Jerusalem, captured amid indiscriminate slaughter in 1099, had been retaken by Saladin in 1187. The last crusader outpost in Palestine had fallen in 1291. Even so, the legacy of the extraordinary marriage of warfare and religious devotion endured. It remained not only in the medieval institution of chivalry and its associated orders of knighthood, but in the structures of Church authority and finance that had supported waves of armed pilgrimages. The recovery of the Holy City might have lost its urgency, but the Pope still sent crusades to fight against Muslims in Spain, pagans in the Baltic lands and heretics at the heart of Europe. These campaigns were united by the foundational principle of crusading: the indulgence. Those who participated – and especially died – on an officially sanctioned expedition could receive forgiveness for all their sins.

As the crusade ideology evolved and fragmented, it could be grafted on to other causes and sentiments. In late-medieval England it supported a growing nationalistic identity. Kings marshalled armies to fight the Scots or the French beneath the fluttering red-and-white banner of Saint George. John Westwyk lived in the midst of the Hundred Years War, when successive English monarchs fought to fulfil their claim to large regions of France. At the same time, from 1378 the Western Church was split between two rival popes: Urban VI in Rome and Clement VII, based in Avignon in southern France. As Europe's kings and princes aligned themselves with either Urban or Clement, political and religious arguments melded to justify and prolong conflict.

Within weeks of the 1378 Church schism, Urban sent proclamations to his supporters in England. He offered a crusade indulgence to anyone who would fight for a year against his rival, Clement, or against supporters of that antipope.[1] Flanders, the historic county occupying half of modern Belgium and parts of France and the Netherlands, was a natural battleground. Its count, Louis de Mâle, was backed by France. Louis and the French both sided with the Avignon antipope Clement. But Flemish cities like Ghent and Bruges had amassed enormous wealth by making cloth with wool imported from England. Their citizens joined England in support of Urban and rejected the authority of Count Louis. The Ghentish leader Philip van Artevelde, who had been brought up in England and had long been in the pay of London, offered to recognise Richard II as both Count of Flanders and King of France if the English sent a fleet and an army to expel the French-backed count.

In 1382, while the teenage King Richard and his council dithered, the French crushed the Ghentish rebellion and imposed a stifling embargo on the wool trade. Many in the Westminster parliament were now desperate to intervene, not least because the wool embargo was crippling English exports. But government finances were in chaos after the Peasants' Revolt of the previous year. Despite selling off some of the king's jewels and taking out a loan from Italian bankers, the total government income between April and September 1382 was a paltry £22,000. Parliament simply could not raise an army.[2]

This is where the Bishop of Norwich came in. The youngest son of a great noble family, Henry Despenser had been well prepared for a career in the Church. He studied law at Oxford before moving to the papal court. There he proved his military credentials by fighting in a crusade against the city of Milan, which was resisting papal control. The grateful Pope soon nominated Despenser to the see of Norwich. As bishop, he saw no need to behave peaceably. The St Albans chronicler described him leading a cavalry charge against the peasants during the 1381 revolt. Fully armed with helmet, breastplate and a double-edged sword, he rode into the midst of the rebels, slashing and stabbing on all sides and 'gnashing his teeth like a boar'.[3] Although clearly energetic, Despenser was no politician; his involvement in the Flanders situation was probably driven by his ambitious chaplain (who later rose to become Archbishop of York). But with the king's council rudderless and parliament paralysed, Despenser saw his chance to use the authority Urban had granted him to launch a crusade.

On 21 December 1382 Despenser set up a large cross in the middle of St Paul's Cathedral in London. There, on the chancel steps, he swore the crusading oath before the Bishop of London. The ceremony for taking the cross had not been performed in living memory, according to one chronicler. The bishop had to search high and low before finally finding instructions for the ritual at Westminster Abbey.[4]

Publicity for the crusade intensified in the new year. Preachers toured the kingdom, promising extravagant spiritual benefits to anyone who participated or helped to fund the expedition. An extra donation, some claimed, could save the souls of the donor's friends or even those already dead. One monk recalled sceptically how preachers promised to summon angels from the skies, which would snatch souls from purgatory and carry them to heaven. Women, he wrote, were especially generous. 'Many people gave more than they could afford,' he complained, 'to obtain absolution for themselves and their close friends. And thus the kingdom's hidden treasury – that kept in the hands of women – was put at risk.'[5]

With some misgivings, the Westminster parliament gave its approval to the venture. Although the main selling-point of Despenser's proposal was that the crusade would largely pay for itself, the parliamentarians agreed to hand over more than £30,000 from the last round of taxation. They were certainly impressed with Despenser's promise to raise an army of 2,500 men-at-arms and 2,500 archers and to keep them fighting the French for a full year. They were less convinced of the bishop's credentials as a general. With evident reluctance, Despenser promised to appoint a royal lieutenant who would take all key military decisions. But he never did.[6]

The recruitment campaign was a triumph. Men signed up from all walks of life: armourers and saddlers, but also fishmongers, tailors, merchants and clerks. Despenser did his best to guarantee the quality of the recruits, commanding that anyone who was not themselves a competent warrior should pay for one to represent them instead. The parliamentary funding and private donations paid for a substantial professional force, including five experienced captains. But the ranks were also swelled by large numbers of churchmen. The muster rolls are replete with chaplains and canons, parsons and prebendaries.[7] And, despite their sworn commitment to the cloister, many monks joined the fighting force.

Their zeal did not impress all their brothers. Houses of silent piety were cast into confusion by men marking themselves with the cross, complained a chronicler at Malmesbury Abbey. 'They deserted divine worship,' he protested, 'saying they were fighting the antipope. But really they were only fighting against chastity.' The St Albans historian Thomas Walsingham agreed:

> The quiet of the cloister displeased them, so they asked the Abbot's permission . . . to turn to warlike deeds and the clash of arms. I will not stay silent about their names. From this monastery John of Bokeden set forth; from the cell of Tynemouth John Westwyk; from the cell of Wymondham William York, from the cell of Binham Roger Beuver and John Bell; from the cell of Hatfield the prior William Eversdon himself . . . and William Sheppey.

In the margin of the manuscript another monk added a further name, Roger Rous, to this list of shame.[8]

The crusade was formally launched at Westminster on 17 April 1383. The bishop raised the banner of the cross in the abbey church. (In a statement that revealed his nationalistic priorities, he had reassured the parliamentarians in nearby Westminster Hall that even if the Clementists converted to the 'true pope', he would still fight for the king's cause.) Followed by a large crowd, he processed two miles along the River Thames to St Paul's Cathedral, to celebrate a solemn Mass. From there, he set off for the south coast to muster the crusading army. While he waited for the troops to gather, he enjoyed the hospitality of St Augustine's, Canterbury, at one of the abbey's manors adjoining the port of Deal.[9]

John Westwyk and his brothers in arms had an arduous journey ahead of them. Knowing how stormy the North Sea could be (to say nothing of the hazards of warfare), it was not one he would have taken lightly. It is tempting to speculate that he was desperate to get away from Tynemouth. But we should resist such easy explanations. For one thing, it would not account for the participation of the prior of the little cell at Hatfield Peverel, whose position already gave him considerable freedom. Nor does it explain why John of Bokeden, a solid citizen whose skills as a mason were celebrated at St Albans, chose to leave the safety of the abbey.[10] What role religious zeal, national pride or a simple sense of adventure played in Westwyk's decision to take the cross, we can only imagine. All we can do is join him on the boat to Calais.

Such voyages over land and sea had been undertaken for thousands of years. They required little navigational science or technology. But scientific development could make travel safer, trade more profitable, invasions more successful. The Middle Ages saw advances in mapping, navigational technology and understanding of oceanic phenomena of tides and currents. Each had a part to play in John's journey.

At first glance, medieval maps look woefully inaccurate. Coastlines are barely recognisable, content unfamiliar. Talking about the 'accuracy' of a map is, however, only a partial assessment at best. Maps are always an answer to a question, a response to a set of priorities. Is clarity more important, or completeness? When you attempt the impossible task of rendering a three-dimensional Earth on a two-dimensional page, do you prioritise consistency of scale, shape or direction? Detail is not always desirable: you would not want a road atlas to be cluttered with all the topological features that appear on hiking maps. Sometimes it even pays to distort: commuters are quite accustomed to using underground rail maps that twist the shape of cities for the sake of simplicity. Medieval maps varied enormously in scale and ambition, precision and content, reflecting a rich visual culture and varied uses. At one extreme are crisp diagrams that proliferate in textbooks like Sacrobosco's *Sphere*, with east–west lines dividing the globe into climatic zones. At the other are unwieldy, immersive visual compendia like the *Mappamundi* at Hereford Cathedral in the west of England.

Made around 1300 from an entire calfskin, almost as large as a double bedsheet, the Hereford world map would have been an imposing presence on the wall of the cathedral. Yet its densely packed inscriptions, with lettering only around three millimetres tall, required close contemplation. It owed something to early medieval schematic 'T-O' diagrams, which split the inhabited world into Asia, Africa and Europe using T-shaped waterways, within a circular frame with east at the top (image 6.1). But it was rich in detail drawn from a range of classical sources, not least the Bible. So while we can find specific detail on the *Mappamundi*, such as the length of Africa or the number of islands in the Orkney archipelago, our eye is drawn to the Tower of Babel, the Garden of Eden and the crucified Christ. At the centre of the map was Jerusalem – 'the navel of the world', as the Pope had supposedly said when he called the First Crusade to capture it.[11] In such visual encyclopaedias, geography was chiefly a framework to organise and display history. Like today's educational atlases that feature a country's wildlife alongside landmark buildings and national dishes, medieval world maps were often stylised and inspiring. Authors could lay them out as any shape. One popular version, devised by the Benedictine monk Ranulf of Higden around 1330, depicted the Earth within an almond-shaped frame.

It is not surprising that monks treasured such providential depictions of a divinely created world. Yet their geographical interests were often much broader. When Henry Despenser delved deep into the Westminster Abbey archives in search of that crusading ritual, he doubtless enlisted the help of brother Richard Exeter, who had just retired as prior. Exeter's own personal book collection included Ranulf Higden's geographical and historical encyclopaedia, bound up with a Latin translation of Marco Polo's *Travels*. After Exeter's death in the winter of 1396–7, the abbey inherited his possessions.

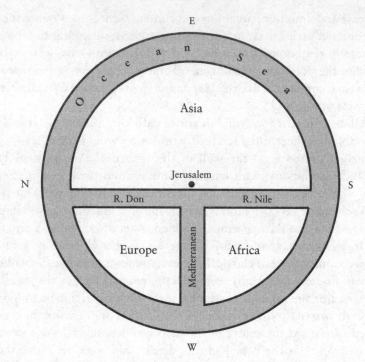

6.1. T-O-style world diagram, popularised by Isidore of Seville, among others.

Alongside a chess set and some fine tableware, he left maps of England and Scotland, and a sea chart.[12]

At St Albans, meanwhile, the chronicler Matthew Paris produced his own maps of Britain, Palestine and the world in the 1250s. His outline of Britain is impressionistically drawn inside a frame punctuated with the names of surrounding lands, including Flanders to the east (image 6.2, plate section). Looking more closely, though, we realise that the map is principally a diagram of an itinerary. Below the Antonine and Hadrian's walls, Matthew's England hangs on a straight spine of towns from Newcastle to Dover. The main route passes by St Albans abbey, of course, as well as its daughter house of Belvoir. The St Albans cells of Binham, Wymondham and Tynemouth feature as detours from the way. Matthew marked the country's principal rivers and used somewhat standardised symbols

for a city, cathedral or mountain. But most of the places he drew
were monastic sites. This was an ideal reference tool for a traveller
like John Westwyk to consult before setting off on a long journey. It
showed all the rest points where he might plan to receive hospitality
on his way from Tynemouth to the south coast.

One obvious absence from these maps are lines of latitude and
longitude. This was not for lack of knowledge of these divisions of
the Earth: as we saw in the previous chapter, John Westwyk and his
contemporaries were perfectly capable of computing and using such
tools of mathematical geography. Tables of the latitudes and longitudes of notable towns and cities abound in medieval manuscripts,
including at St Albans. But mapmakers saw no need to include such
data. Their maps were travel guides – whether that travel was to take
place on the road or in the mind – so it was sufficient to show relative position. On one of his itinerary maps Matthew Paris provided a
scale to show the size of Britain, but on his other maps the names of
the surrounding lands did the same job.[13]

Over the following hundred years, however, maps sprouted a
forest of lines. Mediterranean mariners, who had sketched sailing
directions and port guides for generations, began to compile them
into larger sea-charts in the late thirteenth century. To help lay out
courses when they were out of sight of land, they added radiating
direction – or 'rhumb' – lines, creating what are known as *portolan*
charts. The coastlines on these charts were marked by the thickly
clustered names of harbours, with each word pointing inland from
the shore, while the territories within were left bare of detail.

The flourishing trade of the western Mediterranean made Mallorca a centre of mapmaking, along with the Italian cities of Venice
and Genoa. Master cartographers such as the Catalan Jew Elisha ben
Abraham Cresques produced lavish compendia, blending the practical portolans of Balearic traders with the classical lore of the old
world maps and the latest reports from Europeans who had visited
Asia (image 6.3, plate section). As explorers began to chart the Atlantic islands in the fourteenth century and ventured further down the
coast of West Africa in the fifteenth, these lands, too, were added.

Elisha Cresques and his son Jefuda served the royal family of
Aragon in John Westwyk's day. The 1375 *Catalan Atlas* probably

came from their workshop.[14] The map was originally mounted on eight wooden panels, a rectangle two metres wide in total. It showed the regions from the Atlantic to China, and was accompanied by astrological diagrams and tidal calculators. We owe its survival down the centuries to the fact that it was a presentation copy, never used at sea. But less elaborate versions certainly were risked on board ship. These portolan charts were not necessary for coastal sailing, since experienced pilots knew every inch of their habitual routes; but as competitive traders risked longer journeys out of sight of land, charts became increasingly useful. Such voyages were especially feasible in the Mediterranean, where the tidal flows are too slight to disrupt a well-helmed course.

Why did these charts with their criss-crossing rhumb lines suddenly proliferate in the fourteenth century? The main reason was the increasingly widespread and systematic use of the magnetic compass for navigation. The Catalan Atlas features a compass rose, where the rhumb lines meet at a decorative compass within the chart. This is an early attempt at the device, and few of the lines actually pass through the rose's centre. But compass roses soon became fully integrated with the network of rhumb lines. They made it easy for navigators to plot a basic course from starting point to destination.

While the popularity of portolan charts rocketed, the magnetic compass was the result of more gradual development. The power of lodestones to attract iron had been described in classical Greece and Rome. St Augustine recounted how a fellow North African bishop had seen the magnet's wondrous properties demonstrated at the dinner table. The host, a Roman regional governor, had placed a piece of iron on a silver dish and moved a lodestone underneath it. The guests marvelled to see the iron move, while the dish was unaffected. However, while ancient European philosophers had experimented with various magnetic effects, they made no practical use of them. In China, by contrast, the 'wet' compass, a magnetised needle floating in a bowl of water, was in use by the eighth century (and probably several centuries earlier). In 1088, the government official and polymath Shen Kuo explained how magnetic variation caused such a compass to point a little to the west of true north. A 'dry' compass, whose needle pivots on a narrow pin, was described a few decades later, but was not particularly popular in China.[15]

There is no evidence of any compass in Europe at that time, despite some flimsy myths suggesting it was known to a tenth-century pope or sailors in Salerno.[16] Still, when the compass did eventually appear in Europe, it was most likely an independent invention, since it is not mentioned in Arabic sources until rather later. The first definite Latin reference to a compass comes at the end of the twelfth century, in two works by a St Albans schoolmaster.

Alexander Neckam was born at St Albans on the same night in 1157 as the future King Richard I. His mother, a wet nurse, was said to have fed the prince from her right breast and her own son from the left.[17] While Richard led the Third Crusade against the resurgent forces of Saladin, Neckam was teaching at Oxford. Later he became an Augustinian canon, and eventually abbot, at Cirencester in the west of England. It was there, around the year 1200, that he wrote his most important scientific work, *On the Natures of Things*.

Neckam had first included the compass as an example in a Latin grammar textbook. He wrote that book during a stay in Paris in his early twenties, before returning to teach first in the grammar schools at Dunstable and St Albans, and subsequently at Oxford. The idea of his textbook was to present Latin through the practical vocabulary of daily life. It included a section on sailing equipment, perhaps in memory of his own voyage across the English Channel. Apart from pitch and provisions, oars and an anchor, Neckam had noticed the need for stays and shrouds to support the mast – and also an axe, which, he suggested, might be used to cut the mast down if a storm was brewing. And in case the same bad weather obscured the stars, he wrote, the typical ship had 'a needle mounted on a pivot, which will rotate and revolve . . . so the sailors know where to steer when the Little Bear is not visible'.[18]

Evidently, then, the dry compass was in common usage by the 1180s. Neckam returned to consider it in more detail twenty years later. By then he had had the opportunity to study some of the recently translated works of Aristotle and was full of praise for the Philosopher and his scientific methods. Yet *On the Natures of Things* was intended above all as a moral treatise. Neckam's clearly stated aim, following St Augustine, was to use diverse examples from nature for religious edification. So in his section on the sea he pointed out that all rivers ran towards the coast and briefly outlined some theories of the tides, but lingered at greatest length on the power of the waves and the foolishness of any sailor who thinks he can master them. He closed this section with an account – provided, he assures us, by trusted eyewitnesses – of a sailor who regularly crossed the English Channel with only his dog for crew. Although the dog was trained to pull the ropes with his teeth at his master's command, Neckam thought this a deplorably unwise risk.[19]

In similar fashion, Neckam's detailed account of the maritime compass leads to a lesson for religious leaders: 'a prelate must direct his subjects in this sea [of life]', orienting them by his reason. Before reaching this moral conclusion, though, Neckam put forward a theory to explain the curious behaviour of magnets: they attract iron from any direction, he noted, but they might repel other magnets. He argued that the attractive power found in a lodestone operated most strongly on similar objects – which included iron. However, he suggested, it worked only when a stronger object attracted a weaker one. Thus, despite the obvious similarity of two lodestones, their attractions would cancel each other out. Following this with the story of a statue of the prophet Muhammad that was said to float in mid-air, Neckam pointed out that this marvel could be explained by multiple magnets pulling in opposite directions.[20]

The story of the floating statue is similar to one told six hundred years earlier by Isidore of Seville, and by Pliny before him. As so often with medieval scientific writings, it is hard to be sure how much comes from fresh observation and how much from the recompilation of earlier authorities. Neckam's understanding of the practicalities of navigation was certainly limited, and it seems most likely that his apparently unprecedented insights into the properties of magnets were repeated second-hand, rather than the result of his own experimentation. Nevertheless, writings on the subject accumulated over the later Middle Ages, testifying both to scholarly curiosity about magnetic properties and to the practical popularity of the compass. Another Augustinian canon in the generation after Neckam,

THE BISHOP'S CRUSADE

Jacques de Vitry, asserted that the lodestone was 'truly necessary for navigators at sea'. He had some experience of the matter, having journeyed from his native France to be Bishop of Acre, the last remaining stronghold of the kingdom of Jerusalem, as well as travelling to Egypt to take part in the disastrous Fifth Crusade.[21]

Physicians, too, took an interest in the properties of the lodestone, which some writers thought were weakened (or, in some cases, strengthened) by garlic, onions or goat's blood. One who clearly had performed his own experiments was Jean de Saint-Amand, who worked at Tournai, once part of Flanders, now in Belgium. Jean composed a long commentary on one of the most popular medical textbooks, a pharmaceutical guide known as the *Antidotarium Nicolai*. At the very end, he discussed its recommendation that the flesh of a snake known as the tyrus (which Jacques de Vitry had claimed lived in the Jericho region, not far from Acre) could neutralise poison. How was it, Jean wondered, that such snakeflesh could draw poison out of a patient? How did it overcome the poison's tendency to accumulate within the body, without the venomous snakeflesh itself being drawn inwards? To address this tricky problem, Jean made an analogy with the lodestone's ability to attract iron. It led him into a condensed discussion of magnetic polarity, based on an experiment in which a magnet was placed on a spinning eggshell full of water. He noted how the north pole of a magnet was attracted to south, and pointed out that you could reverse the polarity of a compass needle by rubbing it in a certain way. He also suggested that 'in the magnet is a trace of the world', implying that something in the Earth must underlie the phenomenon.[22] However, it was more commonly believed that the needle pointed to the north pole of the heavens.

By the mid-thirteenth century, magnetism was a routine topic of discussion in works of natural philosophy, mentioned by Thomas Aquinas and Albertus Magnus, among others. Magnets increasingly featured in contemporary poetry, too. Stories spread not only of their marvellous ability to guide sailors but also, conversely, of their power to sink ships by drawing out the nails which secured their timbers. By the time Pierre le Pèlerin – Peter the Pilgrim – a scholar and soldier from northern France, came to compose a treatise on magnetism in 1269, he had an extensive body of knowledge to draw on. But he extended it with substantial new experiments on polarity, described the use of both wet and dry compasses and suggested how magnets might even be used to make a perpetual motion machine. He also summarised the qualities of a good experimenter. For Pierre, mere book-learning was clearly insufficient:

> A master (*artifex*) of this work must understand the natures of things, and should not be ignorant of heavenly motions. He must be skilful at working with his hands, so that he can produce marvellous results using this stone. Indeed through his efforts he may manage to correct some errors, which he could never do with natural philosophy or mathematics alone, if he lacks dexterity. For many people lack practical skill in the hidden arts ... Truly, there are many things which our minds have mastered, but which our hands cannot accomplish.

Pierre's succinct, clearly structured *Letter on the Magnet* was justly popular and unsurpassed throughout the rest of the Middle Ages. We find it alongside works of mathematics, optics and astrology in many university and monastic books, including the Merton Priory manuscript we encountered in earlier chapters.[23]

Geoffrey Chaucer was, as we already know, an astute observer of daily life in John Westwyk's day. He was also, at the time of the Bishop's Crusade, the controller of customs for the busy wool quay at the London docks, where he met countless merchants and mariners. Yet when he outlined his cast of pilgrims in the General Prologue to the *Canterbury Tales*, sketching the wise and worthy Knight with his chainmail and the scholarly Clerk with his books, Chaucer saw no need to include either a compass or a map as emblems of the rugged, tanned Shipman:

A shipman was ther, wonynge fer by weste;	A shipman was there, dwelling far in the west;
For aught I woot, he was of Dertemouthe.	as far as I know, he was from Dartmouth.
He rood upon a rouncy, as he kouthe,	He rode – insofar as he knew how – upon a carthorse,

In a gowne of faldyng to the knee.	in a knee-length woollen gown.
A daggere hangynge on a laas hadde he	He had a dagger hanging on a cord
Aboute his nekke, under his arm adoun;	around his neck, down under his arm;
The hoote somer hadde maad his hewe al broun.	the hot summer had made his hue all brown.
. . .	
But of his craft to rekene wel his tydes,	But of his skill to reckon well his tides,
His stremes, and his daungers hym bisides,	his currents, and the local hazards,
His herberwe, and his moone, his lodemenage,	his pilotage, his Moon, and his navigation,
Ther nas noon swich from Hulle to Cartage.	there was none other such from Hull to Carthage.
. . .	
He knew alle the havenes, as they were,	He knew all the harbours, how they were
Fro Gootlond to the cape of Fynystere,	from Gotland to the Cape of Finisterre,
And every cryke in Britaigne and in Spayne.	and every inlet in Brittany and in Spain.
His barge ycleped was the Maudelayne.[24]	his barge was called the Madeleine.

Whether or not this master navigator had left his compass on his ship, two key qualities distinguished him in Chaucer's view: tidal awareness and local knowledge. Alexander Neckam had admitted that the causes of the tides were 'not yet perfectly resolved'. Even the ancients had been stumped by the question, he pointed out defensively. The issue still troubled philosophers in Galileo's day: the great seventeenth-century astronomer believed that the tides resulted from the Earth's rotation, like sloshing waves of bathwater stirred up by an overexcited child. Yet, as Neckam himself noted, their connection with the Moon had long been obvious to the unlearned. Scholars, too, while musing over their precise mechanisms, were in practice still content to predict the flood and ebb from the age of the Moon. At St Albans around 1250, brother Matthew Paris produced a simple table giving the time of one high tide at London Bridge for each day of the lunar cycle. It was based on the principle that the flood would be forty-eight minutes later with every passing day. A century later, the abbey clockmakers built the same information into a dial on Richard of Wallingford's monumental invention.[25]

Experienced mariners like Chaucer's Shipman relied, above all, on their own local knowledge. Yet despite that character's wide-ranging skill, unsurpassed in time and space from the ancient Mediterranean maritime power of Carthage to the thriving trading port of Hull, Chaucer hints that the Shipman knew some areas better than others. He could navigate 'every creek' on the coasts closest to his home at Dartmouth in the west of England, but over a wider area from the Hanseatic port of Gotland in the Baltic to Finisterre on Spain's Atlantic coast, he knew just the larger harbours. It is likely, therefore, that such a shipman would have carried some sailing instructions to support his memory and assist in less familiar waters. Such instructions might take their details from pictorial charts; in turn, charts were amplified with details from written recollections. The descriptions were rarely intended for long-term preservation – especially as ports grew and declined and coastal sands shifted – but some do survive. One guide, written in Alexander Neckam's day, described all the havens and hazards from the city of York to the eastern Mediterranean, passing first down the River Ouse and out to sea via the Humber estuary, and later through the straits of Gibraltar. We learn, for example, that at Orford on the east coast of England there is 'a good town and a good castle; but the entrance to the port is difficult, because there is a sandbank named "Shinhill" in the middle of it'. The author – possibly a Yorkshire parson named Roger of Howden – showed his wide-ranging knowledge of many harbours on the pilgrimage route to Jerusalem. At Ribadeo on the north coast of Spain, he recalled, 'there is a good, deep port with good holding [for your anchor], but better on the left-hand side'. The details become more sparse as the pilgrimage continues. From Sicily to Alexandria, we learn, is eight days' sailing with good winds. Yet to go on inland from Alexandria to Cairo, only about one-tenth of the distance, is another five days' walk. From Damascus to Baghdad, still only half the distance of the Mediterranean passage, took twenty-six days. It is clear that, for medieval pilgrims and traders, waterborne travel was often much easier than on land; we should think of the seas as highways rather than barriers.[26]

As sailors passed on the benefit of their experience to their successors, they naturally included tidal details, making their instructions rather more useful for passage-planning than a map. Such tidal details appear in the earliest surviving sailing directions in English, written in the early 1400s, just a few decades after John Westwyk went to sea. To cross the English Channel from the Kent coast to Calais, as Westwyk did, the anonymous author tells us of the best time to leave, what course to steer – and, above all, to avoid the notorious sandbank known to centuries of mariners as 'ship-swallower':

> If you are bound to Calais haven and are anchored off the Downs, and the wind is west-south-west, you must raise anchor at a north-north-east Moon, and get the steeple into your weather-vane. Then set your course east-south-east, and afterwards the wind and tide will serve your course. And be sure to seek Calais haven at a south-south-east Moon ... And if you turn in the Downs, come no nearer Goodwin Sands than IX fathoms.[27]

These references to the Moon's cardinal direction may baffle modern navigators accustomed to precisely calculated tide-tables, but they represented ubiquitous medieval convention. The time of high tide at a given location was specified in relation to when the new Moon crossed the meridian. So at Calais, for example, high tide took place 'at a south-south-east Moon', that is, shortly before the new Moon was in the south. (The Catalan Atlas gave this information for fourteen ports on the French and English coasts; the negligible tide in the Mediterranean made it irrelevant much closer to home.) At other times of the month, mariners could adjust this estimate at a rough rate of forty-five minutes – or one point of the thirty-two-point compass – per day. This is why it was essential for Chaucer's shipman to know his Moon. There was also a finger-counting method to find it, much like the manual computus we met earlier. Let us say, for example, that you wish to depart the sheltered anchorage of the Downs, just offshore of the port of Deal, when the Moon is two days old. Adjusting for the Moon's age, you should weigh anchor when you see the Moon two points round from north-north-east: that is, north-east. As well as watching your compass for an east-south-easterly course, you can use the church tower directly behind you, keeping it in line with your stern-mounted weathervane to help maintain the correct bearing.[28]

Experienced navigators, you may have noticed, saw no need to use instruments such as astrolabes to monitor their position during such regular journeys. Only in the fifteenth century, when Portuguese explorers began to colonise the islands of the Atlantic Ocean, did it become necessary to keep track of latitude. Then, as navigators spent many days out of sight of land in unfamiliar seas, they employed well-established methods of measuring the altitude of the Sun or Pole Star. They designed instruments like the cross-staff and the mariner's astrolabe, based on those old principles of astronomy but better suited for use in windy, unsteady conditions. They also had access to fresh geographical texts, not least Ptolemy's *Geography*. Its mathematical maps with lines of latitude and longitude were known to Islamic cartographers by the ninth century but reached European readers only in the fifteenth. Even then, they were more interested in Ptolemy's lists of classical place-names than in his projections. In any case, John Westwyk certainly made his way to Calais with no need for instruments, and probably with no map on board either.

If he suffered from seasickness, he could have followed the recommendation of one Milanese physician: to prepare by drinking a little seawater for a few days beforehand. Rather more blunt advice came from Ibn Sina (Avicenna). Queasy travellers struggling to master their stomachs on lumpy seas could try pomegranate, quince or sour grape juice; but the best thing, said that Persian polymath, was simply to put up with it until you got used to it.[29]

The Bishop of Norwich must have spent a substantial portion of his parliamentary funding on transporting his troops across the narrow strait of the English Channel from the Kent coast to Calais. Early in the Hundred Years War the English commanders had been able to requisition merchant ships for such voyages, but ship-owners protested vociferously about such uncompensated commandeering of their assets. The sailors, too, complained that they were poorly paid and often had to stay in port awaiting orders for months. In any case, the single-masted trading cogs were not well suited for military transport. They were designed for bulk cargo, with deep holds, but the soldiers and their horses – three per man-at-arms – required space on deck. For all but the shortest passages the English government was forced to charter barges, galleys and larger square-rigged carracks, at vast expense, from Germany or Portugal, Genoa or Gascony: whichever European merchants would do business with them at a particular moment. For the half-day hop across the Channel, however, it might still be feasible to ferry the crusaders in multiple trips back and forth. So it is likely that John Westwyk made the uncomfortable trip in a small cargo ship or fishing vessel.[30]

By mid-May 1383 Henry Despenser had eight thousand men at Calais, and was ready to launch the campaign proper. It began with stunning success. The army marched along the coast to Gravelines. 'Our men had the banner of the Holy Cross before their eyes', the St Albans chronicler gushed, 'Keeping their minds fixed on their crusading cause and the absolution of their sins, they considered it glorious to conquer, but gainful to die, for that cause.' They quickly took Gravelines, killing large numbers of the inhabitants and capturing quantities of wine, salted meat and corn, cargo ships and fishing boats. The town held so many horses that the crusaders bought and sold them at only a shilling each; 'thus many of our men who had come as foot-soldiers unexpectedly became cavalrymen.'[31] Next they marched on Dunkirk, which quickly surrendered.

The French army in Flanders had mostly been disbanded the previous December, after its long campaign against the Ghentish uprising. King Charles VI had taken the remaining troops back to Paris, to put down protests against tax rises. He had been forewarned of the English invasion but had done nothing to raise a new army. The defence of Flanders was thus left to veteran troops loyal to the count, alongside some French garrisons and larger numbers of untrained local levies. A few hours after the crusaders had entered Dunkirk, this mixed Flemish army approached from the south. Beneath looming thunderclouds, the nervous crusaders came out to meet them:

> There rectors and vicars, who had been enticed by the gift of absolution, now exposed to danger, appreciated how sweet their own homes were; monks and canons realised how good obedience is; and mendicant friars saw how much easier it is to beg for alms in their own country.

Yet, as lightning flashed overhead, the outnumbered crusaders proved unexpectedly resilient – and lethal:

> Men who were inexperienced in warfare, delicately educated, nurtured in peace and quiet, might have lost heart, if the Spirit of the Lord had not filled them with fortitude . . . Indeed it turned out that some of the monks killed sixteen men in that battle. It was clear that the longer they had matured in the leisure of the cloister, the more they surpassed others in their bravery.

The St Albans chronicler had clearly changed his mind about the crusade. Despite his earlier criticism of John Westwyk and the other brothers who had abandoned the cloister, he concluded that God had blessed the Pope's crusaders. It was, he pointed out, St Urban's Day.[32]

This was the high point of the expedition. Two days later the French began to muster a new army. Meanwhile, as news – and plunder – of the victories reached England, large numbers of untrained, unarmed men rushed to join. Rural peasants, apprentices from towns, and more monks: they dressed up in white hoods and red crosses but brought no equipment or food with them. These new recruits joined Despenser as he laid siege to Ypres, a town of great strategic value – though its inhabitants were mostly already supporters of Pope Urban. The crusaders gained reinforcements from Ghent, but were unable to break through the town's defences. As the siege extended into June and July, the army's provisions and supplies of clean water were stretched thinly across the teeming encampment. Forced to drink fetid water in the summer heat, as the St Albans chronicler explains, 'a deadly plague broke out amongst our men, and every day many died from dysentery'.[33]

Over at Malmesbury Abbey, they viewed events somewhat differently. The crusade's early success had not blunted the chronicler's suspicion of the 'warlike bishop' and the 'armed priests and false religious' who accompanied him. 'He besieged Ypres,' the chronicler recorded grimly, 'and the townspeople defended themselves bravely and killed many men. And God struck them in the backside and they died of bloody flux.'[34] The Benedictine historians, then, agreed that the crusaders suffered from intestinal disease. But was the cause environmental, or divine? Were they killed by God, or by their drinking water?

The short answer is that it could easily be both. In Chapter 5 we saw astrologers discussing whether the Black Death was caused by the stars or mankind's sins, and read Matthew Paris musing whether the drunken Brother William Pigun on the Tynemouth toilet died from cold or by the hand of God. Scholars had long considered such questions. If God did cause disease, was he punishing individuals for their failings, or whole communities, or the entirety of mankind for humanity's original sin in the Garden of Eden? Jesus had, on more than one occasion, been reluctant to blame individuals for their misfortunes, but it was hard to ignore the possibility of punishment, especially at moments when sickness struck an outwardly healthy person with dramatic speed.[35]

Medieval responses to disease were never one-dimensional. On the Seventh Crusade in 1250, for example, the French king Louis IX had such a bad bout of dysentery that the seat had to be cut out of his breeches. His biographer focused on his humiliation, the king riding undignified on a little packhorse, accompanied by only one knight as his army fled by sea, and finally taken prisoner by the Egyptians. In this account Louis's loss of bodily control was inextricably linked to his shameful loss of military control.[36] Yet even if the sick were receiving retribution for their sins, that did not prevent their healthier neighbours treating them with compassion. That is what the St Albans monks did at their two homes for lepers. Although you might pray for a miraculous cure, you could still try practical methods of medicine while you waited for divine intervention.

These varied views of disease gave rise to varied treatments. In the Middle Ages the term 'dysentery' described a collection of symptoms with various possible causes – much as modern microbiologists categorise it as bacterial, viral or parasitic. (Recent enzyme-based tests on crusader latrines in the Levantine city of Acre have confirmed that the amoeba that often causes the condition certainly did afflict the Latin armies.) Common remedies for dysentery included rennet from the stomach of a hare, aged cheese, and spring water blessed by prayer. Yet understanding of intestinal disease, like much of medicine, developed fast in the fourteenth century. In the era of plague, physicians were naturally drawn to investigate the environmental causes of sickness. Meanwhile, some experts showed increasing awareness that, at inappropriate doses, strong medication itself might damage the bowels. They blamed incompetent apothecaries and uneducated healers, who seemed increasingly outdated as medical training became more structured and professionalised in the growing universities.[37]

For a vivid pen-portrait of the state-of-the-art physician, let us introduce another of Chaucer's Canterbury pilgrims:

With us ther was a Doctour of Phisik;	With us there was a Doctor of Medicine;
In al this world ne was ther noon hym lik,	in all this world there was no one like him,
To speke of phisik and of surgerye,	to speak of medicine and surgery,
For he was grounded in astronomye.	for he was trained in astronomy.
He kepte his pacient a ful greet deel	He took great care of his patient
In houres, by his magyk natureel.	at propitious hours, by his natural magic.
Wel koude he fortunen the ascendent	He knew well how to calculate the ascendant
Of his ymages for his pacient.	of his images for his patient.

He knew the cause of everich maladye,	He knew the cause of every malady,
Were it of hoot, or coold, or moyste, or drye,	whether it were hot or cold, or moist or dry,
And where they engendred, and of what humour.	and where they developed, and by what humour.
He was a verray, parfit praktisour:	He was a true, complete practitioner:
...	...
Of his diete mesurable was he,	He had a balanced diet,
For it was of no superfluitee,	for it was of no excess,
But of greet norissyng, and digestible.	but greatly nourishing, and digestible.
His studie was but litel on the Bible.	His study was but little on the Bible.
In sangwyn and in pers he clad was al,	He was dressed all in blood-red and blue,
Lyned with taffata and with sendal.[38]	lined with taffeta and silk.

Clothed in his colourful silk uniform, this is an unmistakable professional. Chaucer introduces the physician by his title in the classical languages: 'doctor' from the Latin meaning 'teacher', signifying that he had the right to lecture in the Faculty of Medicine; the Greek term *physis* alluding to Aristotle's study of change in nature. Eschewing the older English word 'leech', which still commonly meant a medical practitioner, Chaucer emphasised the physician's academic expertise, founded on astrological learning. He understands natural magic, as well as the planetary hours we encountered in Chapter 2. He can calculate the ascendant. He understands 'ymages', which could mean the figures of the constellations but also conjured up the magical talismans of Hermes Trismegistus. Above all, he understands the theory of the four humours.

The planets and stars up above each had their pure elemental qualities, like hot, dry Leo or the cold, wet Moon. Down here below the Moon, though, everything was a mixture of elements. That included the human body. As the celestial influences shifted overhead, and as humans breathed, ate, slept and moved in their environment, the balance of warmth and moisture in the body fluctuated. The physician's task was to maintain a healthy balance.

Since the earliest days of Greek medicine, medical theory had linked that balance to the fluctuations of four fluids: blood, phlegm, red or yellow bile (or choler) and black bile (or melancholy). These were the four humours, based on identifiable fluids within the body but going far beyond them in location, function and complexity. Blood, for example, was the hot, wet humour. But since everything was a mixture, the blood flowing through your veins necessarily contained small amounts of the other three humours too. Those other three each had their seats in particular parts of the human body: phlegm in the cold, moist brain; yellow bile in the gall bladder; and black bile in the spleen. Each person had their own innate configuration – or complexion – of humours. This not only affected your appearance and energy levels but also your temperament. Modern English retains humoral words like 'sanguine' (blood-led) and 'phlegmatic' to describe the personalities that might result.

The humours were essential for the nutrition and maintenance of the body. However, there was considerable debate about their precise functions, as well as how they interacted with the three *virtues*, or powers, in the body. Of these three, psychical virtue controlled the mind and senses, from the brain and nervous system. Natural virtue, through the digestive system and particularly the liver, governed nutrition and growth. Vital virtue kept your pulse going and your lungs breathing, via the organs in your chest. But which had primacy: the brain or the heart? And what about reason or the emotions: could they be pinned down to a particular part of the body? Such questions required physicians like Chaucer's to study widely:

Wel knew he the olde Esculapius,	He well knew the old Asclepius,
And Deyscorides and eek Rufus,	and Dioscorides, and also Rufus,
Olde Ypocras, Haly, and Galyen,	old Hippocrates, Haly Abbas, and Galen,
Serapion, Razis, and Avycen,	Serapion, Rhazes, and Avicenna,
Averrois, Damascien, and Constantyn,	Averroës, John of Damascus, and Constantine,
Bernard, and Gatesden, and Gilbertyn.[39]	Bernard, and Gaddesden, and Gilbertus.

With this lengthy reading list, Chaucer outlined the evolution of medicine down to his – and John Westwyk's – day. The old Asclepius, Greek god of healing with his snake-entwined staff, demonstrated the physician's respect for ancient authority. Knowledge of the broad corpus of writings associated with the ancient Greek Hippocrates was essential, as was adherence to the greatest of all ancient physicians, Galen of Pergamon. Galen was so influential that medieval writers often referred to him just by his initial, 'G'.[40] Yet these authorities were not studied in stagnant isolation: equally important was the work of medical writers from the Islamic world. The tenth-century Persians 'Ali (Haly) ibn al-'Abbas al-Majusi and Muhammad ibn Zakariyya al-Razi (Rhazes), and the Andalusian Muhammad Ibn Rushd (Averroës) were deeply respected by Latin physicians, but it was Ibn Sina that they honoured to almost the same degree as Galen. His *Canon of Medicine* became the standard textbook in the university medical faculties after its translation by Gerard of Cremona.

Other hugely important translations and interpretations had been made by Constantine ('the African'). As we noted in Chapter 3, Constantine brought a library of medical books from Tunisia to the monasteries of southern Italy in the 1060s or 1070s. The Benedictines embraced them with great enthusiasm. The St Albans monks enshrined Hippocrates and Galen, alongside two thirteenth-century Italian surgeons, in the cloister window they dedicated to the art of medicine. And through a succession of monk-physicians trained in the Italian schools, they certainly benefited from Persian medical advances. One Salerno-trained monk, Warin of Cambridge, rose to become abbot in 1183. As well as hiring Alexander Neckam to head the St Albans school, Warin completely reformed the abbey's regime for treatment of the sick and elderly. He built spacious facilities for nuns suffering with leprosy, who had previously had to share space with men at the hospital of St Julian. He brought in new regulations for bloodletting, which was thought the best way to regulate the balance of humours but took an obvious toll on the monks' energy levels. Brothers who had been bled were excused from the midnight office for two days and were permitted to take their meals earlier. Realising the importance of sleep for health, he allowed the monks an additional nap on fast days.[41]

Warin's legacy was cemented by his successor, John of Wallingford, who had studied at Paris and 'could be considered a Galen in
medicine'. The abbey chronicler noted that 'he was an incomparable
judge of urine'. This was significant, since reading a urine sample was,
along with taking a pulse, the most important method of diagnosing
a patient and – more importantly – giving a prognosis for recovery
(or its opposite). The colour, quantity and consistency of urine were
all examined closely for signs of abnormalities in the bodily functions
that produced this most frequent of human outputs. In 1214 the aged
abbot John, gravely ill, wanted to examine his own urine. Bed-bound,
with his eyesight failing, he was unable to inspect it to his satisfaction so asked another medically trained monk to describe it to him.
Hearing the details, he immediately predicted that he had three more
days to live. He was exactly right.[42]

It was not only Benedictine monks who practised the medical
arts in the thirteenth century. Two years after John of Wallingford's
death, an abbot of the Premonstratensian order was called to the
deathbed of a different John: the king of England. King John was
suffering from severe dysentery (though the St Albans historian Matthew Paris blamed him for eating too many peaches and drinking too
much cider). John's counsellors summoned the abbot twenty miles
from his monastery to Newark Castle. There he took the king's confession and eased his suffering. When John passed away, the abbot
dissected his corpse, removing the viscera so that the body would
better withstand the long journey to Worcester for burial in accordance with the king's wishes. He sprinkled large quantities of salt to
preserve the excised entrails, then took them back to his abbey for
an honourable burial. Literally divided between two churches, John
might benefit from prayers at both shrines. In turn, both communities might benefit from the royal connection.[43]

However, already by this time in the early thirteenth century
the monks were retreating from their involvement in medical care.
Monasteries still included infirmary buildings and maintained
their leper hospitals, but treatment was increasingly reserved for
sick and elderly members of their own community. In part this was
because medical study – and the money that might accrue from
it – was a distraction from sacred scholarship. More important,

though, was the rise of a new medical profession, many trained in the new universities. Like Chaucer's doctor, they flaunted their qualifications with uniforms. In some regions they formed organised guilds to represent their interests. Local governments began issuing licences to regulate the profession, as well as employing municipal physicians and operating hospitals, particularly in the wealthy cities of northern Italy. It was in Italy that medical training was most advanced: by 1300, professors there were dissecting corpses to teach anatomy, and autopsies were increasingly performed for forensic purposes.[44]

A professional hierarchy began to emerge; its vestiges can still be seen in the structured branches of medicine today. It consisted, in simple terms, of scholarly physicians at the top, with practical surgeons beneath them. Next were the barbers, who might carry out minor surgical interventions like bloodletting, treating hernias or dentistry; this is why barbers' shops to this day often have a red-and-white striped pole outside them. The apothecaries, making and selling medicaments, complete the traditional picture. In reality, however, the medical marketplace was rather more mixed – as we see from Chaucer's physician, who spoke of both physic *and* surgery. Some budding physicians undertook practical apprenticeships, there were certainly scholarly surgeons, and at all levels women could be found practising, despite some legal restrictions. And away from the university centres, the vast majority of people sought treatment from unlicensed 'empirics', who might practise medicine part-time alongside another craft.[45]

Within the cities, the outward-facing orders of friars, who lived among the people they served, were well placed to tend to the sick. A few years after the bishop's army contracted dysentery at the walls of Ypres, a certain Lady Trussell sought treatment for the same disease in London. She went to a Franciscan friar named William Holme. Holme was renowned for curing the capital's rich and reputable: he had tended to the Duke of York's feet and healed the testicles of a soldier in the queen's household. Some Franciscan colleagues recorded his successful methods as part of a collaborative medical handbook known as *The Slate of Medicine* – an imaginative wiki-style project in which the compilers left space for future corrections and additions. For Lady Trussell, Holme prescribed the following recipe:

> The rind of yellow myrobalan, ½ an ounce
> and of Indian and Chebulic [myrobalan], 2 drachms [⅛ ounce] each
> Dried rhubarb, 1½ ounces
> Mix, and put 1 drachm in 3 spoons of rosewater overnight
> Leave to strain and give to drink.[46]

The marvellous myrobalan fruit was not known to classical experts on medicinal ingredients, like the influential Greek writer Dioscorides. But it became enormously popular after the seventh-century Muslim conquests spread Indian remedies throughout the Mediterranean world. (It is still valued by traditional Indian medicine for its high-tannin, astringent qualities.) In the exceptionally well-documented medical practice of medieval Cairo's Jewish community, it was the most commonly used substance, far outstripping other medicinal plants like saffron, pepper and liquorice. The twelfth-century geographer Muhammad al-Idrisi, a Moroccan Muslim who worked for the Norman King Roger of Sicily (for whom he produced a cutting-edge world map, with south at the top), recorded that myrobalan was traded in the Yemeni port of Aden. The black Chebulic variety, he wrote, grew in the mountains around Kabul, from which it got its name.[47]

The best and most costly variety was yellow myrobalan (*Terminalia citrina*), which came mostly from south-east Asia. But all kinds of myrobalan had their place, according to another friar, the Dominican Henry Daniel. Daniel quantified the intensity of their elemental qualities in his accessible Middle English *Herbal*, written in the 1370s. He noted that they were all cold in the first degree, and dry in the second. Myrobalan was not just 'sovereyn for dissintere', as he put it. 'Sumtime we use it with cassiafistula and tamarindes for to purge colre and blod'. Blood was, of course, the hot, wet humour, while choler was hot and dry. Daniel gave precise instructions for how medicines could be prepared by mixing the powdered fruits with warm water and whey.[48]

When they described the medicine which had cured Lady Trussell's dysentery, the compilers of the *Slate of Medicine* noted that the treatment was recommended not only by William Holme but also by Gilbert, the English physician who appears last on Chaucer's reading list. But a far more thorough study of dysentery was written by a French scholar whose name comes alongside Gilbert's in that list: Bernard of Gordon. Bernard completed his masterwork, the *Lily of Medicine*, in 1305, after twenty-two years of teaching in the pioneering faculty of medicine at Montpellier. It was explicitly aimed at his younger colleagues: he was careful to include elementary details and left out treatments that only the very experienced could use. Concise and neatly arranged to cover every conceivable ailment from head to heel, the *Lily of Medicine* was extremely popular, translated into half a dozen languages before the end of the Middle Ages.[49]

Bernard was keen to disentangle the confusing mess of digestive symptoms and ailments, in which dysentery had been poorly defined against other kinds of intestinal flux (discharge). He drew up a structured survey of seventeen kinds of flux, distinguishing them according to the affected area of the abdomen, internal or external causes, humoral implications, relationship with other organs, and so on. The chapter begins in typically systematic style, categorising causes which, Bernard recognised, could include medicine itself:

> Dysentery is a bloody flux of the abdomen, with excoriation and ulceration of the intestines.
>
> The causes of dysentery and other fluxes are either internal or external. If external, it may be corrupt air; or sharp, intense foods like garlic, onions, vinegar and so on; or medicines like scammony, aloe, colocynth pulp or similar.
>
> And if it arises from internal causes, they are either immediate or remote. If immediate, it may be sharp ulcerating choler, which pierces and stings; or salty phlegm . . . But if the cause is more remote, arising from disease of other organs, then it may come from the head in the form of rheumatism, or from the stomach . . .

Bernard continues with a thorough discussion of how to identify which parts of the intestines are affected. He analyses the intensity and locations of pain, as well as the contents, colour and smell of the patient's stools. From such diagnostic symptoms, he passes to the all-important prognosis:

All abdominal flux and any stool resulting from overheated yellow or black bile, which when thrown on the ground fizzes like vinegar, or if flies avoid it: if that happens at the start of the illness, it is fatal . . . All flux with a wormy and ant-like pulse, which is not eased by consumption of food or medicine, is fatal . . .

The situation was not hopeless, however, for Bernard soon proceeds to a full menu of possible treatments:

First, if the case permits, let blood; then purge according to the state of the humours. And if the cause is choler, purge with yellow myrobalan; and if salty phlegm, with Chebulic myrobalan; if melancholy, with Indian. Prepare it with rainwater containing tragacanth gum, gum arabic, and raisins. If [the cause is] salty phlegm, give very hot vinegar, and with other causes give cold vinegar. Then clean the ulcer with barley water, chickpea broth or with water of salted fish, honey, and rose oil . . .

If the cause is in the upper intestines, the medicines should be taken orally, and applied externally on the affected area. And if the problem is lower down, prepare them as an enema with unsalted goat-kidney fat, and suchlike . . .

In addition to his long list of cures, Bernard recognised the need to maintain the patient's ongoing wellbeing. All the incidental factors that might affect your health were collectively known as *non-naturals*. These conventionally consisted of six types: sleep, air, food and drink, exercise and recovery, fullness and excretion, and the emotions. They were thus only 'non-natural' in the sense of being outside the body's core functions. Conscientious physicians would help maximise their clients' wellbeing by keeping all these aspects of a healthy lifestyle in a holistic balance. So part of Bernard's treatment for dysentery included remedies to aid sleep, such as saffron, opium and egg whites. When it came to nutrition, he recognised that sufferers would be able to digest only small quantities; so he prescribed

foods that nourish in small quantities, like cockerel's testicles, fatty chicken livers, semi-hardened roasted egg yolks, lightly leavened wheat bread, toasted rice with skimmed milk, unrefined astringent bitter wine with cold rainwater.[50]

We are accustomed to modern medicines existing in an entirely different category to food. Foiled packets and folded leaflets from behind the aseptic pharmacist's counter contrast with fresh green apples shining on self-service supermarket shelves. The boundary was not always so clearly demarcated. Chaucer's Physician, remember, was 'mesurable' in his diet, avoiding excess and selecting easily digestible foods. Elsewhere in *The Canterbury Tales* we hear of a wooing clerk chewing liquorice as a breath freshener. Many foodstuffs were prized as much for their contribution to a healthy regimen as for their taste. Simple salt, according to Isidore of Seville, brought out the flavour of sauces, excited the appetite, and by improving the diner's enjoyment might increase happiness. Isidore indeed suggested that the word for 'health' itself, *salus*, was linked to salt. Preserved ginger, meanwhile, was a popular flavouring, but physicians also valued its warming, drying properties and its power to increase sexual desire. Monks in their hushed refectory had their own dietary concerns: it would not do to overindulge in earthly delicacies, but obsessive fasting was good neither for the body nor the soul.[51]

Not all cures, of course, could be accomplished with commonly available foods. After suggesting his readers try boiling lentils in rainwater with milk and a dash of vinegar, Bernard of Gordon could not fail to mention a popular panacea. 'It should be noted', he stressed, 'that theriac is highly effective on fluxes, and most powerful where the cause is cold.' Theriac was a complex concoction including dozens of variable ingredients. Originally a cure for poison and animal bites, its essential component, as Bernard explained, was snake flesh – the same Jericho tyrus snake that had led Jean de Saint-Amand to investigate magnetism. Crucially, though, the specific compound of ingredients had powers greater than the sum of its parts. It was gradually credited with more wide-ranging properties; indeed, the legacy of its almost limitless applications lingers in the English language. The phrase 'snake oil' means a spurious cure-all, and 'theriac' is also at the root of 'treacle', which originally meant something rather more complex and medicinal than plain sugar syrup.[52]

For such compound remedies, a physician might seek the services, or partnership, of a pharmaceutical specialist:

The cause yknowe, and of his harm the roote, / The cause known, and the source of [the patient's] harm,

Anon he yaf the sike man his boote. / he immediately gave the sick man his remedy.

Ful redy hadde he his apothecaries / He had his pharmacists all ready

To sende him drogges and his letuaries, / to send him drugs and his syrups,

For ech of hem made oother for to wynne – / For each of them made the other profit –

Hir frendship nas nat newe to bigynne. / Their friendship was not recently begun.

. . .

And yet he was but esy of dispence; / And yet he was moderate in spending;

He kepte that he wan in pestilence. / he kept what he earned during plagues.

For gold in phisik is a cordial, / For in medicine gold is a heart medication,

Therfore he lovede gold in special.[53] / therefore he loved gold in particular.

Thus, with a wry final nod to the connections between alchemical theory and medicine, Chaucer intimated that this doctor's motives were not always pure. Chaucer conveys some of the difficulty of the Physician's position, in remarking that he profited during periods of plague. He might be accused of exploiting the misfortunes of the sick, but to treat infectious patients put his own health at considerable risk. The alternative, to flee the epidemic, must have been tempting – but he would certainly have been accused of negligence.[54]

Chaucer was far from the only fourteenth-century commentator to suggest that physicians and apothecaries colluded to defraud their suffering patients. His friend John Gower, in the French-language *Mirror of Mankind*, laid out the charges in detail. These are the start of sixty excoriating lines of poetic indictment, completed just a few years before the Bishop's Crusade:

Plus que ne vient a ma resoun	More than is within my power to comprehend
Triche Espiecer deinz sa maisoun	Fraud the Apothecary in his shop
Les gens deçoit ; mais qant avera	deceives people; but when he has
Phisicien au compaignoun,	the Physician as his companion
De tant sanz nul comparisoun	there is no comparison:
Plus a centfoitz deceivera :	he deceives a hundred times more.
L'un la receipte ordeinera	The one writes out the prescription
Et l'autre la componera,	and the other produces it;
Mais la value d'un botoun	but the value of a button
Pour un florin vendu serra :	is sold for a florin [gold coin].
Einsi l'espiecer soufflera	Thus the apothecary whispers
Sa guile en nostre chaperoun.[55]	his guile into our hood.

Should we, then, doubt the motives of Friar William Holme in making yellow myrobalan – the most expensive variety – the main ingredient in his dysentery treatment? Perhaps, but it does seem to have cured Lady Trussell. In the end, it was patient satisfaction that mattered.

The relationship between patient satisfaction and health outcomes was a live issue in the later Middle Ages, when civic and scholarly authorities were compelled to regulate a fast-developing profession. Sometimes arbitration was required. In 1437, for example, two Paris experts clashed about the favourable days for bloodletting and the prescription of laxatives. Although they both explicitly accepted the principles laid down by Ptolemy and Haly, the complexity of astrological interpretation meant that they could not agree which days would be more medically propitious. The university authorities, keen to control the quality of medical interventions, elected two arbitrators: one a master of theology, the other a Benedictine prior. Over fifty pages of written judgement, these two theologians examined the medical almanac Master Laurent had produced and assessed Master Roland's vehement criticism of it. Day by day, they checked the position of the Moon in the zodiac, as well as its relationship to the other planets, justifying their decisions with reference to Albumasar and his Iraqi successor al-Qabisi (Alcabitius). They judged that 3 October, for example, would not be a suitable day, since although the Moon was in the beneficial sign of Sagittarius, it was in an unfavourable aspect to Mercury and Saturn. As far as they could, the arbitrators adopted a balanced position between the disputing masters, explaining their reasons for upholding one view or the other, or ruling that the evidence was inconclusive. On one point, however, they were resolute. Physicians, they stated, must adapt their practices to take account of the complexion of the patient and the elemental qualities of the disease. And for that reason, they ruled:

> all physicians and surgeons must have a full almanac, showing the sign of the Moon on any day, and which planets it relates to, good or bad. And with it they must have an astrolabe, to select – for any day, hour and fractions of hours – the ascendant sign corresponding to the sign where the Moon is, at the hour chosen for bloodletting or laxatives.[56]

Many professionals already possessed an almanac. Several sets of quick-reference tables survive in folding bindings, perfect for hanging from a loop on your belt. They were not only handy to check a celestial configuration but also to certify the doctor's authority – as reassuringly symbolic as the consultant's stethoscope today. Yet such a folding almanac would not satisfy the 1437 arbiters, who insisted on a full almanac for every practitioner.

The arbiters' additional requirement to possess an astrolabe was good news for the instrument-makers of Paris. They were well equipped to supply the demand, since the early 1400s were the heyday of the city's astrolabe workshops. The founder of the most successful and influential of them, Jean Fusoris, had died the previous year, aged over seventy. A canon of Notre Dame and master of medicine, Fusoris had largely foregone therapeutic work in favour of the family metalwork business (his adopted Latin surname means 'the smelter'). He had a rare combination of sublime craft, scientific insight and business acumen, which brought him rapid professional and social success. Not content with manufacturing the finest astrolabes in unheard-of quantities – at least a dozen survive from his workshop – he also wrote treatises on instruments and astrology, drew up mathematical tables, and built an astronomical clock for the cathedral at Bourges. His customers included the Duke of Orleans, the King of Aragon, even the Pope himself.

He may also have been a spy. On 30 August 1415, the garrison of a small coastal town in Normandy arrested a priest. The Hundred Years War was once more aflame, the English army was besieging the port of Harfleur and the priest was carrying letters from the English lines. One letter was from a close counsellor of Henry V – another Bishop of Norwich, Richard Courtenay – to Jean Fusoris. The bishop addressed Fusoris as 'my most excellent friend . . . my dearest fellow and friend'. He supplied details of the English force and inquired about the state of French preparations. The fifty-year-old canon was quickly arrested in Paris and charged with treason. The detailed trial records and statements from numerous witnesses make clear that this medical astronomer had, in his wartime business dealings, been either guilty of the charges or, at best, spectacularly naive.[57]

The canon and the much younger bishop had first met a year earlier, when Courtenay was trying to negotiate a marriage between Henry V and Catherine, the daughter of the French king. Courtenay, a three-time Chancellor of the University of Oxford, was certainly learned in the mathematical arts; he had apparently heard of Fusoris's astrological skill, and asked the canon for a consultation. They began to meet regularly and take walks together. Fusoris had recently completed an innovative seven-piece planetary computer, and the bishop offered him four hundred gold crowns for it – a considerable sum, given that Fusoris would normally sell an astrolabe for thirty crowns at most. Courtenay paid half the price up front, and Fusoris promised to write a set of instructions for the complicated new device.

They met again when Courtenay's mission to Paris resumed in January 1415. Then, according to Fusoris's trial testimony, the bishop was keen to discuss medical matters. He was very overweight and complained of light-headedness, especially when he got out of bed. Fusoris advised him to eat a little toast with spiced wine in the morning. Courtenay remarked that the young King Henry was also unwell, and complained that there were no good doctors in England. He was, he claimed, still unable to pay Fusoris the two hundred crowns he owed him, but urged the anxious instrument-maker to come to England and collect the money. If Henry's marriage to Catherine went ahead, he said, there was every chance that Fusoris could obtain a well-paid post as a royal physician. Then came a test: how likely was the marriage to succeed?, he asked the astrologer. Fusoris answered using Courtenay's own astrolabe and almanac: yes, he said, the marriage would be good for the king and country; but no, the negotiations would not be concluded during this mission. Finally, Courtenay asked Fusoris if he had ever seen an astrological chart for the king's nativity. Fusoris replied that he had not. Then, according to the trial transcript:

> the said bishop immediately led the witness [i.e. Fusoris] to his chamber, and showed him a nativity chart of his king, asking the witness if he could tell from it whether the said king would fall ill in the near future, or would have a long process of recovery. To which the said witness replied to the bishop that he was not sufficiently practised or prepared to know this; and he could not know or do this in less than a year.[58]

This was a sensible reply, since to predict the illness of a king was both theologically and politically risky.

Fusoris applied to join the next French embassy to England, as Courtenay had encouraged. His requests were twice refused. The bishops leading the delegation were concerned that Fusoris sympathised with the Burgundian faction in the simmering civil war between two branches of the French royal family. Even so, Fusoris paid his own way to the English court at Winchester that June. The archbishop and bishop who gave evidence at his trial recalled seeing Fusoris speaking with Englishmen on many occasions around the council chambers. Furthermore, they said, he was often late for meals, or missed them entirely.[59] In Fusoris's own testimony, he was struggling to meet Courtenay – though he also admitted to having a few conversations about the prospects of the marriage alliance and the possible outcomes of renewed war. The bishop did eventually introduce him to King Henry, and he presented the king with an astrolabe and several books on instruments and their uses. The king, who was rather suspicious of astrology, spoke only to thank him in Latin and French. On the last day of the embassy, Fusoris finally received most of the money he was owed, and he set out on the long journey back to Paris.

Six weeks after his return, Fusoris was arrested. He was imprisoned for several months and finally banished from Paris. Whether he was truly guilty, or simply unfortunate to stray from medical astronomy into politics, it was the end of his career in the city. Other men who were also suspected of espionage and faced trial at the same time as him were released without charge. Yet all of them were luckier than Richard Courtenay. In September 1415, just two weeks after Fusoris's arrest, Courtenay fell victim to the dysentery that ravaged the English army at Harfleur. He was only thirty-five. Henry V personally closed the bishop's eyes and sent his body back for burial in the royal tomb in Westminster Abbey. When Henry himself died in 1422, Courtenay's feet had to be amputated and placed under his armpits to make room in the tomb for his friend and master.

Only one of the 1383 monk-crusaders, the prior of Hatfield, died in Flanders. The rest survived that episode of dysentery, and retreated with Despenser's army as the failure of their siege of Ypres became clear and French reinforcements approached. They withdrew first to the strongholds they had captured earlier in the summer, but soon surrendered these in return for modest payments. The news of retreat shocked the young King Richard II. In a panic, he immediately leapt on his horse to ride the seventy-five miles from Daventry to London. He arrived at St Albans in the middle of the night and commandeered the abbot's horse – 'as if the king of France were to be killed that very night', chuckled the abbey chronicler.[60]

The public, which had applauded and contributed to Despenser's venture, shared the king's outrage. There had long been disquiet about the Church's military tendencies: 'Peter preached, but the present Pope fights,' spat John Gower in a Latin poem a few years earlier. And opposition to the long-running Hundred Years War was voiced by the controversial Church reformer John Wyclif and his followers.

The Wycliffites castigated prelates for 'blabbering forth Antichrist's edicts to send Christian men to war with each other'. They reserved particular hatred for the priests who had coerced churchgoers to fund the crusade:

> They will not give the sacraments of the altar, that is, Christ's body, to their parishioners, unless they paid their tithes and offerings; unless they have paid money to a worldly priest to slay Christian men. And if people doubt this, let them inquire truly how it was when the Bishop of Norwich went to Flanders, and killed them in many thousands and made them our enemies.[61]

The Bishop of Norwich himself was impeached when Parliament next met, shortly after the army had limped home that October. In chaotic scenes at Westminster, the bishop was so disturbed and distracted by the abuse hurled at him that he had to beg the Parliamentarians for a second hearing, in the hope of giving better answers to their accusations. Despite his protestations, he was found guilty on all four charges: for failing to raise an army of the size he had promised; for failing to keep them in the field for a full year; for not notifying the king of the names of his captains; and for refusing to share his command with a worthy lieutenant.[62] Stripped of his temporal assets, he withdrew to Norwich and confined himself to diocesan business for much of the next decade.

John Westwyk, too, went underground for ten years. The chronicler Thomas Walsingham recorded that all the monks of the St Albans family – apart from the unfortunate prior of Hatfield – returned to the cloister, where they were received mercifully. 'Never again did they enjoy perfect health,' Walsingham mused, 'but they all experienced the Abbot's unexpected grace.'[63] Two of the monks, John of Bokeden and William Sheppey, had evidently developed a taste for travel, for they soon left again, taking up papal chaplaincies which the cash-strapped Urban VI offered for sale. But if our John had joined them, the chronicler would surely have said so. Most likely he was simply grateful to be readmitted at St Albans; grateful to still be alive. He spent the next decade – like most monks, most of the time – doing nothing to trouble the abbey historians. Yet in that obscurity he was certainly continuing to study science. For when we next encounter him in 1393, ten years on from the crusade, he is making his most notable mark on astronomy: a unique computer he had himself constructed. In the surprising setting of London, and in the fashionable English language, he wrote out the clear, user-friendly instructions for his enormous *Equatorie*.

7

Computer of the Planets

'I will call this circle the limb of my *Equatorie*, which was constructed in the year of Christ 1392, the last midday of December' (image 7.1). A clear voice, announcing its achievement across the centuries, this leapt out at Derek Price on that cold Cambridge morning in 1951. Price was captivated by the thought that it might be the voice, the handwriting, of Geoffrey Chaucer. But we know that it is Brother John Westwyk. We have chased his shadow across England and over the seas. Now, finally, he emerges from a decade's obscurity and proudly introduces us to his new invention.

On the opposite page, in his clear informal script with a deep brown ink, John tells us where he is. 'The year of Christ 1392 complete, the apogee of Saturn was – on the last midday of December at London – I say, the apogee of Saturn in the 9th sphere was 4 double signs, 12 degrees, 7 minutes, 3 seconds etc.'[1] That one disordered, slightly chatty sentence discloses rich details of Westwyk's scientific project. He had made an equatorium – an equation-solver, a computer – and he was calibrating it to give the precise positions of the planets. He could not possibly have constructed it in all its complexity during the few daylight hours of New Year's Eve; that final noon of the year was just a reference point to simplify calculations.

7.1. The first page of John Westwyk's equatorium treatise. 'This cercle wole I clepe the lymbe of myn equatorie / þat was compowned the yer of crist 1392 complet the laste meridie of decembre.'

240

But he had certainly made it for use amid the bustle of England's largest city. Now, in 1393, he was writing instructions for the instrument. He used English, fast developing as a formal written language. He was still sketching out his instructions: between the lines of that one draft sentence he inserted both the word 'last', to make it clear which December midday he meant, and the word 'complete', to clarify that he was speaking in terms of completed years. (I am writing this in June 2019; a medieval scholar might say that the year is either '2019 incomplete', or '2018 complete'.) And with his use of 'double signs', Westwyk was signalling his adoption of the most up-to-date astronomical tables, refined by Parisian experts from ground-breaking work done a century earlier in Spain.

Tables, as we have seen, were an essential tool for astronomers. This manuscript contains John Westwyk's own collection of them, carefully computed for use in London. It is strikingly large, with pages thirty-seven centimetres high – more like a coffee-table book than the compact scientific compilations that filled medieval libraries. The eighty leaves of parchment, albeit only of mediocre quality, would not have come cheap – perhaps a couple of shillings, a week or so's wages for an average worker. But London was the place to obtain them. Paternoster Row, the narrow lane running along the north side of St Paul's Cathedral, was crowded with stationers' stores, jostling with the workshops of scribes, illuminators and bookbinders.[?] There John surely came to buy his parchment. Some of its pages were already filled with pre-packaged tables. Many remained blank for his own creative additions.

London in the 1390s was a busy, noisy city of about forty thousand souls. Its population had halved since the beginning of the century, largely because of repeated outbreaks of plague and other disease. Yet the rising wages and living standards which followed a fall in population meant that the city continued to attract immigrants from across England. In large part it remained within its old Roman walls on the north bank of the Thames, facing across the river to the seething suburb of Southwark. New construction was, to be sure, gradually filling in the space between London and Westminster, two miles upriver with its abbey-palace complex. But the city itself was not yet overcrowded. Gardens were common, and its inhabitants benefited from piped fresh water, channelled from springs north-west of the city to an outlet near St Paul's Cathedral. Just a few minutes' walk away on Broad Street, at the heart of the city on the west-facing slope of Cornhill, was the St Albans inn, the abbey's lodging and office in the city. A family named Westwyk sold candles from a shop on the same street. Whether or not they were relatives of our John, it was surely in this neighbourhood that he made his calculations and observations, drawing, cutting and engraving his enormous equatorium.[3]

The equatorium was designed both to represent the motions of the planets and compute their positions. We shall soon see how it worked. First, though, we must understand how it depended on the astronomical data arrayed in tables. Tables fill most of John Westwyk's book; he copied them carefully on to his large leaves of sheep- or calf-skin. The tables laid out daily and annual changes in the planets' longitudes on their wandering paths through the stars.

Their design was the product of centuries of gradual refinement, by astronomers committed to enhancing their user-friendliness. Ptolemy's *Almagest* had included many astronomical tables, but they had been interspersed throughout his great treatise. That was impractical for astronomers who did not want to thumb through pages of theoretical arguments and proofs. Realising this, Ptolemy issued a revised, slimmed-down collection: the *Handy Tables*. Their format inspired later Muslim astronomers, who also incorporated Indian models for their sets of tables, known as *zijes*.* The tables of eastern Muslims like al-Khwarizmi and al-Battani influenced their Spanish counterparts in turn. The most important of these western Muslim astronomers was the 'blue-eyed' al-Zarqali (Arzachel). Working with colleagues in Toledo in the 1060s and 1070s, al-Zarqali put together a fluid, evolving set of materials that became famous as, simply, the 'Toledan Tables'. These tables, as much as any wordy treatise, brought Islamic astronomy – and Christian respect for Islamic science more generally – to the heart of Europe. It is no surprise that John Westwyk cited Arzachel explicitly in his own personal compilation.[4]

The Toledan Tables were enough of a household name by the

* The Arabic word *zīj* (plural *azyāj*) derives from Persian. It initially meant a thread. By extension from the criss-crossing threads of a fabric to rows and columns, it came to be used first for a single table, and later a set of tables.

fourteenth century for Chaucer to drop them into his *Canterbury Tales*. *The Franklin's Tale* features a French astrologer:

His tables Tolletanes forth he brought,	He brought forth his Toledan Tables
Ful wel corrected, ne ther lakked nought,	fully updated, nor did they lack anything:
Neither his collect ne his expans yeeris,	neither the grouped or single years,
Ne his rootes, ne his othere geeris.[5]	nor the baseline positions, nor any other gear.

In fact, by Chaucer's day the original Toledan Tables were outdated technology. Chaucer surely knew that, so he may have been purposely painting the astrologer as old-fashioned. Or perhaps it was the narrating Franklin (a middling landowner), who protests his ignorance of technical terminology, that was supposed to be old-fashioned. Either way, a team of astronomers working for the Castilian King Alfonso X, 'the Wise', led by two Jews, had overhauled Arzachel's work in the 1270s. This team wrote in their patron's Castilian Spanish, but their tables were tweaked and translated into Latin in Paris soon after 1320. These revised 'Parisian Alfonsine Tables' quickly spread all over Europe. We get a sense of the vibrant international scientific networks that transmitted them from the fact that one popular Latin version of these Judeo-Spanish tables, produced by an astronomer from Amiens in northern France, was dedicated to an Italian churchman who was dean of the Scottish city of Glasgow.[6] Copernicus was still using them for his revolutionary theories more than two hundred years later.

One innovative feature of the Parisian Alfonsine Tables was their rejection of the multi-table format described by Chaucer's Franklin. Traditional tables had laid out the motions of the planets in single 'expanded' years and 'collected' groups of twenty or twenty-four years, as well as in months, days and hours. The new-style tables showed only the motion in days, from 1 to 60. This simplified daily presentation enabled astronomers to work not only with the 365¼-day Christian calendar starting from the Incarnation of Jesus, but also with the 354-day Islamic calendar, whose era started from the Hijra (the migration of Muhammad in 622 CE), and even with the Persian Yazdijird era. Since the new layout showed only days, it did require a lot of laborious multiplication and division, compared with the old sets of tables that needed only a few easy additions.

To help users with this, the table-makers made a small but significant change. They divided the 360-degree zodiac into six signs of 60 degrees, rather than the traditional twelve signs of 30 degrees. John Westwyk was emphasising his use of those 60-degree segments when he gave the position of Saturn's apogee in 'dowble signes'. This simple change meant that every column was sixty times the one next to it: one sign was 60 degrees, just as one degree was sixty minutes and one minute was sixty seconds. Astronomers could now compute large multiples and tiny fractions just by switching sexagesimal columns. If, for example, you know that the mean Sun moves 0;59,8° – a little less than a full degree along the ecliptic – in one day, it must move 0;0,59,8° in one-sixtieth of a day, which is twenty-four minutes.* For a certain number of hours, you can multiply that number using the multiplication tables that often came with the planetary predictors. John Westwyk's pre-packaged set had some tables that could multiply any number up to 60 x 60. They gave the result in sexagesimal format, saving the user considerable effort.[7]

Image 7.2 shows a new Parisian-style table in John Westwyk's handwriting, with days numbered 1 to 30 down the left and 31 to 59 on the right. He could afford to miss out the value for sixty days, of course, because it was the same as the value for one day, only shifted one sexagesimal column to the left. It is worth taking a closer look at this page, as it provides a rare insight not only into the intricate mathematical methods of medieval science but also into the mindset of practitioners like John Westwyk. Let us examine a little of his meticulous copying and computation in the busy city.

The tables allow you to find astronomical data for any moment in time: past, present or future. All you have to do is calculate

---

* A reminder of the standard sexagesimal notation used by historians: the degrees are followed by a semicolon, and further fractions are separated by commas.

7.2. Table of the mean daily motion of apogees and fixed stars. Rows of days, 1 to 60. Columns of signs (S), degrees (D), minutes (M), seconds (2nds), and so on up to sexagesimal ninths. The value for one day is 0,0;0,0,4,20,41,17,12,26,37° (see below for why that is fractionally larger than it should be).

a multiple – or a fraction – of the daily motion and add it to a baseline reference point. The Franklin clearly understood the importance of these baseline 'root' values. John Westwyk knew them by the Latin term *radix*. The Alfonsine Tables provided root values of all the main planetary motions, for eras ranging from the Flood (Thursday, 17 February, 3102 BC) to the 1252 coronation of King Alfonso, via Alexander the Great, the Hijra and the Christian epoch. We have just heard John state with crystal clarity that he was using AD 1, 'the year of Christ'. So if, for example, he wished to find the mean Sun on New Year's Eve 1392, he had to start by working out the number of days in the 1,392 years since Christ's Incarnation. Next he multiplied that number by the Sun's daily mean motion. Finally, he added the result to its baseline radix for 1 January AD 1.

That calculation would give you only the *mean* Sun, not its true position. This is because, as we saw in Chapter 4, the Sun does not move at a constant rate around the zodiac. The Whipple Museum astrolabe has an off-centred calendar, because the Sun was thought to move on an eccentric circle. When it is furthest from us, at its apogee, it is moving slowly through the summer signs. It is closest to us, at perigee, in the winter.

Every planet had its own eccentric circle, displaced in the direction of its apogee (image 7.3). (They are a pretty good first approximation to the elliptical orbits of modern astronomy.) This direction was essential to any planetary calculation. That is why John Westwyk informed us of the longitude of Saturn's apogee. When he made his equatorium, he marked the centre of Saturn's eccentric circle on a line from the Earth to that longitude. He did the same for all the other planets, including the Sun.

Yet that line to the apogee, which John carefully engraved 'with a sharp instrument', could not be fixed permanently. For, as John makes clear, the planets' apogees themselves moved over time. This was a result of precession – the slow drift of the constellations across the framework of celestial equator and ecliptic. Chaucer's Franklin shows off his knowledge of precession when he remarks that the French astrologer 'knew full well how far [the star] Alnath was shove / from the head of that fixed Aries above'. This stellar drift

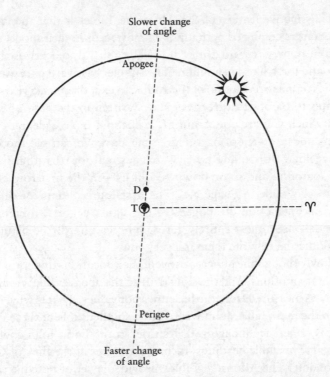

7.3. The Sun's eccentric circle, with its centre (D) removed from the Earth (T) along the *line of apsides*. It appears to move more slowly when it is near its apogee. The line of apsides slowly rotates. Its position is measured from the vernal equinox or 'Head of Aries' (♈). Every planet has its own eccentric circle, with the centre displaced from Earth in a particular direction. All except the Sun have at least one more circle in their model.

was practically impossible to observe, since the fixed stars and apogees took nearly a hundred years to move just one degree. It is not surprising, then, that astronomers disagreed on the precise nature of this motion (which is now understood to be a result of the Earth's axis wobbling as it turns, like an unsteadily spinning top). Ptolemy had argued that precession was a simple uniform motion, but the observations of later astronomers indicated that it was slowing down and might ultimately change direction.

A distinctive feature of the Alfonsine Tables is that the Parisian astronomers combined both kinds of movement in their model of precession: a slow eastward drift, adjusted by a slightly quicker oscillation, back and forth like a pendulum. The simpler eastward part was fixed at one complete revolution of the zodiac in exactly 49,000 years. That equates to 0.00002 degrees per day. Written in the base-60 system with which you are now familiar, it is about 0;0,0,4 degrees – that is, 4 ÷ (60 x 60 x 60) of a degree per day. We can see those four sexagesimal thirds, and twenty sexagesimal fourths (that is, 20 ÷ 12,960,000) – and so on down to ninths – in the first row of John Westwyk's table in image 7.2. The first four columns of this very slow 'mean motion of apogees' – the signs, degrees, minutes and seconds – are almost entirely zeros. John can hardly be blamed for not bothering to write them in every row.[8]

Slowly though the apogees moved, their locations underpinned all planetary motions. And the slow drift of the apogees, like everything else, was measured from the baseline root values. That is why, underneath the main table, John wrote out a smaller table of these radices for easy reference (image 7.4). Its title, in the Latin John continued to use for his table headings, is 'mean apogees at the time of Christ, at London'. The Alfonsine Tables provided a set of baseline radices for the Christian epoch, of course, but they had been drawn up for Toledo.[9] And midday at Toledo was not the same thing as midday in London.

The time difference between London and Toledo was tricky to

7.4 (expanded from image 7.2). *Auges medie ad tempus Christi London.* Radices of the mean apogees of Saturn, Jupiter, Mars, the Sun and Venus, and Mercury.

measure. One St Albans list of almost two hundred global locations stated that the two cities were separated by a longitude of 5 degrees, but medieval estimates ranged from 4 degrees (the true value) to ten. Whoever calculated the London radices in John's little table adjusted them from the standard Toledo data by 8° 26′ of longitude. That is equivalent to a time difference of thirty-three minutes and forty-four seconds. Since midday in London was half an hour before midday in Toledo, the radices were reduced – by a fraction of the daily motion equal to 8;26 ÷ 360. The daily mean motion of the apogees is line one in image 7.2: 0;0,0,4,20,41,17,12,26,37° per day. A rather laborious calculation of 8;26° of that yields 0;0,0,0,6,6,24,41,51,9°.[10]

Bear with me, for now the medieval mathematics get revealing. What happens when we lay out the required subtraction?

	S	D	M	2nd	3rd	4th	5th	6th	7th	8th	9th
Mean apogee of Saturn (Alfonsine)	3	53	23	42	4						
minus adjustment for 8° 26′ of time						6	6	24	41	51	9
TOTAL	3	53	23	42	3	53	53	35	18	8	51

This is the total we find for Saturn in the first line of John Westwyk's little table. Now, it is obvious from the briefest glance at image 7.4 that all the planets share the same values at the right-hand end of the table. We can see how that happened. The Toledo values were rounded to sexagesimal thirds, and the London adjustment was much less than one sexagesimal third. Subtracting the same tiny number from a range of much larger, rounded numbers means that the conscientious medieval scholar was effectively subtracting from zero in every column. His half-hour adjustment from Toledo to London was meaningless.*

You might well wonder why any astronomer would care to

* The creators of the Parisian Alfonsine Tables did something similar: all the AD 1 radices for the apogees end in that same 4.

calculate down to the level of sexagesimal ninths. These are unimaginably tiny fractions. The 37 that appears in the column of ninths for one day's motion of the apogees (image 7.2, line one) is equal to one 98,000,000,000,000,000th part of a complete circle. It would take approximately 750 billion years for these daily 37s to accumulate to even a degree's difference in the longitudes of the apogees. Such precision clearly does not reflect observational accuracy, but it came from calculations carried out by standard methods in accordance with Ptolemaic theory. An astronomer would have to be brave, even arrogant, to discard the painstaking work of his forebears. And it required considerable mathematical sophistication to reject the opportunity for precision. I am guilty of the same lack of sophistication whenever I bake a cake: I measure out the sugar to the last half-gram, even though I am well aware that the eggs may vary widely in weight. So John Westwyk was left with a little table of apogees that was very precise but not necessarily very accurate.

Such obsessive calculation, so often carried out by monks, may well have been an arithmetical exercise, even a meditative activity. In this case, though, the calculations were probably not John's own work. At the bottom of the column of eighths, second from the right, we can see two 4s instead of 8s (image 7.4). That is a clear copying error: an indication that John transcribed this table from another source. If the mistake was already in his source, he failed to spot it.

A few pages earlier, though, we can catch John himself in the act of calculating. He was working on another table of the mean motion of the apogees. This one gives motion in years rather than days. It has an unusual layout (image 7.5). The astronomer who compiled it wrote out rows for 1, 2, 3, 4 years, then switched to intervals of four years, up to 56, and finally wrote out 1, 2, 3 years again. If we examine the rows closely, we discover why. The first three years are years of 365 days each, while all the rest, including the last three, are years of 365¼ days. This clever layout enabled users to factor in the right number of leap years, no matter where they were in the cycle.

A striking feature of the table is that, below the first three rows, two columns in the middle of the table are almost empty – suspiciously so. This, it turns out, is because this *annual* table of apogees was calculated by multiplying the *daily* motion (from the big table we

7.5. Mean motion of the apogees in years. Years (in leftmost column) in sequence 1, 2, 3 (of 365 days), 4, 8, 12 . . . 56, 1, 2, 3 (of 365¼ days).

were just looking at) by 365 or 365¼, as appropriate. You may think that that sounds perfectly sensible. But wait: it turns out that those daily values, so precise down to the last thirty-seven ninths, are not the ones we find in other manuscripts. They are fractionally larger than they should be. The reason is that they were themselves generated from a *rounded* annual value. If we examine this table's value for one year of 365¼ days (marked with an arrow in image 7.5), we find that a rounded figure of 0;0,26,26,56,20°, divided by 365¼ to a precision of sexagesimal ninths, and then multiplied out again to a full nine sexagesimal places, is precisely the 0;0,26,26,56,20,0,0,1,44° that John wrote.[11]

This is an unusual insight, as the rough workings of medieval arithmetic almost never survive. Parchment was precious, so calculations and jottings were chalked on a piece of slate and soon erased.

It is only because these tables were computed to so many sexagesimal places that we have been given this rare glimpse of the practical methods of medieval science.

Here, again, John could have been copying another astronomer's arithmetical exercise. But he was not done yet. After he had written it out he noticed an inconsistency. Seemingly just as he was outlining it with vibrant red ink, sitting amid the hubbub of the abbot's London inn, he spotted that the final column did not add up. The figure for four years, near the top of the table, ends with a 57, but the figure for one year, towards the bottom, ends in 44. As John apparently realised, 44 x 4 should end in 56, not 57.* Luckily, there was a simple solution: he quickly split the difference into quarters and squeezed 15, 30 and 45 into the bottom-right-hand corner of the table. With this cramped last-minute addition, he made the table internally consistent – and superficially enhanced its precision to an incredible tenth sexagesimal place. We might find such medieval precision pointless, but we cannot fault John Westwyk's attention to detail.

John's exercises with these tables were not only mathematical. On the page facing his table of the daily motions of the apogees, there is another table giving the same information in a slightly different format. Beside it, we find a block of cryptic signs and symbols (image 7.6). This, as Derek Price immediately realised from the familiar arrangement of text and spaces, was a passage of coded text. For each letter of the alphabet, John substituted a symbol from his own personal cipher scheme. In this corner of the page – and on four other pages of his manuscript – he wrote ciphered messages, each up to fifty words long. The code was not very complicated. As soon as Price spotted the recurrence of the symbols U60 and guessed they might represent the letters THE, he was able to decipher the rest of the script in a matter of minutes.[12]

The mysterious symbols turned out to say, in Middle English, 'if you wish to know the true apogees of planets for years, months or days, add these mean motions to the roots of the true apogees for year 1392; and take there the true apogee of your desired planet. Add a year for a year or a day for a day.' John was simply stating how the tables

* The number 176, written sexagesimally, is 2,56.

7.6. Ciphered text: 'if the liketh to knowe the verre auges of planetes for yeris or montis or daies adde thise mene motes to the rotes of the verre auges of a.1392. & tak ther the verre aux of thi planete desired adde a yer for a yer or a dai for a day'.

of motions worked with the radices. In fact, as we shall see, he had not got it quite right. Leaving that aside, though, this basic instruction hardly seems like the sort of information that is worth encoding. Certainly, if you were expecting the revelation of deep medieval secrets, you would be disappointed. Yet the use of such ciphers was not uncommon among medieval scholars. Like any puzzle, they were an intellectual exercise and a challenge for both setter and solver; a way for astronomers like John Westwyk to educate and amuse themselves.

*

Among all this mathematical and coded complexity, one part at least seems simple: working out the number of days in 1,392 complete years. That number appears, written in base-60, on an early page of the manuscript. Next to it, as Derek Price found when he persuaded the Peterhouse librarian to cut the pages out of their tight binding, was the Latin label *radix Chaucer* – Chaucer's baseline. The discovery of this name led Price to claim a connection between the equatorium completed in 1392 and the poet who had written *A Treatise on the Astrolabe* in 1391. John North, the historian whose ground-breaking study gave Richard of Wallingford his rightful place in the pantheon of medieval science, initially disagreed with Price. But the *radix Chaucer* note made North change his mind. Finding the number of days in 1,392 years, North thought, was 'a trifling matter' for any astronomer capable of using these tables. Why would anyone cite the source of such simple data? he asked. It would only make sense, North argued, if the citer was Chaucer himself, or someone who had a close working relationship with him. If it was the latter, that person would have to have 'moderately advanced astronomical knowledge' but not be a top-ranking university astronomer. Since no person fitting that description could be identified in London in this period, North concluded, it must have been Chaucer.[13]

On this point John North was – highly unusually – wrong. In the first place, these tables are full of citations. Apart from the reference to the Toledan astronomer al-Zarqali, they also acknowledge the Provençal Jew Profatius, the English friar John Somer and the scientific reformer Roger Bacon.[14] More importantly, there are enough mistakes here to suggest that our astronomer did not find such matters completely 'trifling'. Already in this chapter we have spotted him making (or at least failing to amend) a blatant copying error, and watched him fiddle a little pointlessly, albeit conscientiously, with some tiny fractions. Moreover, his carefully coded instructions for the table of apogees are incorrect. That table was not actually sufficient to find the true apogees, because it gave only the linear component of precession and did not take into account the other component: the oscillation of the apogees and fixed stars. And that was not the only mistake in the manuscript, as we shall soon see.

We are back, then, with North's other explanation: that this manuscript belonged to an astronomer who had some sort of working relationship to Chaucer. Even then, either the year 1392 must have been linked to Chaucer in some way, or the writer must have had some other reason for associating himself with Chaucer. Now we know that the handwriting is John Westwyk's. It is conceivable that Chaucer, who was writing the *Treatise on the Astrolabe* around this time, provided John Westwyk with the tables we have been looking at, and a grateful Westwyk acknowledged his source. But a more important reason why Westwyk wished to cite Chaucer, I believe, was the early success of his *Astrolabe* manual, and Chaucer's pioneering use of English for science. In this manuscript, sometime before September 1393, John Westwyk adopted Chaucer's data and, as we shall see below, he adopted Chaucer's scientific English. It seems he saw himself as an astronomical apprentice to the great London writer.

An acquaintance between the monk and the poet is not as unlikely as it may appear. Chaucer spent much of his working life in London, as a customs officer on the riverside and later as Clerk of the Works, managing major building projects. Between 1389 and 1391 he was busy building a new wharf for the Tower of London. A more temporary project was the construction of an arena for the great tournament held at Smithfield, just outside the city walls, in 1390. Although London is curiously absent from Chaucer's poetry, he certainly moved within a broad intellectual network in the city, including lawyers, knights, civil servants and other writers. This network supported wide-ranging learning.

One of Chaucer's literary apprentices was Thomas Hoccleve. Hoccleve was a government clerk and a regular in London's taverns. In one witty poem, he confesses his immoderate eating and drinking, and his fondness for the women who

At Poules Heed me maden ofte appere
To talke of mirthe and to disporte and pleye.

often summoned me to Paul's Head [near the cathedral]
to talk of mirth and fool around and play.[15]

Elsewhere, he writes more mournfully of his sore back and eyes, from years spent stooping over sheep-skin parchment. But Hoccleve's arduous commitments as a scribe of the Privy Seal did not stop him also writing a guide to princely behaviour for the future king Henry V. It begins during a sleepless night in his lodging at the Bishop of Chester's Inn, which occupied a prime riverside plot between London and Westminster:

Musynge upon the restlees bysynesse
Which that this troubly world hath ay on honde,
That other thyng than fruyt of bittirnesse
Ne yildith naght, as I can undirstonde,
At Chestres In, right faste by the Stronde
As I lay in my bed upon a nyght
Thoght me byrefte of sleep the force and might.

Musing on the restless busyness
which this troubled world always has at hand,
which besides the fruit of bitterness
yields nothing, as far I can understand,
at Chester's Inn, very near to the Strand
as I lay in my bed one night
Anxiety took away my power of sleep.[16]

Hoccleve's poem belongs to a genre, the 'Mirror for Princes', which was hugely popular in this period. Princely guidance was one aim of John Gower's *Lover's Confession*; Gower's verses of astronomical and magical lore sit within a syllabus designed, he claimed, for Alexander the Great, and written up for Richard II. Gower divided the education of a king into three parts: theory, rhetoric – 'to telle a tale in juggement / so wel can no man speke as he' – and practice. Practice was most important: a ruler, Gower wrote, must display truth, generosity, justice, pity and chastity. But such regal behaviour had to be underpinned by a comprehensive theoretical education, ranging from theology to astrology.

Monarchs throughout the Middle Ages took such matters very seriously. Charlemagne, for example, maintained a courtly correspondence with scholarly advisers, monks from England and

Ireland. They diligently answered the emperor's earnest queries about eclipses and other celestial phenomena he had observed at court and during military campaigns in the years around 800.[17] At the other extremity of the period, seven hundred years later the Tudor court of Henry VII boasted at least three astrologers. One, the Welsh astronomer-physician Lewis of Caerleon, treated three queens in the 1480s and 1490s. As a trusted adviser to leading figures on both sides of the War of the Roses, Lewis helped to broker the peace-making marriage of the Lancastrian Henry with Elizabeth of York. Such involvement in politics earned him a cell in the Tower of London, mere months after the young Princes in the Tower had met their fate there. Yet Lewis was able to spend his incarceration drawing up supremely detailed tables of eclipses, comparing the calculation methods of Richard of Wallingford and al-Battani and developing his own procedures. After his release, he hired a professional scribe to write up his results into at least three presentation manuscripts. The most sumptuous of these may well have been presented to the king himself.[18]

In the later Middle Ages, a king's education could draw on the increasingly fashionable genre of encyclopaedic 'mirror' literature. These mirror texts surveyed human knowledge, above all for moral edification. Today the most prestigious journal of medieval history is named *Speculum* – Latin for 'mirror'. In its first issue, in 1926, the editor wrote that the name 'suggests the multitudinous mirrors in which the people of the Middle Ages liked to gaze at themselves and other folk – mirrors of history and doctrine and morals, mirrors of princes and lovers and fools'. Using the metaphor of sight for understanding, and the ray of light as the vehicle of knowledge, medieval mirrors reflected almost infinite interpretations. Your rational mind might be a mirror of haphazard sensory input; monks could check their behaviour in the mirror provided by rulebooks like St Benedict's; and Nature was a mirror-image of God's plan and mankind's place in it. The French mystical theologian and poet Alain de Lille, like his twelfth-century contemporary Alexander Neckam, thought nature had clear lessons to teach us:

Omnis mundi creatura,	Everything created in the world
Quasi liber, et pictura	is like a book, and a picture,
Nobis est, et speculum.	and a mirror for us.
Nostrae vitae, nostrae mortis,	Of our life, our death,
Nostri status, nostrae sortis	our position, our fate
Fidele signaculum.	it is a reliable sign.

The subsequent verses of Alain's hymn highlight how the rose begins to wither as soon as it blooms. It thus mirrors mankind's transient existence.[19]

Although Aristotle had written lengthy studies of animals and plants, those natural books of his received little attention in the medieval universities. But cultured readers in John Westwyk's time took enormous interest in natural history, in part as a mirror of mankind. Abridged and translated versions of Latin encyclopaedias provided accessible education and entertainment for aristocratic families. In illustrated bestiaries, for example, nobles could read – or have shown to them – the marvellous attributes and behaviours of animals from all over the world. Some of those animal descriptions were accurate, others were utterly fanciful; but all conveyed a moral lesson to the reader. For this reason, bestiaries were also popular among preachers. On the virtue of chastity, for instance, the actions of the beaver were exemplary. This rare animal, according to bestiaries, has fur like an otter and a tail like a fish, and its testicles produce an oil of great medicinal power. Knowing instinctively that that is why it is hunted, when a beaver finds itself in danger it will bite off its own testicles, throw them to the hunter and make its escape. If pursued a second time, it will rear up on its hind legs and show the hunter that he is wasting his efforts. This ability to self-castrate was, it seemed, the source of its Latin name *castor*.* In one bestiary, produced for a house of the Dominican preaching friars, readers could marvel at a graphic illustration of the amazing animal in the act of self-mutilation, chased by a hunter dressed in vivid green, blowing his horn and carrying a large club. Beneath the

---

* Although plant-based castor oil has been used since antiquity, it acquired that name only in the seventeenth century, probably because traders confused its source, *Ricinus communis*, with another plant, *Vitex agnus-castus*.

vibrant painting, readers were advised that 'every man who inclines towards the commandment of God and wants to live chastely must cut himself off from all vices and all indecent acts – and must throw them in the Devil's face'.[20]

Thomas Hoccleve's *Regement* [conduct] *of Princes* includes some similar moral and educational material, though it also digresses in various directions, not least to praise Chaucer several times. Hoccleve acclaims his deceased idol as 'flower of eloquence', 'universal father in science', 'heir in philosophy to Aristotle, in our language'. The poetry of Chaucer, Gower and Hoccleve evokes a vibrant intellectual life in London, a world in which pleasurable and useful knowledge was traded freely and lowly clerks might mingle with nobility.

Nor were monks excluded from this world. St Albans had long cultivated relationships with worthy friends and benefactors, including members of the royal family. Richard of Wallingford produced a little manual of astrology for the queen in around 1330, and the young John Westwyk might have witnessed the magnificent funeral procession of Blanche, Duchess of Lancaster, which paused for an elaborate requiem Mass at St Albans abbey on its way to London in 1369. Thomas de la Mare, the abbot who hosted the honoured royal cortege, re-established the abbey's confraternity. This was a supporters' association of influential laypeople: patrons and scholars. Many of them would have visited the abbot's inn on Broad Street.[21]

The inn was much more than a convenient place for the abbot to stay the night when he was in the city on parliamentary or monastic business. It was an office, warehouse and showroom for goods grown on the abbey's estates. It was a sizeable parcel of urban real estate that generated income through rents and the produce of its own garden. And it was a social, cultural and entertainment centre. In the early thirteenth century the abbey had paid a hundred marks to buy the site, and half as much again to expand and enclose the buildings. 'It stretched out like a great palace', wrote the chronicler Matthew Paris, so that the abbot, 'and all his successors, and any monks who wished, could stay there in comfort and privacy.' Apart from bedrooms and kitchens, it had a chapel, a garden and an orchard, a courtyard and a well, and the all-important stables – because stabling horses was as much of a problem in the city as car parking is today. Later abbots expanded and refurbished the inn, buying up surrounding properties and raising the rents. In this way St Albans projected its authority and influence in the economic heart of England. Still, they could not control everything, as abbot John of Wheathampstead discovered in 1430. His move to block up three windows of a property overlooking the inn caused a memorable neighbourly dispute.[22]

Just as Thomas Hoccleve's government lodging, rented from the Bishop of Chester, gave him opportunities to see different slices of life, so John Westwyk could well have met Chaucer around the abbot of St Albans's inn. By the 1390s, Cornhill was an intellectual centre with growing educational opportunities for children of both sexes.[23] It was just the sort of place where Chaucer might find a readership for his child's guide to the astrolabe – and where John Westwyk would have had an opportunity to read it.

The *Treatise on the Astrolabe*, as we have it, consists of two parts: a description of the instrument and instructions for its use. But in his prologue to the book, Chaucer proposed three more parts. He promised 'Little Lewis' 'diverse tables', followed by a 'theorike to declare the moevying of the celestiall bodies'. Finally, he planned an introduction to the rules of astrology.

There are striking parallels between Chaucer's blueprint and John Westwyk's production. Westwyk's tables are not exactly what the poet promised – though both men cite the Carmelite friar John Somer as an influential table-maker. But the *Equatorie* fits the description of a 'theoric' rather well. If that word is unfamiliar to you, it was to Chaucer's readers too: this use in his prologue is its first appearance in the English language. Its Latin source, *theorica*, meant a model in both senses of the word: a theoretical description and its physical reproduction. (In modern English, think of an 'economic model' and a child's 'model aircraft'.) It could be ambiguous. When medieval astronomers wrote *theoricae* textbooks – which they did a lot – sometimes they were describing pure geometrical theory, albeit with diagrams representing the three-dimensional movements of the heavens; sometimes they were describing tangible instruments in wood and brass; and sometimes it was something in between. We can read very convincing descriptions and even drawings of apparent instruments which turn out to be thought-experiments. Their authors invented devices that they never meant to make.

John Westwyk, though, certainly did intend his device to be made – he had already done it himself. If Chaucer's *Astrolabe* contains the first use of the word 'theoric' in the English language, Westwyk's *Equatorie* has its second. It was just one example of the influence Chaucer's language had on John's writing. John even cited 'the Tretis of the Astrelabie' explicitly as his source for the name of one part of his equatorium. But to understand John's other influences, and to grasp just how his equatorium worked, we need to go back to what his 'theorike of the celestiall bodies' was trying to achieve.

John's equatorium treatise comes at the end of his book, occupying just fourteen pages – after more than 140 pages of tables. The very last of those tables was an incomplete list of stars. Above the column giving the stellar altitudes as seen from London, John wrote a reassuring statement in Latin: 'I have tested these.' Not all astronomers carried out such checks on their tables. John was evidently proud of having done so, though he wrote down London altitudes for only seven of the forty-three stars in his list. For the first of them, Aldebaran, he recorded an accurate maximum altitude of 53° 36′. Next to its Arabic-influenced name, he also wrote a description: 'the heart or eye of the bull'. This was the bright, reddish star which, as John Gower had rhymed, was magically associated with rubies. It was easy to spot, twinkling in Taurus' right cheek beneath the outstretched horns of the constellation. But if John was watching the sky at the end of July 1392, when Aldebaran was visible for a few hours before each dawn, he would have seen it outshone by an even redder star that did not twinkle. Tracking through the fixed stars, already brighter than the bull's eye and getting brighter with each passing week, was Mars (image 7.7).[24]

Since long before Ptolemy, a pressing problem for astronomers had been to make sense of planetary motions. It is easy to understand why these celestial anomalies aroused such fascination. The Sun, as we have seen, gradually travelled around the ecliptic circle in the course of a year. But the planets did much more than that. Not only did they zigzag either side of the ecliptic; sometimes they seemed to stop their motion through the stars, and even went into reverse, for

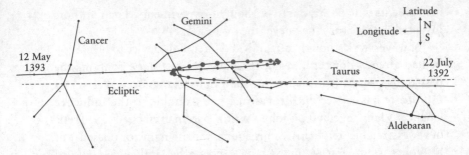

7.7. Motion of Mars, July 1392 to May 1393. There is one week between each dot; where the dots are larger, the planet looks larger in the sky.

periods of weeks or months. When Mars passed a few degrees north of Aldebaran on 29 July, the planet was moving normally through the constellations – increasing its longitude by a little more than 4 degrees, about the width of three fingers, each week. But as it crossed the ecliptic and moved into Gemini, its westerly course curved more sharply north and began to slow. In the first week of November it stood stock-still by the midriff of the southern stellar stick-man of the Twins. Then it slowly started moving retrograde. In mid-December, when it was diametrically opposite the Sun on the far side of the ecliptic, it was at its largest and brightest. Finally, in late January 1393, it slowed to a halt once more. It resumed its normal westerly trajectory, repeating its passage through Gemini and gradually returning south towards the line of the ecliptic.

Ancient astronomers used all their ingenuity in devising geometrical models to explain these curious clashes with the normal purity of celestial harmony. Aristotle outlined one explanation devised by his contemporary Eudoxus.[25] Eudoxus suggested that each planet might be carried on a system of four concentric spheres, each nested one inside the other (like Russian dolls), rotating at different angles around the Earth. A planet like Mars would need its first sphere to rise and set each day with the rest of the heavens; its second sphere to keep it moving around the ecliptic with the Sun and the other planets; its third sphere to reproduce its zigzag motion north and south of the ecliptic; and its fourth to create that characteristic retrograde motion. This was an attractive solution. Philosophers loved the clever geometry by which Eudoxus had managed to replicate direct and retrograde motion with nothing more than pure homocentric circles. But it had obvious drawbacks. It could not explain, for example, the planets' dramatic changes in size. Both Mars and Venus, in particular, can more than quadruple their apparent diameter during the retrograde phase of their cycles.*

One way of modelling such changes in size was to use a circle mounted on another circle: an *epicycle*. A Greek astronomer named Apollonius, who lived in the third century BCE, had proved that an epicycle by itself could produce the same changes in size and speed as an eccentric circle. We only know about Apollonius' work from a summary that Ptolemy wrote almost four hundred years later. But it is clear that Apollonius did enough to show that the combination of a planet travelling around an epicycle, with the whole epicycle itself travelling around the Earth, could create retrograde motion. While the planet revolved on its circular epicycle, the centre of the epicycle itself revolved on a larger 'carrying' (or *deferent*) circle (image 7.8). The relative sizes of the epicycle and deferent would control the size of those retrograde loops.

There was still a problem: the planets' retrograde loops are visibly irregular. They not only vary in length but also in their spacing in different parts of the zodiac. If you made the deferent circle eccentric, you could match either the changing sizes of the loops, or their spacings – but not both at the same time. For a while this did not concern the Greeks too much. As long as their models replicated the general shapes and patterns of planetary motion, that was good enough for them. But astronomers, led by Hipparchus in the second century BCE, gradually began to demand theories that were not only geometrically plausible but numerically precise. This was, in part, because of the growth of astrology. Astrologers, ever conscientious and competitive, demanded accurate predictions of the planets' positions in the signs and houses. Yet although Hipparchus rigorously

---

* We now know that retrograde motion occurs as a planet nearer the Sun overtakes one that orbits further out, catching it up and passing it closely on the inside. This is why planets outside the Earth, like Mars, are retrograde when they are opposite the Sun; the planet appears to move backwards as we pass between it and the Sun.

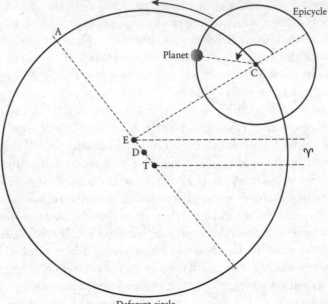

7.8. The deferent-epicycle-equant model for a single planet. From the point of view of an observer on earth (T), the planet will be retrograde, and will appear larger in the sky, when it is on the inside ('lower') part of its epicycle. The epicycle moves around the deferent at a constant rate with respect to the equant point (E). The equant and deferent centre (D) are displaced from Earth along the line to the apogee (A), the *line of apsides*. Since the 'Head of Aries' (♈) is effectively infinitely far away, its angle from E and T is the same. (The epicycle must be correctly sized to produce retrograde motion: the Moon has an epicycle but does not move retrograde.)

pointed out the problems with existing planetary theories, he was not able to improve upon them.

The last big step was taken by Ptolemy, around 150 CE. He added one more detail to the deferent-epicycle model, making it predict the planets' positions with remarkable accuracy. Just as the planet moved around the epicycle at a uniform speed, so the epicycle moved around the deferent circle at a uniform speed. But that 'uniform' speed results in a uniformly changing angle only when observed from a particular point – like the way the Sun seems to move more slowly when it is near its apogee. Ptolemy's masterstroke was to suggest that the epicycle's speed around the deferent circle was not uniform from the point of view of an observer on Earth; nor from the centre of the deferent; but from a third point. That point was called the *equant*. With this final addition, Ptolemy produced a reliable and durable model of planetary motions. An astrologer who needed to know the true longitude of each planet at a certain time did not have to worry about their complex looping motions. He only had to concern himself with two pieces of data, which both changed uniformly: the angle of the planet around the epicycle (known as the *mean anomaly*) and the angle of the epicycle centre around the deferent (normally known as the *mean longitude*, depending on where you measure it from).

Ptolemy realised that, with his innovative equant, he might be accused of violating ancient principles. Plato and Aristotle had both argued that motion in the heavens must be uniform and circular, and astronomers all accepted this – at least in principle. The equant seemed to pull the system out of shape. Writing the *Almagest*, Ptolemy took a step back from terse mathematics to defend himself against such accusations. 'If we are compelled', he pleaded, 'by the nature of our subject to use a procedure not in strict accordance with theory . . . or to make some basic assumptions which we arrived at not from some readily apparent principle, but from a long period of trial and application . . . we may accede [to this compulsion] . . . provided only that [the assumptions] are found to be in agreement with the phenomena.' In the long run, the lack of physical evidence for the equant would cast a shadow over Ptolemy's theory. But for centuries the best proof of its existence was simply that it worked. Astronomers did not trouble themselves about the aesthetic appeal of the model. They preferred to work on refining parameters such as the relative sizes of the deferents and epicycles, and on putting the theory into practice in their astrological predictions. And they built physical models to set Ptolemy's diagrams in motion.[26]

When John Westwyk came to cut out the circles for his prototype equatorium, he was not just inspired by Chaucer; he was building on the foundations laid by centuries of practical astronomers. As they worked to turn the epicycle and deferent into a calibrated geometrical computer, they grappled with two big challenges. The first was that the theory was slightly different for each planet. Not only did all their circles vary in size, but there were additional tweaks in some cases. Mercury and the Moon each had a whole extra circle to account for their exceptional motions. The second challenge was the fact that the epicycle's motion was measured from the equant, but the planet's actual location in the zodiac – its celestial longitude – was measured from Earth. That meant that an extra scale might be needed, possibly for each planet, since every equant was offset in a different direction.

The obvious solution was to make a whole new instrument for each planet. This is what several inventors working in Islamic Spain did, including the Toledan table-maker al-Zarqali (though he did manage to squeeze them all on to two sides of a single plate). The first Latin Christian scholar to write on the subject, an Italian canon named Giovanni Campano da Novara, likewise designed a compendium of seven separate instruments. Campanus' *Theoric of the Planets*, in which he described his seven-in-one equatorium as well as painstakingly calculating the distances of all the planetary orbits, was hugely influential. Roger Bacon, who was memorably rude about most of his contemporaries – claiming in one polemic that Latin scholars had not produced a single original work in theology or the sciences – praised Campanus as an outstanding mathematician, alongside the magnetic experimenter Pierre Pèlerin.[27]

However, not everyone was uniformly enthusiastic. Jean de Lignières, the Amiens astronomer who had dedicated his Alfonsine Tables to the dean of Glasgow, admired Campanus, but had reservations about the Italian's equatorium:

> Recently a certain good and God-blessed man named Campanus designed a certain very necessary instrument. With it the true places of the planets are known – and their stationary points, forward motions and retrogradations. But its construction is extremely tedious, owing to the multitude of plates contained in this instrument, with their various cavities. And also, because of the size of this instrument, it cannot easily be moved from place to place.[28]

Jean did what every motivated medieval astronomer longed to do – he turned his ingenuity to designing a better instrument. His improved model was a simplified version of al-Zarqali's invention. All the deferents were now on the same side of the main plate. But the planets still needed extra smaller plates, and anyone wanting to follow Jean's instructions had to do quite a lot of careful engraving and dividing of circles.

Here was John Westwyk's chance to make his mark. No text in the Middle Ages was wholly original: just as Chaucer's *Treatise on the Astrolabe* was an adaptation of a Latin text, itself based on Arabic forebears, so John's English equatorium manual was probably partly a translation. Even so, paying due respect to one's predecessors left plenty of room for personal creativity. He began dutifully enough by quoting a certain 'Leyk'. No one has been able to identify this person, but it may have been John Loukyn, the St Albans sub-sacrist responsible for maintaining Richard of Wallingford's clock at this time. Westwyk called on the authority of this 'Leyk' to challenge any craftsman who valued portability. 'The larger you make this instrument', he pointed out, 'the larger will be its divisions.' Wider-spaced divisions on the instrument meant more precise readings, down to fractions of degrees. 'And the smaller the fractions', John emphasised, 'the nearer the truth your calculations will be.' Taking a leaf out of Chaucer's treatise with this direct decree to the reader, he suggested just what size you should build his invention for precise enough measurement. 'Take therefore a plate of metal, or else a board that be smooth shaved, levelled, and evenly polished, of which . . . the whole diameter shall contain 72 large inches, or else 6 foot.' This was no portable astrolabe. But if, as John suggested, the six-foot disc should be 'bownde with a plate of yren in maner of a karte whel' (bound with iron like a cartwheel), London was the place to get the work done.[29]

The medieval city resounded with the clangour of blacksmiths' workshops. Many metalworkers had their forges outside the city centre, but those that remained within the walls still created quite a disturbance. One sleep-deprived citizen a little after John Westwyk's time was driven to compose explosively alliterative verse on the subject:

Swarte smekyd smethes smateryd with smoke	Soot-blackened smiths smutted with smoke
Dryve me to deth wyth den of here dyntes;	drive me to death with the din of their dints;
Swech noys on nyghtes ne herd men nevere,	such noise at night heard no man, never,
What knavene cry and clateryng of knockes,	what knavish crying and clattering of knocks,
The cammede kongons cryen after col! col!	the snub-nosed rascals cry for 'Coal! Coal!'
And blowen here bellewys that al here brayn brestes.	and blow their bellows till their brains all burst.[30]

The noisy urban forges were not all bad. They might even inspire medieval scholars. Remember the Paris Condemnations of 1277 and the turf war between theologians and physicists over whether the creation of a vacuum was possible even for God? The Paris University master Jean Buridan used his knowledge of the metalworkers' giant bellows to strike a hammer blow in the debate. He pointed out that it was impossible to pull the sides apart when the nozzle was stopped up. This, said Buridan, proved that a vacuum could not exist. The bellows led him to a further conclusion. Although the hot smithy air could not be further compressed if the bellows were stopped up, Buridan knew that the air would fill a smaller space if it were cooled, demonstrating – in his eyes at least – that the matter of the air (its substance) was something separate from its quantitative form (its magnitude). Even if Buridan had not got personally smutted with soot – and he was not the only Paris philosopher who used this example – he clearly had enough familiarity with such crafts to draw on them for complex and convincing scientific arguments.[31]

Monks, too, were willing to get hands-on with metal and wood. Richard of Wallingford, after all, had been brought up in a smithy, and his clockmaking shows that he had not fully forgotten the heat of the furnace and the weight of the hammer. And the first version of John Westwyk's equatorium was made by none other than John himself. His instructions are dotted not only with rueful references to the mistakes he made in his first attempt but also little suggestions to help his reader avoid repeating them. One key innovation of his instrument was that it combined all the planets' deferent circles – we shall soon see it in action – but first it was essential to calibrate those combined parts perfectly. 'I advise you,' he warns, 'not to write in the names of Signs until you have checked that your common deferent centre is correctly and accurately placed' (image 7.9, plate section). If it was not, he had a solution ready. 'If you mishap in this case, I shall teach you a remedy: knock your deferent centre in or out until it stands exactly on . . . the limb of your equatorium.'[32]

John's first model was only a mini mock-up. He did not say precisely what size it was but lamented that Mercury's special extra circle 'hath but 24 holes on my instrument'. Those twenty-four must have been the most he could squeeze around its circumference, since he had earlier instructed that 'this little circle shall be pierced full of small holes . . . in 360 holes if it be possible or in 180 or in 90 at least'. He did not record what materials he had made it from either. He could well have used wood, or even parchment. Although John's manuscript is now bound in brown leather-backed boards, added after Derek Price took it apart in the 1950s, it was once wrapped in thick parchment. Part of that parchment wrapping survives; it still bears the scored scars and crimson-inked curves of an experimental astrolabe projection.[33]

John Westwyk's instructions were not written in the international language of science – Latin – but in the Middle English of a craftsman. This was a period when English was developing rapidly, mingling freely with Latin and French. When we examine scientific manuscripts from this period, we find them written in a combination of English and Latin (and sometimes French too), far more often than in any single one of those languages. At times it is hard even to say which language a given word belongs to, such is the flexibility with which vocabulary was blended and re-forged. In September 1392, for example, when a London jeweller named John Pyncheon made his will, he expressed his wish to distribute his money to the poor with polyglot panache: 'Ieo volle que la moneye soit despendu . . . to the pore men.' Nonetheless, with the Hundred Years War constantly simmering, an increasingly patriotic political class promoted the earthy English of the common man as a symbol of national unity. Literacy in Latin and French began a slow decline. When Geoffrey Chaucer wrote his *Astrolabe* treatise in the vernacular, it was not just because his ten-year-old son was yet to master Latin (though the poet did remind 'Little Lewis', somewhat bluntly, of his deficiency in that regard). It was also in English for 'the King, that is lorde of this langage'. Chaucer championed the blunt clarity of English's 'naked words', against the opaque sophistry of classical grammar. English, he maintained, was as good for science as Greek, Arabic, Hebrew or Latin, 'just as different paths lead different folk the right way to Rome'. Chaucer's English instructions taught his childlike readers only the uses of an astrolabe, but John Westwyk also covered construction in his instrument manual. An English-speaking craftsman could read the instructions – or have them read to him – and follow the step-by-step method, cutting a six-foot ring of slender brass and attaching silken threads to smooth-shaved wood to make the new equatorium.[34]

Chaucer's pioneering use of English for an astronomical manual not only inspired John Westwyk; it also provided him with some stocks of technical vocabulary. A careful reading of John's draft reveals at least seven words that appear in Chaucer's *Astrolabe* but nowhere else before this time. They include components of the instrument like 'riet' (rete), and 'label', for the revolving rule on the front. But Westwyk was not writing about an astrolabe; for his own instrument, he needed to use his own words. He took great care to define and explain them for his reader. 'This little hole that is no wider than a small needle,' he pointed out, 'shall be named the common deferent centre of planets', thus making sure we notice that combined component, which is his instrument's defining feature.[35]

More than twenty words or phrases have their first English appearance in John Westwyk's handwriting, in this manuscript. Most of them were astronomical terms or parts of his instrument, and he took pains to make their meaning absolutely clear. Some expressed more general science, such as the instruction to 'drawe out', meaning to subtract, or 'remnaunt', to denote what was left over after that subtraction. Some we still use unchanged today, like 'geometrical'. In using all these words, John might give a creative tweak to the meaning of existing English words, or he might simply borrow terms from Latin.[36]

Other words he used, like 'aryn' or 'alhudda', came from further afield. Those two are both Arabic in origin, but again John slightly customised their meaning. *Arim* was the name many medieval geographers gave to the centre of the habitable Earth – usually zero or 90 degrees longitude. It is easy to see why John might have chosen this name for the centre-point of his equatorium – the position where we stand observing the planets wheeling in their endless circles. *Alhudda*, meanwhile, was what John called the line running from that centre-point to the top of the instrument. This word is unique to this single manuscript, but a similar Arabic term, *alucha*, was very occasionally used to denote the equivalent part of an astrolabe. The way that John picked up and repurposed such Arabic terms reveals how medieval science helped languages blend and develop.

It does not, however, prove that John himself knew Arabic. Such terminology had long been absorbed into Latin. Reading manuals like John's, we even get a sense that these buzzwords had exotic allure. That may be why the very first words of the *Equatorie* are 'In the name of God, pitos [compassionate] and merciable [merciful]' – a direct translation of a phrase ubiquitous in Arabic: *bismillahi r-rahmani r-rahim*. This Muslim invocation, which could preface any Qur'an reading, prayer or other action requiring God's blessing, had become fashionable among Christian astronomers, such was the prestige of Islamic sciences.

All the same, John could not allow his fondness for foreignisms to confuse his reader. Once again, he defined each term carefully. And what makes his writing stand out, far more than his vocabulary, is the way he addresses us directly. His words reach out from every page, making direct contact through phrases such as 'I counsel thee', 'I say consider', 'work as I taught thee'. Reading the *Equatorie* feels like overhearing an astronomy class, a master coaching a pupil. At times we catch moments of self-deprecation: in one aside, John excuses a perfectly serviceable diagram of his instrument with 'I know it is roughly drawn.' Self-deprecation was a popular literary technique at the time, but it does give the feel of a genuine personal relationship between the monk and his reader – every bit as believable as that between Chaucer and his supposed son, Lewis. John Westwyk moulded his language to inform and impress, warn and recommend, to tell a story and to motivate his reader. In short, he was teaching.

Just what was he teaching? The equatorium was a tool for learning astronomical concepts, but its primary purpose was to pinpoint planetary positions. No astrologer, and few physicians, could even begin their work without knowing precisely where the planets were. The challenge for any instrument designer, as we saw with the rectangulus in Chapter 4, was to balance two competing priorities: ease of construction and ease of use. Within each of those priorities John had to make tough decisions. Under the first priority, modifications that reduced the cost of construction materials might demand more expert craftsmanship. Within the second, meanwhile, simplifying the task of finding a longitude might obscure the instrument's clarity in demonstrating Ptolemy's diagrams of deferent and epicycle. These conflicting priorities challenged medieval astronomers – but gave them ample opportunities for creative geometry. It is worth taking a look at some of their innovations, to get a sense of how ingenious they could be.

Any inventor who wanted to improve on Campanus' equatorium design had to figure out a way to fit all the planets' deferent circles on a single plate or face of the instrument. The problem was, these deferent circles were all different sizes. This was not because the planets' spheres were of varying sizes – though of course they were, with the Moon adjoining the sphere of fire while cold Saturn was almost out among the fixed stars. What mattered to most astronomers, though, was the angular geometry of planetary theory. This demanded that each planet's deferent and epicycle were in precise proportion, in order to correctly model its orbit and the length and frequency of its retrograde loops. However, astute medieval astronomers realised that those proportions were only relative. You could make the deferent circle any size you wanted, as long as the epicycle was adjusted to maintain the right ratio.

Even Campanus had seen the possibility of adjusting the epicycle size. He was trying to overcome a different practical problem: the epicycles of Jupiter and Saturn were so small that it was impossible to mark a useful scale of degrees on their circumference. His solution was to engrave two concentric circles for each planet. He kept each distant planet's tiny epicycle the right size, but without any scale of degrees; then he added a larger circle outside it, which was graduated with degrees. If he fixed a thread at their centre, he could then read the position of the planet by stretching the thread out to the scale on the larger circle. Where the thread crossed the smaller circle was the planet's true position on its epicycle.[37] Later astronomers, including Jean de Lignières, soon saw that a single epicycle could stand for all the planets using this method. Rather than drawing circles for each of them, though, Jean marked a revolving pointer with the radius of each planet. As the pointer turned, it traced out each epicycle (image 7.10).

The order of the planets on that pointer was set according to the size of their epicycles, relative to a given size of deferent circle. Now, if you are standardising the size of all the planets' deferent circles, you can eliminate them altogether. This is what one anonymous fourteenth-century designer realised. The epicycle centre had to revolve around the deferent centre, at a constant distance from it. So why not change the deferent from a circle into a simple straight line? A craftsman could easily make a straight metal bar. One end would be fixed at the deferent centre, while the other end revolved around it, with the epicycle attached to it (image 7.11a). As the end of the bar revolved like a clock hand, it traced out the deferent circle. This 'epicycle tail' (as one astronomer in the generation after John Westwyk dubbed it) was used to replace the deferent circle in several different equatorium designs. One large brass instrument made according to this specification around 1350 survives in the medieval library of Merton College, Oxford, though its epicycle is lost.[38]

The next step, however, is unique to John Westwyk's design. He must have realised that the combined epicycle would not need a tail at all, if he just expanded it to have the same radius as the deferent (image 7.11b). He could then pin the rim of that combined epicycle to the deferent centre of any planet. If attached fairly loosely, it could still turn, allowing him to position the epicycle centre in the correct direction from the equant. At one point on the rim John made that hole, 'no wider than a small needle', to hold a pin. That hole was the *common deferent centre*.

This improvement, from an 'epicycle tail' to the common deferent

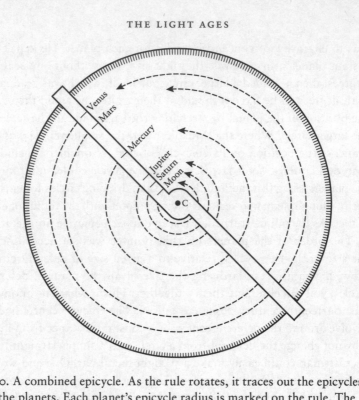

7.10. A combined epicycle. As the rule rotates, it traces out the epicycles of the planets. Each planet's epicycle radius is marked on the rule. The planet's *mean anomaly* is read on the scale on the outside.

centre, may look like no improvement at all. For one thing, if the whole epicycle is pinned down by a particular point on its rim, any scale you engrave on that rim becomes rather less usable, since you can no longer rotate the whole epicycle to 'zero' it. More obviously, the enlarged epicycle uses much more metal than the previous version. Yet John clearly felt that these problems with his design were outweighed by its two advantages. First, as he emphasised, a larger epicycle meant more precise measurement of each planet's mean anomaly around that epicycle. More importantly, though, it meant that the combined epicycle was exactly the same size as the main face of the instrument – six feet in diameter. This radically simplified the most difficult task in its construction: dividing both those six-foot circles into degrees and minutes. Accurately dividing a circumference into 360 equal parts was a notoriously difficult problem for

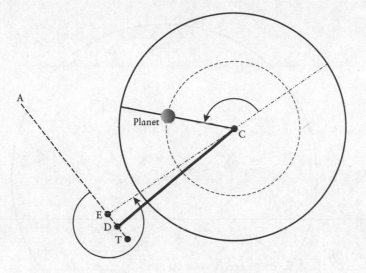

7.11a The 'epicycle tail' model of equatorium. The thick line represents the deferent radius. It pivots around the deferent centre (D). The radius keeps the epicycle centre (C) at a constant distance from D, on the circumference of the – now eliminated – deferent. The position of the epicycle centre may be measured at the equant E as an arc from the apogee (A). The planet's longitude will be measured from Earth (T).

instrument-makers. John clearly felt that, if it meant he had to do the job only once, the cost of the extra metal required for the enlarged epicycle would be worth it.

Once we have divided these circles according to John's instructions and marked the equant point and deferent centre of each planet in their precise positions on the face of the instrument, we are ready to find a planet. First – with help, for it is far too bulky for one person to lift – we lay the wooden face of the equatorium flat on a table-top. We may take a moment to admire its polished grain, as well as the shining brass plate in the middle that bears the engraved planetary centres. Before using it, though, we must look up the mean longitude and mean anomaly in the tables which Westwyk himself copied, adapted and conveniently packaged with his treatise. Longitude is measured around the zodiac, starting from the equinox, or *Head of Aries*, where the ecliptic intersects the celestial equator. So

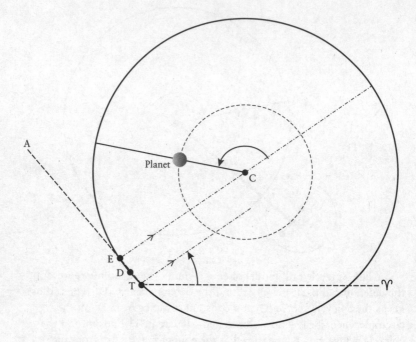

7.11b John Westwyk's equatorium design. The combined epicycle
now has the same radius as the standardised deferent. The *common
deferent centre* (a point on the circumference of the epicycle) is held in
place at D. The position of the epicycle centre (C) is measured as an arc
(the *mean longitude*) from the 'Head of Aries' (♈). Since the scale of
mean longitude is centred on Earth (T), the angle is transferred to the
equant (E) using parallel threads. The planet's position on the epicycle
is the end of an arc (the *mean anomaly*) measured from the thread
that crosses C.

7.12. Steps in finding the longitude of a planet using John Westwyk's
equatorium. The small round plate (dark grey) can be rotated to adjust
the deferent centre and equant point for precession. The mean longitude
($\bar{\lambda}$) is measured from the Head of Aries (♈) and the mean anomaly ($\bar{\alpha}$) is
measured from the end of the white thread. The true longitude (λ) is read
where the black thread, stretched to the planet's mark on the rotating
pointer, crosses the scale engraved on the limb of the equatorium.

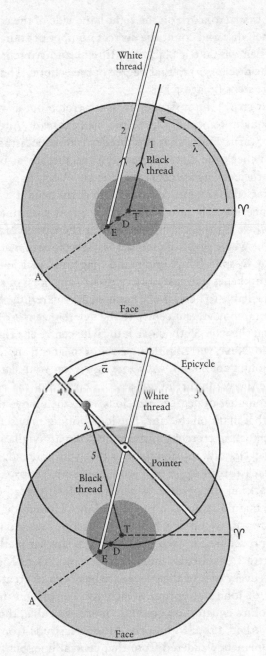

we start at that zero mark on the right-hand side of the equatorium's solid wooden disc and count the correct number of zodiac signs and degrees. Then we take a black silk thread and stretch it from the centre of the disc (*aryn*) to that degree of the ecliptic. That is the first of five short steps (image 7.12).

The next step, John instructs us, is to stretch out a white thread from the equant point, parallel to the black thread. 'And proeve by a compas', warns John, 'that thy thredes lyen equedistant.' Using a compass to check the distance between the threads at both ends is indeed an efficient way to ensure they are absolutely parallel. Now we have transferred the longitude from Earth to the equant, we can bring the combined epicycle into play. Taking great care not to disturb the threads, we lift the big brass ring over the wooden face (step 3). Although the whole ring is six feet across, it is made of metal only two inches wide. Despite John's command that 'this epicicle m[us]t have suffisaunt thikkenesse to sust[ai]ne hymself', any brass of such dimensions will be liable to bend if not handled very carefully. We start by fixing its common deferent centre to the deferent centre of the planet we seek, and then we gently move it until its centre sits right over the white thread. Next, we remind ourselves of the mean anomaly which, following John's command, we wrote down 'on your slate'. Starting from where the white thread crosses the far rim of the epicycle, we count that number of degrees anticlockwise, then move the rotating pointer to that mark on the rim. The long pointer may scrape noisily over its supporting crossbar and we must remain vigilant to ensure that the epicycle stays set exactly over that white thread. Finally, we take the black thread again and stretch it from 'centre aryn' to the planet's mark on the pointer. Where that black thread crosses the limb of the main disc, we can read the planet's longitude.[39]

In order to draw up a complete horoscope, we would need to repeat the procedure for each planet, with a simpler method for the Sun but extra steps for the more complex models of Mercury and the Moon. (Every part of the procedure has been re-created in a virtual model of John's invention, which you can try out for yourself online.)[40] John's equatorium could do more than this, though. In its upper half it had a tool to find the Moon's latitude from the ecliptic. All the planets wandered from that central line, but the Moon's latitude was particularly important to astronomers because eclipses could occur only when the latitude was zero. The lower half of the equatorium, meanwhile, had another tool, this time to calculate the oscillating part of the slowly drifting precession. John did not bother to explain that tool in full – unlike the lunar latitude, where he wrote out three worked examples – perhaps because it gave only an approximate result. Over time it would become useful, indeed necessary, as later users could shift the central brass plate to reposition the deferent centres and equants. But for John himself the main goal was to find the planets in 1393. This his weighty device could do in just a few minutes, giving each planet's longitude to a high level of accuracy and a precision of around two minutes of arc.

John Westwyk probably never made his equatorium at the scale he specified. As he himself hinted, hammering out brass of such dimensions would have been a challenge for any craftsman. Still, he certainly enjoyed experimenting, making his own mock-up, trying out different tables that substituted for some of its processes, and refining his instructions. Forensic study of his manuscript reveals that he continually crossed out or scraped away words, correcting and improving his text. Here he adds a Latin gloss to clarify his English; there he changes 'blak thred' to 'white'; in another place he alters the dimensions of his invention. On two pages he struck out whole sections of text. 'This instruction is wrong', he lamented above them.[41] He was evidently still figuring out his design as he drafted – and sometimes the model frustratingly failed to perform as he hoped.

We can imagine him at work in the St Albans inn. Despite the peace the abbots had procured by purchasing surrounding properties, he could not shut out the buzz of Broad Street. It was not only craftsmen that caused disturbances, but animals too. Just behind the inn was the Hospital of St Anthony: an alms-house offering free accommodation to two priests, a schoolmaster and twelve poor men. According to local law, the only pigs permitted to roam freely were those belonging to St Anthony's hospital. Pigs were destructive, and sometimes killed small children, so any found unfenced were liable to be slaughtered. Their neglectful owners could be fined fourpence per day. But St Anthony's pigs were exempt. According to local tradition, pigs too small to sell were donated to the hospital. As they trotted through the streets, Londoners fed them up from runts to valuable livestock, in small but frequent gestures of civic charity. The hospital marked its porcine property with bells to prevent their confiscation and deter theft. For John Westwyk, though, the grunting and clanging from the street cannot have aided his attempts to comprehend Ptolemaic planetary theory.[42]

We should not be surprised that John was still learning, even while teaching his reader how to make and use the equatorium. The belief that teaching and learning go hand in hand is popular in today's schools, but it has an ancient pedigree. Westwyk could easily have encountered it in the writings of Seneca, very fashionable in his day. One well-known saying of the Stoic philosopher was *homines dum docent discunt* – 'while men teach, they learn'.[43] Equatoria were not, after all, purely practical instruments. To be sure, they were valued for their capacity to save users the effort of time-consuming calculation and error-prone references to endless tables. But they were also teaching devices. That is why they continued to feature in astronomy guides long after the Middle Ages. Printing technology allowed rotating volvelles, which had been a rare feature of manuscript books, to be easily mass-produced. Publishers produced textbooks with paper parts that readers could cut out and assemble themselves – they needed only to supply their own silken strings. The most expensive printed books might be sold with hand-painted instruments ready-assembled, and little seed pearls sliding up and down the silk threads to serve as indicators. These revolving paper diagrams gave their privileged readers hands-on tuition.[44]

Yet no quantity of bejewelled pointers or intricately woodcut dragons representing the lunar nodes could overcome the problems with Ptolemaic planetary theory. Even as astronomers acknowledged the astonishing predictive power of the models, they bemoaned their inconsistencies. At the finest level of detail, they wanted to know, for example, how the planets' apogees could oscillate slowly backwards and forwards without affecting the obliquity of the ecliptic. But there were more general questions too. If God had created the heavens with no wasted space and no overlap, could all the deferents and epicycles be precisely nested inside each other in a way that matched observations of the Sun and the Moon? Most fundamentally of all, were the planets really moving in perfect circles, if their circles were centred on different points?

The frustration of medieval scholars was well captured by Moses Maimonides. The great Jewish philosopher – who was honoured, like Aristotle, Galen and St Paul, in the windows of the St Albans abbey cloister – wrote his *Guide for the Perplexed* in Egypt between 1185 and 1190. In it, he tackled some of the most troubling problems in law, science and theology. The question of whether the universe was eternal or created led him to ask how Aristotle could be reconciled with the *Almagest*:

> If what Aristotle has stated with regard to natural science is true, there are no epicycles or eccentric circles and everything revolves round the centre of the earth. But in that case how can the various motions of the stars come about? Is it in any way possible that motion should be on the one hand circular, uniform, and perfect, and that on the other hand the things that are observable are . . . accounted for by one of the two principles [i.e. epicycles and eccentrics], or both of them? This consideration is all the stronger because of the fact that if one accepts everything stated by Ptolemy concerning the epicycle of the Moon . . . it will be found that what is calculated on the hypothesis of the two principles is not at fault by even a minute. The truth of this is attested by the correctness of the calculations . . .
>
> Furthermore, how can one conceive the retrogradation of a star, together with its other motions, without assuming the existence of an epicycle? On the other hand, how can one imagine a rolling motion in the heavens or a motion around a centre that is not immobile? This is the true perplexity.[45]

Maimonides ducked the question slightly, stating that the models (which he calls hypotheses) do not need to be literally true; astronomers care only about whether they produce accurate results. In this he was echoing Ptolemy's excuse for introducing the equant. Yet Ptolemy did believe that his planetary models were physically real. We can see that, for example, in his discussion of the order of the planets. It was universally accepted that the planets must be ordered according to their orbital periods, with the slowest planets, like Saturn, having the largest spheres. Aristotle even supplied a physical explanation: the outermost planets were slower because they were closest to the sphere of fixed stars. Their steady progress eastwards round the ecliptic was dragged back, he suggested, by the daily rotation of the heavens in the opposite direction. But what was the order of Venus and Mercury, which, staying close to the Sun, both completed their circuit of the sky in one year? It made sense to put the Sun above them, since it would be in prime position in the middle of the seven planets. (In that position, the Sun would also separate off the outer planets Mars, Jupiter and Saturn, whose retrogradations all followed the same pattern, from the other planets that behaved differently.) For the order of Venus and Mercury, though, Ptolemy followed Aristotle's lead and resolved the question with physical arguments. He reasoned that the more complex model of Mercury proved that its sphere was below that of Venus. Like the Moon one step below, he wrote, it was disturbed by the nearby spheres of fire and air.[46] The extra circles and mobile deferent centres of Mercury and the Moon were not abstract theoretical devices but real physical phenomena.

Late-medieval astronomers paid close attention to the physical theories Ptolemy had set out in his *Planetary Hypotheses*. When the Austrian scholar Georg von Peuerbach rewrote the standard university handbook of planetary astronomy in the 1450s, his *New Theories* [or *Theorics*] *of the Planets* was filled with deep, almost three-dimensional woodcut diagrams of the deferents and epicycles. Set in neatly nested black-and-white shells, their motions were visibly constrained, like a bobsleigh on its narrow iced track (image 7.13). Peuerbach's German student Johannes Regiomontanus continued his teacher's work, completing a comprehensive commentary known as the *Epitome of the Almagest* in 1463. Both astronomers wrote very lucidly. As they updated ancient works, they also explained the complex Ptolemaic theories with greater clarity than ever before.

But the main reason for the stunning success of their textbooks was the rise of printing in the late fifteenth century. In 1471 Regiomontanus moved to Nuremberg, the centre of Europe's commerce and communication. There he set up his own printing press – the world's first dedicated scientific publisher. The first book he printed was his late mentor's *New Theories of the Planets*. The first printed edition of Sacrobosco's *Sphere* came out around the same time, in

octauę spherę sup axe z polis ecliptice mouent. Sed orbis epicy/
clū deferens super axe suo axem zodiaci secante secundū successio/
né signoꝛ mouet: z poli eius distant a polis zodiaci distantia non
ęquali. Quare fit vt auges coꝛ eccentricoꝛ nunꝗ eclipticā ptran/
seant sed semper ab ea versus aquilonem z opposita versus austꝛ
mancāt: ita vt auges sez deferentiū epicyclos similit opposita at/
ꝗ cétra z poli deferentiū eccentricoꝛ circumferentias supficiei ecli
ptice virtute mot⁹ octauę spherę describāt equidistantes. vn eriaꝛ
in illis supficies eccentricoꝛ a supficie ecclipticę inęꝗliter secabunt

Theorica Trium superioꝛū z Veneris.

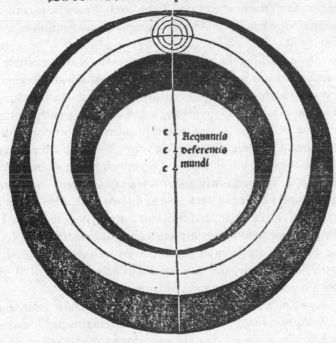

7.13. The theory of the three superior planets and Venus, from Georg
von Peuerbach's *Theoricae novae planetarum*. In a compilation edition
with Sacrobosco's *Sphere*, printed by the German pioneer Erhard Ratdolt
at Venice, 6 July 1482.

the Italian university city of Ferrara. Printing not only meant that scientific books could be produced and read in much larger quantities, spreading ideas quickly; it also allowed complex diagrams to be copied more accurately, astronomical almanacs to be cheaply mass-produced. To be sure, we have seen John Westwyk filling blank spaces with accurate diagrams long before the age of print, and the early printed editions were not free from typos. And one-off, handwritten astronomy continued for some time after Regiomontanus.[47] But printing spread scientific ideas and data far more efficiently than Westwyk could have imagined. If you spotted a mistake in a treatise you were reading, you could – at least in theory – inform its publisher for correction in the next edition. When John Westwyk was thrown by an apparent error, he could only express his frustration in manuscript margins.

Apart from making the astronomy of the *Almagest* accessible to a wide readership, one impact of the work of Peuerbach and Regiomontanus was to highlight the physical inconsistencies in the complex planetary models. In the late fifteenth and early sixteenth centuries, a succession of printed books noted problems with existing theories. Nicolaus Copernicus soon joined the chorus of complaints. In the letter to the Pope which prefaced his epoch-making book *On the Revolutions of the Heavenly Spheres* (1543), Copernicus lamented that astronomers 'have not been able to discover the main thing, that is, the form of the universe, and the clear symmetry of its parts. They are like someone who draws a picture, and includes hands, feet, a head and other limbs from different places, well-drawn indeed, but not based on one body, and not matching each other at all. A monster, not a man, would be produced from them.'[48]

Just how Copernicus came to the conclusion that the solution was to put the Sun at the centre of the universe remains much disputed among historians. It was certainly a conclusion that many of his contemporaries found scientifically unjustifiable (though some medieval philosophers, such as Jean Buridan, had discussed the separate possibility that the Earth rotates). Copernicus had been encouraged, he told the Pope, by learning that several ancient Pythagorean philosophers believed in a moving Earth. He was impressed, too, to read that the Roman educationalist Martianus Capella had theorised that

Venus and Mercury orbited the Sun. Still, his ability to make such a system work depended on the carefully constructed geometry of medieval astronomers, many from the Islamic world. Having grown up in a wealthy Polish family, Copernicus' access to these wide-ranging theories and data was first gained during university studies in Krakow, Bologna, Padua and Ferrara. The thriving medieval universities, and the printed editions of texts and tables that began to multiply at this time, were essential for Copernicus to compare different theories and refine his own.[49]

Copernicus was keen to eliminate the troublesome equant and restore Plato's principle of uniform, circular motion. Luckily for him, Islamic astronomers since Ibn al Haytham in the 1020s had been experimenting with different geometrical tools that might overcome the problems with Ptolemy. Like Eudoxus back in ancient Greece, they were attracted by homocentric models.[50] The right combination of circles turning around the same centre but at different angles, they thought, might provide predictions as accurately as the dominant eccentric devices but be more physically plausible. Several significant figures laid the unglamorous geometric foundations for Copernicus' heliocentric system. But the most important of them was the Persian polymath Nasir al-Din al-Tusi.

Tusi came from a respected family of Shi'ite scholars in the Khurasan province of what is now north-eastern Iran. He studied both Islamic law and the mathematical sciences, latterly in the Iraqi city of Mosul. Around 1235, aged in his mid thirties, he won the patronage of an Isma'ili ruler, first in his own province and later in the northern hills. Whether he converted willingly to the esoteric doctrines of Isma'ilism or, as he later claimed, 'had fallen into the power of the heretics', their support for his astronomy was clearly beneficial.[51] He produced much of his most original work over the following twenty years. Even so, he might well have remained a peripheral scholar, but for the capture of the Isma'ilis' mountaintop redoubt of Alamut by the Mongols in 1256. The Mongol ruler Hulagu, grandson of Genghis Khan, was a supporter of science, and evidently a sound judge of character too. He immediately employed Tusi as an astrological advisor. Tusi quickly persuaded him to fund the building of a substantial observatory at Maragha, in the far north-west of Iran.

Maragha was not the first observatory in the Islamic world, but it was unprecedented in its scale and ambition. From the outset, Tusi warned his patron that this would be a costly project. After some early financial uncertainties, Hulagu gave him access to a Muslim charitable endowment fund, a *waqf*, and the astronomer channelled this into constructing, staffing and outfitting a state-of-the-art research complex.[52] Work began in 1259, on a flattened hilltop about 400 metres long and 150 metres wide. As well as the huge main building with a library of thousands of volumes, there was a domed observatory, a mosque, and a residence for Hulagu. The complex was immensely impressive to visitors. One contemporary was inspired to express his admiration in Arabic verse:

bina'un la-'umri mithlu banihi mu'jizun –	By my life! A building that, like its builder, is inimitable.
tuqarribuhu l-alhazu wa-nnafsu tubhaju	Glances hasten to it, and it delights the soul.
sa-yablughu asbaba ssama'i bi-sarhihi –	He will ascend along the heavenly paths with his lofty edifice,
yunaghi ki'aba zzuhri minha tabarraju	and whisper to the bright stars like dice adorning the zodiac.
aqulu wa-qad shada l-bina'u bi-dhikrihi –	I speak mere words – but the edifice itself sings his praises!
wa shayyada qasran lam yashid-hu mutawwaju.	He has built a castle that no crowned king has ever built.[53]

Tusi assembled an international team of astronomers and planners. From the Syrian desert came the engineer Mu'ayyad al-Din al-'Urdi; from the verdant southern city of Shiraz came the young geometer and avid chess-player Qutb al-Din al-Shirazi. At least one member of the observatory, Fao Munji, was Chinese. Many others came to study, to undertake research in the peerless library and, of course, to carry out observations. They benefited from an array of instruments, many made at monumental scale for precise measurement.

The proudest production of the observatory was a completely revised set of tables, known as the Ilkhani Tables after the great Khan Hulagu. He did not live to see their completion in 1272, but the observatory was sufficiently well established that it outlasted his patronage – and that of five other Mongol rulers. Even decades after it was ruined, the Timurid astronomer-king Ulugh Beg visited Maragha and was inspired to construct an even larger observatory at Samarkand in the 1420s.

The work of Tusi and his followers had an enormous influence over the following centuries. Their creative use of geometry to provide physically plausible alternatives to Ptolemy's planetary models was taken up by a number of astronomers who never visited the hilltop observatory. One such was Ibn al-Shatir, who was official timekeeper at the great Umayyad Mosque in Damascus when John Westwyk was growing up. Another was Ali Qushji, whose father was Ulugh Beg's falconer but who rose to head the Samarkand observatory in the 1440s and later worked in Istanbul.

When Copernicus came to work out the mathematics of his Sun-centred astronomy, he was indebted to the work of the 'Maragha School' and their successors. His models for the latitudes of the planets, as they weaved slightly north or south of the ecliptic, were based on those of al-'Urdi and al-Shirazi. His model for the complex motion of the Moon was very similar to Ibn al-Shatir's. And in order to do away with the philosophically objectionable equant and show that the planets could move in eccentric circles around the Sun, he made use of theorems by Tusi and Qushji.[54] Copernicus' new system, in the end, was no simpler than the one it replaced. But he believed it was true, and he made it work.

How Copernicus came to hear of the theories of these Muslim scholars is a question that has long vexed historians. Central Asian science, to be sure, was justly famous: only a few years after the foundation of the Maragha observatory, the English friar Roger Bacon was praising the Mongols' commitment to astronomy. Some of the detailed theories Copernicus required may have come through networks of Jewish scholars, who communicated freely with both Islamic and Christian communities across the Mediterranean around 1500. They were certainly active in Padua, where Copernicus learned much of his astronomy. Much of the essential geometry, though, came through his reading of Regiomontanus.[55]

Both Regiomontanus and his mentor Peuerbach had been supported by a powerful Byzantine émigré scholar and cardinal named Basileos Bessarion. Bessarion was a key figure of the blossoming Renaissance. If the Renaissance was a rebirth of ancient learning, as its partici-pants claimed, they needed access to ancient texts. In fact, respect for ancient learning and study of ancient texts was nothing new in the fifteenth century – it was a key feature of the Middle Ages. John Westwyk himself proves that. But quite apart from the change in out-look the Renaissance represented in fields such as the visual arts, it did see an acceleration of efforts to find, study and translate works of classical Greek and Latin. Communication with Constantinople was essential to those efforts. Even before the historic centre and haven of Greek culture was captured by the Ottoman Turks in 1453, educated Byzantines like Bessarion were travelling to Italy, bringing with them Greek works previously unknown to Western philos-ophers. Bessarion himself worked tirelessly to support Greek refugees and to introduce Greek learning to Latin-speaking scholars like Regio-montanus. Before he died he donated his entire library, comprising more than eight hundred mostly Greek manuscripts, to the Venetian Senate. Bessarion was far from the only cultural go-between. It was through such studies, translations and donations that the legacy of centuries of Islamic astronomy sowed a seed for modern European science. Copernicus was unmistakeably a Renaissance man – his masterwork *On the Revolutions of the Heavenly Spheres* was struc-tured and written in explicit and intimate dialogue with Ptolemy's *Almagest*. But he has also been called 'the most noted follower of the "Maragha School" '.[56]

Beside this international litany of astronomers, little John Westwyk at his London inn may seem unimportant. Yet Westwyk was part of a noisy conversation through which astronomical theories were com-municated, challenged and refined. As he tried out different versions of the Alfonsine Tables, or computed the slowly oscillating apogees of the planets, he was working at the coalface of astronomy. As he built his models, he was showing their physicality and helping pop-ularise their procedures. As he taught, translated and learned, he was spreading precise scholarship in multiple languages. John may have been drawn to astrology, but he contributed in his own small way to modelling that changed science. His equatorium was designed to find the positions of the planets, but it also helped its users understand their place in the universe.

Epilogue

The Mystery Instrument

At half past three on a blustery Tuesday afternoon in 2012, I arrived at the back door of the Whipple Museum of the History of Science in Cambridge. I rang the doorbell and waited by the cluttered bicycle racks and humming refrigeration units of the Cambridge University New Museums Site. This six-acre block near the city centre embodies the history of the university founded in 1209. Occupied by Augustinian friars for much of the Middle Ages, it became home to a carefully curated botanic garden in the 1760s. One hundred years later, when Cambridge's scientific profile was rapidly expanding, the university developed the then New Museums. Museums in that era were sites of intense scientific activity, where scholars painstakingly investigated zoological or geological specimens sent from all over the world. As science gradually moved into purpose-designed research laboratories, these too were constructed on the site. Many notable scientific discoveries were made in the complex of buildings on the imperceptible gradient of Peas Hill. These included the electron and neutron, and the structure of DNA, in the iconic Cavendish Laboratory. Yet as modern science continued its inexorable growth, such laboratories withdrew in turn to more spacious sites outside the city centre. The vacant mish-mash of buildings met the evolving needs of the modern university: enhanced lecture halls and libraries; audio-visual support; student services. The Whipple Museum was founded soon after the Second World War, when striking demonstrations of the power of science spurred interest in its history. The museum has sat snug in the site since 1959, though the massive stone block above the main entrance still bears the carved name of its former occupant: Laboratory of Physical Chemistry.

I was at the Whipple Museum on that autumnal day in pursuit of another mystery: not a manuscript this time, but a model. While Derek Price was studying what he thought was Chaucer's equatorium, he had volunteered at the newly founded museum, helping to build up its collections. When, much later, Price's supervisor wrote his recollections of those exciting early years, he mentioned that a large replica of the equatorium had once hung on the museum wall.[1] No such 'equatorium' was listed in the museum's catalogue, but I had come to see if any trace could be found of it.

Inside the cramped museum office beneath the galleries, I outlined my quest to two patient curators. I explained what the equatorium was and what the replica might have looked like: a six-foot wooden circle with a similarly sized brass ring attached. There was a pause as they exchanged a meaningful look. 'Do you think', said one, who was seated at a computer with the museum database open, 'that it might look a bit like this?' On the screen they called up a database record. A thumbnail photograph showed an object that I had never seen – but recognised instantly.

We quickly went to see it in the museum's storeroom. Wheeling it out from its resting-place behind a large cupboard, I came face to face with Price's replica: dusty and a little scratched, but unmistakeable. With space on the site always at a premium, the bulky equatorium had languished in offsite storage for many years. By the time it was formally accessioned into the museum, its origins had been forgotten. So when in the 1990s the Whipple installed a new electronic catalogue, which required each object to have a name, the cataloguer used the nickname this large wooden disc had acquired: King Arthur's Table.

John Westwyk's equatorium was made in the Cavendish Laboratory. Forcefully energetic and scientifically trained, Derek Price had gained the patronage of the Cavendish Professor himself, the Nobel Prize-winning physicist Lawrence Bragg. When Bragg had reorganised the Cavendish Laboratory after the Second World War he had been particularly concerned to ensure that there was a surplus of workshops: it was better, he argued, to have a few spare tools than to keep researchers waiting for vital experimental apparatus. So it was that in March 1952, just a few months after Price's discovery of the *Equatorie* manuscript, the Cavendish technicians began work on an unusual project. A BBC Radio reporter was there to witness it. It was startling, the reporter noted, that 'this instrument, designed more than five hundred years ago, will have first been made in a laboratory famous for atomic research'.[2] The fact that Westwyk's instructions could be successfully followed, so long after they were written, is testament to his communicative skill. Perhaps it was that which had ensured the survival of his draft manuscript in the library of Cambridge's oldest college, bound with sections of a commentary on St Augustine and a Roman manual of military strategy, for Price to rediscover and reconstruct. Just eleven months later, in one of the same workshops, Francis Crick and James Watson built their celebrated model of DNA.

A line runs from the Middle Ages to modern science. It is not an unbroken line, of course, and certainly not straight. But if you struggled with any of the trigonometry in earlier chapters, you will admit that medieval people – who carried out such painstaking calculations without the help of any electronics – were not stupid. Throughout this book we have learned what we owe to medieval monks and scholars. It was the Middle Ages that saw systematic translation of classical and Arabic works and gave us the universities that became centres of their study. It was the Middle Ages where intense interest in astronomy – and, yes, astrology – made people look outwards to the heavens, testing predictions, compiling tables and refining theories that ultimately led to the reorganisation of the universe. It was the Middle Ages when, to regulate their religious routines, monks designed mechanical clocks and challenged calendrical orthodoxy. It was the Middle Ages when Christians adopted Hindu-Arabic numerals; when Europeans experimented with marvellous medicaments from across the world; when theories of sight and light competed to explain human understanding; when alchemists developed practical techniques still used in modern chemistry; when mathematics was inspired by the miracle of transubstantiation. It was in the Middle Ages that Europeans began exploring over the oceans, aided by new technologies of mapping and the magnetic compass. And it was in the Middle Ages that they built complex instruments to model their divinely ordered cosmos. When Isaac Newton, the hero of the Scientific Revolution, wrote with false modesty that he was 'standing on the shoulders of giants', he was not only more right than he realised, he was making use of a medieval metaphor.[3]

We have seen, too, that religion was no impediment to scientific progress. Time and again we have witnessed medieval Christians respecting and absorbing learning from other faiths without prejudice. Pious faith motivated investigation of the natural world; institutions from individual monasteries to the papal monarchy itself instigated and supported science. To be sure, when novel visions of creation were promoted, disagreements could arise. But where those erupted into conflict, they were primarily fuelled by political or personal factors. We saw this among the masters of thirteenth-century Paris. After the Middle Ages, the celebrated cases of Giordano Bruno and Galileo Galilei, often held to be emblematic of the mortal incompatibility of faith and reason, owe much to the particular beliefs and circumstances of two provocative individuals, as well as to the violent fragmentation of the Holy Roman Empire amid Reformation and Counter-Reformation.[4]

Why, then, do we persist in belittling the Middle Ages? In part it is certainly to exalt ourselves. When prominent present-day scientists assert that Copernicus 'dethroned' the Earth from a proud pedestal at the centre of the universe, they are implicitly boasting of the modesty of the moderns.[5] As it happens, medieval thinkers often pictured the Earth at the bottom, rather than the centre, of the vast universe; as far as possible from the perfection of the heavens was hardly a desirable place to be. That is why, in Galileo's *Dialogue on the Two Chief World Systems, Ptolemaic and Copernican*, the Florentine astronomer had his spokesman, Salviati, assert that 'we are trying to make [the Earth] more noble and more perfect . . . and in a sense to place it in heaven, from which your philosophers have banished it'.[6] Nevertheless, the tale of the Earth's demotion is often framed as a blow to medieval arrogance; and modernity, by contrast, is supposed to have succeeded through the enlightened modesty of scientists. Neil deGrasse Tyson, astrophysicist and self-proclaimed successor to Carl Sagan, has written that, when he sees the tiny Earth in a planetarium show, 'I . . . feel large, knowing that the goings-on within the three-pound human brain are what enabled us to figure out our place in the universe.'[7]

Yes, the Middle Ages stumbled into some scientific dead ends. But so will we. The Roman farmer Palladius was aware that lead is poisonous – yet we continued to pump it through our cars and into the air we breathed until the very end of the twentieth century. And if, as Bernard of Gordon recognised, much medieval disease was caused by medicine itself, this problem has not gone away either. Of course, modern science has made our lives longer and more comfortable in ways medieval people could only imagine. But the biggest barrier blocking further progress may be our own complacency. The doctrine of 'scientism', the belief that an infallible scientific method is the only route to reliable knowledge, is, in its own way, as dangerous as blind religious faith. As long as science is a human activity, it will have human flaws. In this respect, perhaps the many mistakes of the Middle Ages can teach us some helpful humility, and motivate us to identify opportunities for improvement in our own day. Studying medieval scholars' errors, as well as their magnificent achievements, helps us to appreciate human endeavour in all its fascinating complexity.

And yet success and failure are, in an important sense, utterly irrelevant. The fact that the thought processes – and scientific abilities – of medieval scholars were not as exotically alien from us as we might have imagined could tempt us to compare them with the present day. But we should not award the Middle Ages points for how much they manage to be like us, for two clear reasons. In the first place, they were not trying to be like us. Medieval science was not trying to understand the workings of a coldly mechanistic natural world but a living cosmos endowed by God. Even when, as we have seen, they saw the universe functioning as a predictable machine, they were less interested in *how* it worked than *why*. We may casually say that science explains 'why' nature is the way it is, but we often confuse 'how?' for 'why?' As any parent of a four-year-old knows, beyond every 'why?' lies another one. Medieval people hoped to follow that trail of 'why?'s back to the mysteries of Creation, and mankind's place in it. We don't think of ourselves as backward, even though we – like the medievals – are well aware that there are questions we haven't yet answered. And we would not like future generations to belittle us for failing to answer those we hadn't – we couldn't possibly have – posed.

Secondly, even if medieval scholars scored no points at all – even if they failed utterly to be like us – even if they really had believed the world was flat – they would still be interesting. This book has sought to tell the story of medieval science, less as part of the long history of science from antiquity to the present day and more as an integral part of medieval life and culture. We have seen the place of science in literature, in art, in music and religion. Again and again medieval people have shown us that their science was not the separate cultural sphere it so often is today but was intimately engaged with other forms of thought and behaviour. When a monk observed the stars silently rising above the smooth arc of a Romanesque window, or when an urban craftsman hammered brass into the curved canine tongue of Sirius, we see science in everyday medieval life. That is why it has been so important to observe the fine grain of scientific practices, from the colour of a patient's urine to the ninth sexagesimal place in a table of slowly moving apogees. When we leaf through a beautifully decorated bestiary and see an elephant giving birth in water to avoid the danger of dragons, we can read that this symbolises Eve fleeing the subversive serpent, or we can simply marvel at the power of the medieval imagination and the creative skill of scribes and artists. Either way, study of the natural world was a fundamental part of medieval life. (It is a fact we can easily fall to notice, if the only histories we read are of kings and battles.) Even when monks, closeted away in their cells, meditated through their divine reading, they had to start from something concrete. Grounding your position in space and time could be the gateway to transcendence.

But what of *our* monk? As he drafted the instructions for his equatorium, it seems John Westwyk's journey was nearing its end. For one final trace of his existence, let us enter the hushed Reading Room of the Vatican Secret Archives. (Or, alternatively, we can complete the photoreproduction order form, pay a small fee and receive some scanned pages via email.) On the second day before the Ides of May, 1397, a registrar at St Peter's made an entry in a papal register. He signed his name, Nicolas of Benevento, and acknowledged receipt of thirty *grossi*, the silver fourpence groat coins named for their substantial size. Pope Boniface, he wrote, sent formal greetings to his 'beloved son John Westwyk, monk of the monastery of Saint Alban, of the Order of Saint Benedict, in the diocese of Lincoln'. John's entry in the register joins another St Albans monk on the same day – and more than five hundred other clergy and laypeople, men and women from all over England, in a year's worth of such records over ninety pages of the papal register. To each the Pope confirmed the right for a confessor of their choice to grant them, being duly penitent, full remission of their sins. For some this voucher could be used only once, at the hour of death; for others, including Westwyk, it could be used 'as often as you please'. Nevertheless, such licences were often granted to people nearing the end of their life, and this may well have been the case with John.[8]

The previous year, the great abbot Thomas de la Mare had died and the monks had gathered to choose his successor. The abbey chronicler listed the monks present at that election in the autumn of 1396; John had not been among them. He may have returned to the abbey between that October and the following May, but he was not there when the brothers next elected an abbot, in 1401.[9] Certainly, if he was dying, he would have found excellent care at St Albans. It was common for older monks in larger monasteries to retire to the well-appointed infirmary, where they received a sustaining diet and appropriate medical treatment. As their final hours drew near a young monk might be assigned to offer them constant companionship, and they would be perfectly placed to receive the last rites.[10] The customs recorded in rule-books and narrative accounts only hint at the feelings of the community when one of their members passed away, but it is clear that, alongside the prayers and hopes of salvation, there was sadness at the loss of a beloved brother. The emotional energies of mourning were channelled into a sequence of commendatory prayers, exequies of the dead, requiem Masses.

Such ceremonies were recorded only on exceptional occasions, when the passing of the most senior monks was marked with due solemnity. For John Westwyk, we know none of this. At his death, as in so much of his life, he eludes us. Perhaps, despite what it says in the papal register, he preferred to stay among the vibrant comings-and-goings of the abbot's inn in London. More likely, though, is that he chose to live out his final days at St Albans. The papal licence, which allowed

John to choose his own confessor, was sometimes used by monks to avoid having to deal too closely with their own abbot or other senior monks.[11] After his misadventures at Tynemouth and on Crusade, John may well have wanted to maintain some distance from his St Albans superiors. For a thirty-groat fee he could safeguard this small measure of independence, while still benefiting from the care of the infirmary and enjoying the proximity of his birthplace at Westwick.

Westwick itself (or Gorhambury, as it remained known after the twelfth-century abbot who had given it away) was to achieve greater fame in the history of science, as the home of the Jacobean statesman and philosopher of scientific method Francis Bacon. But by Bacon's day the medieval abbey of St Albans was no more. Its cloisters and clock were destroyed, and its precious library ransacked, at the Dissolution of the Monasteries. The abbey church still stands proudly as the city's cathedral.

Such cathedrals, towering over so many European cities, are testament to the achievements of the Middle Ages. The chimes of their innovative clocks surely tell us that it is time to redefine the word 'medieval'. Rather than a synonym for backwardness, it should stand for a rounded university education, for careful and critical reading of all kinds of texts, for openness to ideas from all over the world, for a healthy respect for the mysterious and unknown.

And, of course, it stands for modesty. John Westwyk did not claim credit for his astronomical efforts; he hardly left his name at all. But he left us his anonymous work. Perhaps it was his communicative skill that ensured his draft equatorium manuscript somehow survived. This is an authentic medieval gift; it is why it seemed truer to the spirit of the Middle Ages to focus this study of science on an unknown, flawed monk, rather than a famous figure. The Oxford scholar and royal administrator Richard of Bury, whom we met in Chapter 3, did find fame as Bishop of Durham. But, in his heartfelt hymn to *The Love of Books*, completed on his fifty-eighth birthday in 1344, he recognised the limitations of his own achievements:

> Alexander, the conqueror of the world; Julius, invader of Rome and of the world, who was the first to unite dominance in both war and arts within a single person . . . would not now be remembered, without

the aid of books. Towers have been torn down; cities overthrown; triumphal arches have succumbed to decay; nor can either pope or king find a better way to bestow the privilege of perpetuity than books. The book he has made renders its author this service in return: that so long as the book survives, its author remains immortal and cannot die, as Ptolemy testifies in the prologue to the *Almagest*: He is not dead, he says, who has brought science to life.[12]

Ptolemy never quite said that. But he did recognise, in that *Almagest* prologue, that studying and teaching the perfect beauty and symmetry of the ordered universe was the surest way to approach the divine.[13] Perhaps, then, through close contemplation of John Westwyk's lifelong labours, we have honoured his memory in just the way that a medieval monk would have wanted.

Further Reading

This is a brief, selective guide. It is intended primarily for non-specialist readers so is slanted towards accessible (and reasonably priced) books, and websites, in English. However, I have included academic works where I found them particularly important – or, as on many topics, where they are all that exist – and especially if they are available free of charge online. Other sources I have used are cited with full bibliographic details in the endnotes. Web addresses were correct as of November 2019. A full bibliography is available at sebfalk.com.

PRIMARY SOURCES

The original materials of medieval science are more widely accessible than ever before. Many instruments are displayed in museums, so you can visit and see astrolabes and equatoria for yourself. The astrolabe that takes centre stage in Chapter 4 is displayed at the Whipple Museum of the History of Science (Cambridge). The Whipple is also home to 'King Arthur's Table'. Other particularly notable collections of medieval instruments are held at the History of Science Museum (Oxford), the Adler Planetarium (Chicago) and the Museo Galileo (Florence). However, if you look out for them, you will spot an astrolabe on display at many national and regional museums.

Medieval manuscripts are harder to access, but many can now be viewed online. I urge you to do so; the carefully pricked parchment and flowing handwriting will transport you back to the candlelit scriptoria of the Middle Ages. Notable collections that have been substantially digitised include:

Bibliothèque Nationale de France	https://gallica.bnf.fr
Bodleian Library, Oxford	https://digital.bodleian.ox.ac.uk and https://medieval.bodleian.ox.ac.uk
British Library	http://www.bl.uk/manuscripts
Corpus Christi College, Cambridge	https://parker.stanford.edu
Trinity College, Cambridge	https://www.trin.cam.ac.uk/library/wren-digital-library

Many of the manuscripts featured in this book can be viewed at the above websites simply by searching for the classmark I give in the endnotes. Of course, other libraries and archives have digitised their collections; where I have cited those, I have included the web address in the endnotes.

Many of the ancient and medieval texts that monks like John Westwyk would have read are available online, often in translation. Two good examples are the Latin text of Isidore of Seville's *Etymologies* (http://penelope.uchicago.edu/Thayer/E/Roman/Texts/Isidore) and the 1855 English translation of Pliny the Elder's *Natural History* (http://www.perseus.tufts.edu/hopper/text?doc=Perseus:text:1999.02.0137). Many translations of Boethius' *The Consolation of Philosophy* are available online (or in print); it is not only a window into the medieval mind but remains powerful and insightful today. Online editions can often be found with a quick search, or via the 'References' or 'External links' section of many pages on Wikipedia (https://wikipedia.org – for all its imperfections, Wikipedia is an extraordinarily useful resource). In addition, books published more than seventy years ago may be available to download via the Internet Archive (https://archive.org/). These include early printed editions of medieval texts, such as the Alfonsine Tables or Bernard of Gordon's *Lily of Medicine*, as well as early translations into English. Both Wikipedia and the Internet Archive are managed by non-profit foundations; if you benefit from them, I encourage you to make a donation.

Excerpts from a wide range of medieval scientific texts are collected in *A Source Book in Medieval Science*, ed. Edward Grant (Cambridge, MA, 1974). It contains many translated sources that were not previously available in English but is, unfortunately, hard to get hold of. Much medieval literature is online, including works of John Gower and Thomas Hoccleve in the splendid TEAMS series (https://d.lib. rochester.edu/teams/text-online) and many of Geoffrey Chaucer's works with interlinear modern English translation (http://sites.fas. harvard.edu/~chaucer).

GENERAL WORKS

The best general overview of early science, scholarly but still readable, is David Lindberg, *The Beginnings of Western Science: The European Scientific Tradition in Philosophical, Religious, and Institutional Context, Prehistory to A.D. 1450* (Chicago, 2nd edition, 2007). A more swashbuckling story is provided by James Hannam, *God's Philosophers: How the Medieval World Laid the Foundations of Modern Science* (London, 2009). Jim al-Khalili, *Pathfinders: The Golden Age of Arabic Science* (London, 2010), is an evocative account of science in the medieval Islamic world. For those seeking more detail, *The Cambridge History of Science, Volume 2: Medieval Science*, ed. David C. Lindberg and Michael H. Shank, is an indispensable collection of essays. Lynn Thorndike's *A History of Magic and Experimental Science*, vols. 1 to 4 (New York, 1923–34) put ancient and medieval science into their contexts of magic and wonder, with detail and erudition that remain unparalleled. Scholarly reference works have also been essential for me, particularly the *Complete Dictionary of Scientific Biography*, the *Biographical Encyclopedia of Astronomers* and the *Oxford Dictionary of National Biography*.

An absorbing introduction to John Westwyk's England is Miri Rubin, *The Hollow Crown: A History of Britain in the Late Middle Ages* (London, 2005), valuable for its own detailed guide to further reading as well as its attention to all levels of society. Ian Mortimer, *The Time Traveller's Guide to Medieval England: A Handbook for Visitors to the Fourteenth Century* (London, 2009), is rich with enjoyable information on all aspects of daily life. A broader view of the vibrant European Middle Ages is provided by Chris Wickham, *Medieval Europe* (New Haven, 2016).

More than fifty years after it was first published, C. S. Lewis, *The Discarded Image: An Introduction to Medieval and Renaissance Literature* (Cambridge, 1964), remains an outstanding guide to medieval ideas of nature and the universe. From the same era, still pithy and powerful, is Lynn White, *Medieval Technology and Social Change* (Oxford, 1962). Umberto Eco, *The Name of the Rose* (1980; tr. William Weaver, New York, 1983), is a novel, but it evokes the atmosphere of medieval monastic scholarship as no non-fiction work can.

An excellent example of a fourteenth-century 'micro-history', a book which reconstructs the details of a specific story as a gateway into past cultures, is Robert Bartlett, *The Hanged Man: A Story of Miracle, Memory, and Colonialism in the Middle Ages* (Princeton, 2004). Eileen Power pioneered popular social history; her *Medieval People* (London, 1924) is a series of six richly textured portraits of lively medieval characters. A seminal biographical microhistory is Carlo Ginzburg, *The Cheese and the Worms* (1976, tr. John and Anne Tedeschi, Baltimore, 1980).

THE 'DARK AGES' AND DEREK PRICE

Derek Price has been, in several ways, an inspiration for this book. Price told the story of his discovery of the 'Chaucer' equatorium manuscript in a typically ambitious work, *Science Since Babylon* (New Haven, enlarged edition, 1975); that book and a series of recorded lectures Price gave at Yale University in 1976 are available on a website maintained by his family, http://derekdesollaprice.org. Equally foundational for my project was a magnificent article by Kari Anne Rand, 'The Authorship of *The Equatorie of the Planetis* Revisited', *Studia Neophilologica* 87 (2015): 15–35. Price's experiences at Cambridge, and his production of 'King Arthur's Table', are recounted in Seb Falk, 'The Scholar as Craftsman: Derek de Solla Price and the Reconstruction of a Medieval Instrument', *Notes and Records of the Royal Society* 68 (2014): 111–34 (freely available via https://doi.org/10.1098/rsnr.2013.0062).

Views of the Middle Ages in both academic and popular culture are thoughtfully dissected in David Matthews, *Medievalism:*

A Critical History (Woodbridge, 2015). A useful discussion of the term 'Dark Ages' is Janet L. Nelson, 'The Dark Ages', *History Workshop Journal* 63 (2007): 191–201. Peter Harrison, *The Territories of Science and Religion* (Chicago, 2015), dismantles the myths of a perennial conflict between science and religion. *Galileo Goes to Jail, and Other Myths about Science and Religion*, ed. Ronald L. Numbers (Cambridge, MA, 2009) and its sequel, *Newton's Apple, and Other Myths about Science*, ed. Numbers and Kostas Kampourakis (Cambridge, MA, 2015), amusingly and effectively debunk a wide range of common misconceptions about the history of science.

WESTWICK, LEARNING AND ARITHMETIC

David S. Neal, Angela Wardle and Jonathan Hunn, *Excavation of the Iron Age, Roman, and Medieval Settlement at Gorhambury, St. Albans* (London, 1990), is an excellent example of what archaeology can add to our understanding of life in the Middle Ages – specifically, in this case, the farmland and fishponds of John Westwyk's birthplace. On John's education, Nicholas Orme, *English Schools in the Middle Ages* (London, 1973), remains a good starting point, and Roger Bowers, 'The Almonry Schools of the English Monasteries, c.1265–1540', in *Monasteries and Society in Medieval Britain*, ed. Benjamin Thompson, Harlaxton Medieval Studies NS 6 (Stamford, 1999): 177–222, is an important article. Mary Carruthers, *The Book of Memory: A Study of Memory in Medieval Culture* (Cambridge, 2008), is an essential account of the methods and status of memory in medieval learning (and includes translations of medieval memory techniques that still work well today).

The folk astronomy of sunrises, seasons and constellations is brilliantly and accessibly covered in Stephen C. McCluskey, *Astronomies and Cultures in Early Medieval Europe* (Cambridge, 1998). To understand the astronomy, I found the free computer programs *Planetary, Lunar, and Stellar Visibility* (http://www.alcyone.de) and Stellarium (http://stellarium.org) very helpful. Otto Neugebauer, *The Exact Sciences in Antiquity* (New York, 1962), is a succinct introduction to early mathematics and astronomy. On their Babylonian foundations, Eleanor Robson, *Mathematics in Ancient Iraq: A Social History* (Princeton, 2009), is ground-breaking. The work of Charles Burnett at the coalface of the medieval sciences is indispensable for scholars; his *Numerals and Arithmetic in the Middle Ages* (Farnham, 2010) is an original and erudite collection of essays. An accessible yet scholarly blog on all aspects of medieval and early modern mathematics and astronomy is Thony Christie, *The Renaissance Mathematicus*; a typically informative post, on Hindu-Arabic numerals, is at https://thonyc.wordpress.com/2018/05/03/as-easy-as-123. For monastic mathematics and its use in the calendar and wider sciences, an essential website is *The Calendar and the Cloister* (http://digital.library.mcgill.ca/ms-17), with high-quality images of a twelfth-century manuscript and commentary by Faith Wallis.

ST ALBANS, TIME AND CALENDARS

Mark Freeman, *St Albans: A History* (Lancaster, 2008), is a well-illustrated introduction to the cathedral city and its rich history. Eileen Roberts, *The Hill of the Martyr: An Architectural History of St. Albans Abbey* (Dunstable, 1993), is also valuable. There are archaeological riches to be found in the *Transactions of the St Albans & Hertfordshire Architectural and Archaeological Society*, whose early publications are freely available online (https://www.stalbanshistory.org); see, for example, Ernest Woolley on 'The Wooden Watching Loft in St. Albans Abbey Church' (1929): 246–54. Michelle Still, *The Abbot and the Rule: Religious Life at St Alban's, 1290–1349* (Aldershot, 2002), mines the gems of monastic life as revealed in the abbey chronicle. A new translation of Thomas Walsingham's chronicle, *The Deeds of the Abbots of St Albans*, tr. David Preest (Woodbridge, 2019), was published just as I was completing this book. It is edited by James G. Clark, whose *A Monastic Renaissance at St. Albans: Thomas Walsingham and His Circle, c.1350–1440* (Oxford, 2004) is an important work on intellectual life at the abbey.

David Knowles, *The Religious Orders in England*, vols. 1 and 2 (Cambridge, 1948–55), remains a thorough and valuable guide. Joan Greatrex, *The English Benedictine Cathedral Priories: Rule and Practice, c.1270–c.1420* (Oxford, 2011), does not cover St Albans specifically but is rich with details of life in the most wealthy and influential monasteries. *The Rule of St Benedict* is essential reading for an understanding of monastic life; it is available in several modern translations.

On early timekeeping, *Time and Cosmos in Greco-Roman Antiquity*, ed. Alexander R. Jones (Princeton, 2017), is a beautifully illustrated exhibition catalogue with informative essays. John North, *God's Clockmaker: Richard of Wallingford and the Invention of Time* (London, 2005), is masterly not only on Wallingford's clock (and other instruments) but on the development of timekeeping across the Middle Ages. Jean Gimpel, *The Medieval Machine: The Industrial Revolution of the Middle Ages* (London, 2nd edition, 1988) is a wide-ranging introduction to medieval technology, including clock-making. E. R. Truitt, *Medieval Robots: Mechanism, Magic, Nature, and Art* (Philadelphia, 2015), is a beautifully evocative guide to the place of mechanical arts in medieval cultures.

The work of Philipp Nothaft is fast revolutionising our understanding of the way medieval people understood time and used calendars. His *Scandalous Error: Calendar Reform and Calendrical Astronomy in Medieval Europe* (Oxford, 2018) is essential reading on this subject. Danielle B. Joyner, *Painting the* Hortus deliciarum: *Medieval Women, Wisdom and Time* (University Park, PA, 2016), is a fascinating, beautifully illustrated introduction to Herrad of Hohenburg and medieval attitudes to time. An important collection of essays on the *Gregorian Reform of the Calendar: Proceedings of the Vatican Conference to Commemorate Its 400th Anniversary (1582–1982)*, ed. G. V. Coyne, M. A. Hoskin and O. Pedersen (Vatican City, 1983), is free to download from archive. org. Bonnie Blackburn and Leofranc Holford-Strevens, *The Oxford Companion to the Year: An Exploration of Calendar Customs and Time-reckoning* (Oxford, 1999), is a valuable reference work and a joy to dip into.

OXFORD, THE MEDIEVAL UNIVERSITIES AND NATURAL PHILOSOPHY

The development of the medieval universities in general, and Oxford in particular, has been the subject of intense study. *The History of the University of Oxford*, vol. 1, ed. Jeremy Catto, and vol. 2, ed. Jeremy Catto and Ralph Evans (Oxford, 1984–92), are indispensable collections of essays. *A History of the University in Europe*, vol. 1, ed. Hilde de Ridder-Symoens (Cambridge, 1992), is equally important for the wider European scene. Edward Grant, *The Foundations of Modern Science in the Middle Ages: Their Religious, Institutional, and Intellectual Contexts* (Cambridge, 1996), focuses on what was studied in the universities. Here the work of James Weisheipl has been fundamental: see, for example, 'Curriculum of the Faculty of Arts at Oxford in the Early Fourteenth Century', *Mediaeval Studies* 26 (1964): 143–85, and a follow-up article in vol. 28 (pp. 151–75) of the same journal. Weisheipl also edited an important collection of essays on *Albertus Magnus and the Sciences* (Toronto, 1980). An equivalent volume, *Roger Bacon and the Sciences: Commemorative Essays*, ed. Jeremiah Hackett (Leiden, 1997), also repays careful reading. While there is no equivalent recent volume for Robert Grosseteste's science, the work of the *Ordered Universe* project (https://ordered-universe. com) is doing much to bring it to clearer light. On the scientific differences between Franciscans and Dominicans, see Roger French and Andrew Cunningham, *Before Science: The Invention of the Friars' Natural Philosophy* (Aldershot, 1996). Carl B. Boyer, *The Rainbow: From Myth to Mathematics* (Princeton, 1987), is fascinating reading on the history of attempts to understand that mind-bending natural phenomenon.

Benedictines in Oxford, ed. Henry Wansbrough and Anthony Marett-Crosby (London, 1997), is a useful collection of short essays on monks' university lives. Raymond Clemens and Timothy Graham, *Introduction to Manuscript Studies* (Ithaca, NY, 2007), is an informative and well-illustrated guide for students. Christopher de Hamel, *Meetings with Remarkable Manuscripts* (London, 2016), paints a dazzling portrait of a dozen medieval books, with fascinating detail on their histories. The British Library has a selection of short online videos on manuscript production (https://www.bl.uk/medieval-english-french-manuscripts/videos).

Many of the philosophers who filled the medieval universities are catalogued in *A Companion to Philosophy in the Middle Ages*, ed. Jorge J. E. Gracia and Timothy B. Noone (Oxford, 2006), which also has useful articles on topics such as scholasticism and the Parisian condemnations. For a truly accessible, entertaining introduction to medieval philosophy (in the broadest possible sense), I recommend Peter Adamson's *History of Philosophy* podcast (https://historyof philosophy.net); the medieval episodes have recently been published in book form (Oxford, 2019).

John of Sacrobosco's staggeringly successful *Sphere* textbook is available in Lynn Thorndike's translation, in *The Sphere of Sacrobosco and Its Commentators* (Chicago, 1949); online at http://www.esotericarchives.com/solomon/sphere.htm. Olaf Pedersen, 'In Quest of Sacrobosco', *Journal for the History of Astronomy* 16 (1985): 175–220, available via https://ui.adsabs.harvard.edu, is an essential introduction to the mysterious man himself. On the myth of a flat medieval Earth, see Jeffrey Burton Russell, *Inventing the Flat Earth: Columbus and Modern Historians* (New York, 1991). James Evans, *The History and Practice of Ancient Astronomy* (New York, 1998), contains a clear explanation of Eratosthenes' methods and is, in general, a wonderful, inspiring how-to guide to the subject.

ASTRONOMICAL INSTRUMENTS

The best explanation of how an astrolabe works is in J. D. North, *Chaucer's Universe* (Oxford, 1988), which is a fascinating and exhaustive study of astronomy and astrology in the work of Geoffrey Chaucer. Templates for producing your own astrolabe are also available on various websites. Chaucer's *Treatise on the Astrolabe* can be read with a (passable) parallel modern English translation at http://www.chirurgeon.org/treatise.html. On stereographic projection and many other issues in Ptolemaic astronomy, Otto Neugebauer, *A History of Ancient*

Mathematical Astronomy (Heidelberg, 1975), remains a touchstone, but it is not for the faint-hearted. The succinct, well-illustrated essays in *Astronomy before the Telescope*, ed. Christopher Walker (London, 1996), provide a stimulating introduction to the subject.

Many museums have produced illustrated catalogues of their medieval scientific instruments, and increasingly these are online. For example, *Western Astrolabes*, ed. Roderick Webster and Marjorie Webster (Chicago, 1998), and *Eastern Astrolabes*, ed. David Pingree (Chicago, 2009), are exemplary guides to the collections at the Adler Planetarium. Oxford's History of Science Museum has an excellent online catalogue (http://www.mhs.ox.ac.uk/astrolabe), which allows the world's largest collection of astrolabes to be browsed and searched. S. R. Sarma has recently produced an exhaustive *Catalogue of Indian Astronomical Instruments*, which is freely available at https://srsarma.in.

The Whipple Museum of the History of Science: Objects and investigations, to celebrate the 75th anniversary of R. S. Whipple's gift to the University of Cambridge, ed. Joshua Nall, Liba Taub and Frances Willmoth (Cambridge, 2019), is a new essay collection representing the range of history of science museums, from brass to silicon. It includes an essay – by me – on the astrolabe we met in Chapter 4, and others on medieval sundials and modern fakes, and is freely available via https://www.cambridge.org/core.

On Richard of Wallingford's life and *Albion*, see North, *God's Clockmaker* (cited above). For full editions, translations and technical explanations of Wallingford's works, see J. D. North, *Richard of Wallingford* (Oxford, 1976). On Geoffrey of Monmouth and the Albina myth, see Jeffrey Jerome Cohen, *Of Giants: Sex, Monsters, and the Middle Ages* (Minneapolis, 1999).

TYNEMOUTH, TRIGONOMETRY, ASTROLOGY AND MAGIC

Many of the most important sources for the history of Tynemouth Priory are collected in H. H. E. Craster, *A History of Northumberland, Volume VIII: The Parish of Tynemouth*

(Newcastle, 1907). Those interested in the archaeology and architecture of the priory will enjoy consulting past volumes of the
journal of the Society of Antiquaries of Newcastle, *Archaeologia
Aeliana*, available online via the Archaeology Data Service (https://
doi.org/10.5284/1053682); articles in vols. 4:13 and 4:14 (1936–7)
are replete with details of the monastic buildings. On the status of
such priories, see Martin Heale, *The Dependent Priories of Medieval English Monasteries* (Woodbridge, 2004).

Ptolemy's *Almagest* is available in English translation (tr. G. J.
Toomer, London, 1984), but it is famously difficult. It is made somewhat easier by Olaf Pedersen, *A Survey of the Almagest* (revised
edition, ed. Alexander Jones, New York, 2011). For a practical introduction to spherical trigonometry in historical context, see Glen Van
Brummelen, *Heavenly Mathematics: The Forgotten Art of Spherical Trigonometry* (Princeton, 2013), and also his *The Mathematics
of the Heavens and the Earth* (Princeton, 2009) for fuller historical
background to the theories. James Evans, *The History and Practice
of Ancient Astronomy* (New York, 1998), will get you working flu-
ently with oblique ascensions.

Medieval astrology is the object of renewed, wide-ranging academic interest. Sophie Page, *Astrology in Medieval Manuscripts*
(London, 2002), is a brief, well-illustrated introduction. Nicholas
Campion, *A History of Western Astrology, Volume II: The Medieval and Modern Worlds* (London, 2009), is a narrative of astrology's
place in society. J. D. North, *Horoscopes and History* (London,
1986) thoroughly unpicks the mathematics essential for medieval
astrologers. The work of Charles Burnett has been foundational,
editing and translating previously unpublished source materials,
and co-ordinating and supporting research; *From Māshā'allāh to
Kepler: Theory and Practice in Medieval and Renaissance Astrology*, ed. Burnett and Dorian Gieseler Greenbaum (Ceredigion, 2015),
is one useful collection of scholarly essays. For a recent academic
reassessment of a range of medieval writings, see H. Darrel Rutkin,
*Sapientia Astrologica: Astrology, Magic and Natural Knowledge,
ca.1250–1800, Vol. 1: Medieval Structures* (Cham, 2019). 'Celestial
Influence – the Major Premiss of Astrology' is among many erudite
and original articles by J. D. North collected in *Stars, Minds and*

Fate: Essays in Ancient and Medieval Cosmology (London, 1989). Hilary M. Carey, *Courting Disaster: Astrology at the English Court and University in the Later Middle Ages* (Basingstoke, 1992), is that rare combination: scholarly, succinct and very enjoyable to read.

The Routledge History of Medieval Magic, ed. Sophie Page and Catherine Rider (London: 2019), is a magnificent collection of short essays by all the main scholars in this field. Monks practising magic are fascinatingly covered in Sophie Page, *Magic in the Cloister: Pious Motives, Illicit Interests, and Occult Approaches to the Medieval Universe* (University Park, 2013).

THE CRUSADES, TRAVEL AND MEDICINE

The Crusades have been the subject of endless fascination and frequent historical treatment. Christopher Tyerman, *God's War: A New History of the Crusades* (London, 2006), is one accessible yet sensitive attempt to understand this bizarre chapter of human history. On the Hundred Years War, the series by Jonathan Sumption (London, 1990–2015; four volumes of a planned five have been published so far) is an immense yet elegant work of scholarly writing. *Volume III: Divided Houses* (2009) covers the 1383 Bishop's Crusade.

The History of Cartography (Chicago, 1991–2015) is a monumental project. Volume 1, on ancient and medieval Europe (ed. J. B. Harley and David Woodward), and volume 3, on Renaissance Europe (ed. David Woodward), were useful to me. Both these well-illustrated, scholarly volumes are freely available at https://www.press.uchicago.edu/books/HOC. Kenneth Nebenzahl, *Mapping the Silk Road and Beyond: 2,000 Years of Exploring the East* (London, 2004), is a beautifully illustrated introduction to the development of European map-making. Julian Smith, 'Precursors to Peregrinus: The Early History of Magnetism and the Mariner's Compass in Europe', *Journal of Medieval History* 18 (1992): 21–74, is an excellent survey of the first European writings on the magnetic compass. Felipe Fernández-Armesto, *Pathfinders: A Global History of*

Exploration (Oxford, 2006), is an ambitious yet accessible introductory work.

Chaucer's portraits of the Shipman and Doctor of Physic (as well as the monk, cited in Chapter 2) are sensitively analysed in essays in *Historians on Chaucer: The 'General Prologue' to the Canterbury Tales*, ed. S. H. Rigby and A. J. Minnis (Oxford, 2014). There are many good introductory works on medieval medicine; Nancy Siraisi, *Medieval and Early Renaissance Medicine: An Introduction to Knowledge and Practice* (Chicago, 1990), and Carole Rawcliffe, *Medicine & Society in Later Medieval England* (Stroud, 1995), are particularly clear. Luke Demaitre, *Medieval Medicine: The Art of Healing, from Head to Toe* (Santa Barbara, CA, 2013), is an excellent recent survey, which draws on the author's long engagement with the writings of Bernard of Gordon and includes an especially impressive chapter on digestive illness. *Practical Medicine from Salerno to the Black Death*, ed. Luis García Ballester et al. (Cambridge, 1994), is an important collection of scholarly essays. A beautiful, thought-provoking recent book, Jack Hartnell, *Medieval Bodies: Life, Death and Art in the Middle Ages* (London, 2018), uses the human body as a starting point for a fascinating, wide-ranging exploration of medieval cultures.

LONDON, THE *EQUATORIE*, BESTIARIES AND RENAISSANCE ASTRONOMY

Urban life in the later Middle Ages is effectively explored in Caroline Barron, *London in the Later Middle Ages: Government and People, 1200–1500* (Oxford, 2004); the work of Martha Carlin on medieval Southwark is also important. For an atmospheric novelistic portrayal of the medieval city, authored by an academic historian, try Bruce Holsinger, *A Burnable Book* (London, 2014).

The definitive account of the Alfonsine Tables is José Chabás and Bernard R. Goldstein, *The Alfonsine Tables of Toledo* (Dordrecht, 2003). In their more recent work these two scholars have greatly enhanced our understanding of astronomical tables and their uses.

A solid overview of the importance of tables is provided in John North's monumental and magisterial *Cosmos: An Illustrated History of Astronomy and Cosmology* (Chicago, 2008).

The starting-point for understanding John Westwyk's equatorium remains Derek J. Price, *The Equatorie of the Planetis* (Cambridge, 1955, reissued 2012). My PhD thesis on the equatorium, 'Improving Instruments: Equatoria, Astrolabes, and the Practices of Monastic Astronomy in Late Medieval England' (Cambridge, 2016), can be downloaded via https://doi.org/10.17863/CAM.87. John Westwyk's manuscript is displayed at the Cambridge University Digital Library, https://cudl.lib.cam.ac.uk/view/MS-PETERHOUSE-00075-00001. There, alongside high-resolution images of the manuscript and a full transcription and translation, you can try out a virtual model of Westwyk's planetary computer. Scholars seeking a full understanding of such planetary instruments must consult Emmanuel Poulle, *Les Instruments de la théorie des planètes selon Ptolémée: équatoires et horlogerie planétaire du XIIIe au XVIe siècle* (Geneva, 1980).

Bestiaries and the marvels of the wide world are brilliantly explored in Lorraine Daston and Katharine Park, *Wonders and the Order of Nature, 1150–1750* (New York, 1998). Lisa Jardine, *Worldly Goods: A New History of the Renaissance* (London, 1996), is captivating on the visual wonder of this period and gives ample attention to changing scientific ideas and communication practices. Elizabeth Eisenstein, *The Printing Revolution in Early Modern Europe* (Cambridge, 2nd edition, 2005), is a seminal account of the transition from manuscript to printing.

Michael J. Crowe, *Theories of the World from Antiquity to the Copernican Revolution* (Mineola, 2nd edition, 2001) is a succinct textbook full of valuable excerpts from key texts in the history of astronomy. Owen Gingerich, *The Book Nobody Read* (New York, 2004), recounts the author's exciting worldwide quest to see every surviving copy of the first two editions of Copernicus' *On the Revolutions of the Heavenly Spheres*; along the way we learn a great deal about that 1543 masterwork and the charged atmosphere of sixteenth-century astronomy. Thony Christie, *The Renaissance Mathematicus*, has a thorough series of blog posts on this subject: https://thonyc.wordpress.com/the-emergence-of-modern-astronomy-a-complex-mosaic.

Before Copernicus: The Cultures and Contexts of Scientific Learning in the Fifteenth Century, ed. Rivka Feldhay and F. Jamil Ragep (Montreal, 2017), is a series of essays that effectively lay out the multicultural foundations for the heliocentric revolution. *Knowledge in Translation: Global Patterns of Scientific Exchange, 1000–1800 CE*, ed. Patrick Manning and Abigail Owen (Pittsburgh, 2018), has rich examples of medieval scientific transmission, with essays on Tusi and the Catalan Atlas, among other subjects.

Notes

PROLOGUE

1. These are the words of the Nobel Prize winner Lawrence Bragg (quoting his wife, Alice), in a letter to Professor Nevill Mott, 4 May 1962, Royal Institution MS WLB 55F/89. Bragg also told a colleague that 'there is nothing wrong with the man himself; it is his background'. (Letter to C. Singer, 15 Dec. 1955, RI MS WLB 55F/47.

2. Price's application is amply documented in his Cambridge University file: CU Archives, Records of the Board of Graduate Studies, 1, 1953–4, Price D.J.; S. Falk, 'The Scholar as Craftsman: Derek de Solla Price and the Reconstruction of a Medieval Instrument', *Notes and Records of the Royal Society* 68 (2014), 111–34.

3. M.R. James, *A Descriptive Catalogue of the Manuscripts in the Library of Peterhouse* (Cambridge, 1899), 94.

4. D. de S. Price, *Science since Babylon*, enlarged edition (New Haven, 1975), 26–7.

5. D.J. Price, 'In Quest of Chaucer – Astronomer', *Cambridge Review*, 30 Oct. 1954, 123–4.

6. 'Chaucer Holograph Found in Library', *Varsity*, 23 Feb. 1952; 'Possible Chaucer Manuscript: Discovery at Cambridge, *The Times*, 28 Feb. 1952; e.g. 'Skrev Chaucer Bog om Astronomisk Regnemetode?', *B.T.* (Copenhagen), 27 Feb. 1952; 'Was Chaucer a Scientist Too?', *The Hindu* (Madras), 6 April 1952.

7. C. Sagan, *Cosmos* (New York, 1980), 335.

8. D. Wootton, *The Invention of Science: A New History of the Scientific Revolution* (London, 2015). The first sentence of Wootton's impressive polemical tome qualifies the claim implicit in the title: '*Modern* science was invented between 1572 ... and 1704' (1, my emphasis); but in arguing for the importance of the Scientific Revo-

lution, he is quite explicit that he does not consider the science of previous centuries worthy of the name (see esp. 573–5).

9. J. Gribbin, *Science: A History* (London, 2002).

10. W. Camden, 'Certaine Poemes, or Poesies, Epigrammes, Rythmes, and Epitaphs of the English Nation in Former Times', in *Remaines of a Greater Worke* (London, 1605), 2.

11. E. Gibbon, *The History of the Decline and Fall of the Roman Empire*, vol. 6 (London, 1788), 519.

12. e.g. the description by the then UK Deputy Prime Minister Nick Clegg of 'these barbaric, medieval types in Isil', A. Chakelian, 'Nick Clegg: "It's Not Obvious" What the UK Can Do Legally on New Terror Powers', *New Statesman*, 2 Sept. 2014, https://www.newstatesman.com/politics/2014/09/nick-clegg-it-s-not-obvious-what-uk-can-do-legally-new-terror-powers; S. Javid, Twitter, 8 March 2019, https://twitter.com/sajidjavid/status/1104054288064675840; B. Moor, 'Residents Frustrated at "Medieval" Cellphone Coverage in the Far North', *Stuff*, 29 Aug. 2018, https://www.stuff.co.nz/auckland/local-news/northland/106654790/residents-frustrated-at-medieval-cellphone-coverage-in-the-far-north.

13. J. Swan, 'White House Review Nears End: Officials Expect Bannon Firing', *Axios*, 18 Aug. 2017, https://www.axios.com/white-house-review-nears-end-officials-expect-bannon-firing-1513304936-a38d90a8-131a-4198-9e77-5a28042b5c58.html; D. Snow, tweets at 9:47 a.m. and 10:31 a.m., 19/8/2017, https://twitter.com/thehistoryguy/status/898828840197345280 and 898839891949256704.

14. I. Newton, *Philosophiae naturalis principia mathematica*, 3rd edition (London, 1726), 529. The 'General Scholium' afterword was first written for the second edition in 1713, but there Newton used the phrase 'experimental philosophy' (*philosophiam experimentalem*), tr. A. Motte, 1729 (*The Mathematical Principles of Natural Philosophy*, 2 vols., 2:391–2). See A. Cunningham, 'How the "Principia" Got Its Name: Or, Taking Natural Philosophy Seriously', *History of Science* 29 (1991), 377–92, and a subsequent debate between Cunningham and another historian, Edward Grant, in an issue of *Early Science and Medicine* 5:3 (2000); M.H. Shank and D.C. Lindberg, 'Introduction', in *The Cambridge History of Science, Volume 2: Medieval Science*, ed. D.C. Lindberg and M.H. Shank (Cambridge, 2013), 1–26.

15. D.J. Price, *The Equatorie of the Planetis* (Cambridge, 1955), 149; Chaucer, 'A Treatise on the Astrolabe', II.4; see, for example, F.N.

Robinson, Preface to 2nd edn of *The Works of Geoffrey Chaucer* (London, 1957), ix; D. Pearsall, *The Life of Geoffrey Chaucer: A Critical Biography* (Oxford, 1992), 218–19. The *Equatorie* is included in J.H. Fisher, ed., *The Complete Poetry and Prose of Geoffrey Chaucer* (New York, 1977 and subsequent editions).

16. C.J. Singer, letters to R. F. Holmes (8 May 1959) and A. W. Skempton (27 March 1959), London, Wellcome Collection PP/CJS/A.47; D. J. de S. Price, letter to C.J. Singer, 22 Dec. 1959, London, Wellcome Collection PP/CJS/A.47.

17. A. Liversidge, 'Interview: Derek de Solla Price', *OMNI* (Dec. 1982), 89–102 and 136, at 89.

18. K.A. Rand Schmidt, *The Authorship of the Equatorie of the Planetis* (Cambridge, 1993); K.A. Rand, 'The Authorship of The Equatorie of the Planetis Revisited', *Studia Neophilologica* 87 (2015), 15–35.

19. John Gower, *Confessio Amantis*, VII.625–32.

20. L.P. Hartley, *The Go-Between* (London, 1953), 5.

CHAPTER 1: WESTWYK AND WESTWICK

1. St Albans Book of Benefactors, London, British Library Cotton MS Nero D.VII, ff. 81v–83v. Ed. W. Dugdale, *Monasticon Anglicanum*, new edn, ed. J. Caley, H. Ellis and B. Bandinel (London, 1819), 2:209n.

2. D. Knowles, *The Religious Orders in England, Volume II: The End of the Middle Ages* (Cambridge, 1955), 231–2; R.B. Dobson, *Durham Priory, 1400–1450* (Cambridge, 1973), 56–7.

3. Book of Benefactors (note 1), f. 83r.

4. E. Woolley, 'The Wooden Watching Loft in St. Albans Abbey Church', *Transactions of St. Albans and Herts Architectural and Archaeological Society* (1929), 246–54.

5. Matthew Paris, *Chronica Majora*, ed. H. Luard, Rolls Series (London, 1880), 5:669.

6. British Library Cotton MS Tiberius E.VI, f. 236v; Inquisition post mortem Alphonsus de Veer (1328), National Archives C 135/10/12; D.S. Neal, A. Wardle and J. Hunn, *Excavation of the Iron Age, Roman, and Medieval Settlement at Gorhambury, St. Albans* (London, 1990), 102–3.

7. Dobson, *Durham Priory* (note 2), 57–9; N. Orme, *English Schools in the Middle Ages* (London, 1973), 50–51; A.E. Levett, *Studies in*

Manorial History, ed. H.M. Cam, M. Coate and L.S. Sutherland (Oxford, 1938), 292–3.

8. Islam may have chosen a purely lunar calendar to distinguish itself from the older Judaism and Christianity. C.L.N. Ruggles and N.J. Saunders, 'The Study of Cultural Astronomy', in *Astronomies and Cultures*, ed. Ruggles and Saunders (Niwot, 1993), 1–31; J. North, *Cosmos: An Illustrated History of Astronomy and Cosmology* (Chicago, 2008), 185.

9. Oxford, Bodleian Library MS Digby 88, f. 97v.

10. 'Houses of Benedictine Nuns: St Mary de Pre Priory, St Albans', in *A History of the County of Hertford: Volume 4*, ed. W. Page (London, 1971), 428–32; Ver Valley Society, 'Mills', http://www.riverver.co.uk/mills/.

11. S.C. McCluskey, *Astronomies and Cultures in Early Medieval Europe* (Cambridge, 1998), 13; North, *Cosmos* (note 8), 11–12.

12. Cambridge, Pembroke College MS 180; Cambridge, Trinity College MS B.2.19; British Library Royal MS 2 A X. See T.A.M. Bishop, 'Notes on Cambridge Manuscripts', in *Transactions of the Cambridge Bibliographical Society* 1 (1953), 432–41, at 435.

13. Cambridge, Emmanuel College MS 244. Translation adapted from Rutilius Taurus Aemilianus Palladius, *The Work of Farming (Opus Agriculturae) and Poem on Grafting*, tr. J.G. Fitch (Totnes, 2013), 177. Latin text ed. J.C. Schmitt (Leipzig, 1898), http://www.forum-romanum.org/literature/palladius/agr.html.

14. John 11:9; J.D. North, 'Monastic Time', in *The Culture of Medieval English Monasticism*, ed. J.G. Clark (Woodbridge, 2007), 203–11, at 208.

15. Cambridge, Peterhouse MS 75.I, f. 64r.

16. J.R. Harris, 'On the Locality to Which the Treatise of Palladius De Agricultura Must be Assigned', *American Journal of Philology* 3 (1882): 411–21.

17. See, for example, *The Kalendarium of Nicholas of Lynn*, ed. S. Eisner (Athens, GA, 1980); J.D. North, *Chaucer's Universe* (Oxford, 1988), 104–9.

18. Cambridge, Trinity College MS O.3.43; Oxford, Bodl. MS Auct F.5.23; D. Wakelin, *Humanism, Reading, and English Literature 1430–1530* (Oxford, 2007), 43–5.

19. Palladius, *De re rustica/On Husbondrie*, VII.60–63. Two editions are based on two different manuscripts of the fifteenth-century translation: *The Middle-English Translation of Palladius De re*

rustica, ed. M. Liddell (Berlin, 1896), 181, and *Palladius on Husbondrie, from the unique MS. of about 1420 A.D. in Colchester Castle*, ed. B. Lodge (London, 1879), 160.

20. See, for example, Cambridge, Corpus Christi College MS 297, ff. 23r–91v.

21. Virgil, *The Georgics*, I. 208–11.

22. Virgil, *The Georgics*, I. 220–21.

23. Accounts of the precise definition of 'dog days' vary, and as precession has delayed the rising of Sirius, its link with the hottest part of the summer has weakened. See B. Blackburn and L. Holford-Strevens, *The Oxford Companion to the Year* (Oxford, 1999), 595–6.

24. N. Sivin, *Granting the Seasons: The Chinese Astronomical Reform of 1280* (New York, 2008).

25. *Gesta Abbatum monasterii Sancti Albani (GASA)*, ed. H. Riley (London, 1867), 1:73–95.

26. C.B.C. Thomas, 'The Miracle Play at Dunstable', *Modern Language Notes* 32 (1917): 337–44.

27. *GASA* (note 25), 1:73.

28. Cambridge University Library MS Ee.4.20, f. 68v; J.G. Clark, *The Benedictines in the Middle Ages* (Woodbridge, 2011), 70–71; M.T. Clanchy, *From Memory to Written Record: England 1066–1307*, 2nd edn (Oxford, 1993), 13; Orme, *English Schools in the Middle Ages* (note 7), 49–50.

29. *GASA* (note 25), 1:196. See R. Bowers, 'The Almonry Schools of the English Monasteries, c.1265–1540', in *Monasteries and Society in Medieval Britain: Proceedings of the 1994 Harlaxton Symposium*, ed. B. Thompson (Stamford, 1999): 177–222, at 191–2.

30. Statutes in British Library Lansdowne MS 375, ff. 97–105, edited in *Registrum Abbatiae Johannis Whethamstede* (London, 1873), 2: 305–15.

31. See, for example, the popular *Carmen de Algorismo* of Alexander Villedieu. Edited in J.O. Halliwell, *Rara Mathematica* (London, 1841), 73–83. On the development of decimal numerals in fifth-to seventh-century India, especially the vital contribution of Brahmagupta (*c.* 598–*c.*668), see K. Plofker, 'Mathematics in India', in *The Mathematics of Egypt, Mesopotamia, China, India, and Islam: A Sourcebook*, ed. V.J. Katz (Princeton, 2007), 385–514.

32. Cambridge University Library MS Ii.6.5, fol. 104r. Facsimile and transcription in K. Vogel, *Mohammed ibn Musa Alchwarizmi's*

Algorismus: das früheste Lehrbuch zum Rechnen mit indischen Ziffern (Aalen, 1963); a translation is available in J.N. Crossley and A.S. Henry, 'Thus Spake Al-Khwārizmī: A Translation of the Text of Cambridge University Library Ms. Ii.vi.5', *Historia Mathematica* 17 (1990): 103–31.

33. J.N. Crossley, 'Old-Fashioned versus Newfangled: Reading and Writing Numbers, 1200–1500', *Studies in Medieval and Renaissance History* 10 (2013): 79–109.

34. University of Aberdeen MS 123, ff. 66r–67v.

35. Cambridge, Corpus Christi College MS 7, f. 98r. Edited in *GASA* (note 25), 3:399–400, 454–7. See also F. Madden, B. Bandinel and J.G. Nichols, eds., *Collectanea Topographica et Genealogica* (London, 1838), 5:194–7.

36. Bede, *The Reckoning of Time*, ed. and tr. F. Wallis (Liverpool, 2004), 9.

37. See, e.g., *The Crafte of Nombryng*, from British Library Egerton MS 2622, ed. R. Steele, *The Earliest Arithmetics in English*, EETS ES 118 (London: 1922): 3–32, at 5.

38. e.g. Oxford, St John's College MS 17, ff. 41v–42r (Thorney, 12th century), http://digital.library.mcgill.ca/ms-17.

39. Doubt has been cast on the extent (if any) of Alfred's involvement in the translation attributed to him; see J. Bately, 'Did King Alfred Actually Translate Anything? The Integrity of the Alfredian Canon Revisited', *Medium Ævum* 78 (2009): 189–215. Elizabeth's 1593 translation survives in her own handwriting (Kew, National Archives SP 12/289).

40. T. Kojima, *The Japanese Abacus: Its Use and Theory*, quoted in L. Fernandes, 'The Abacus vs. the Electric Calculator', https://www.ee.ryerson.ca/~elf/abacus/abacus-contest.html.

41. e.g. St John's College MS 17 (note 38), ff. 41v–42r http://digital.library.mcgill.ca/ms-17.

42. D. Knowles, *The Religious Orders in England* (Cambridge, 1957), 1:285.

43. John Gower, *Confessio Amantis*, Prologue, ll. 27–30 (Kalamazoo, 2006), http://d.lib.rochester.edu/teams/publication/peck-confessio-amantis-volume-1.

44. For Westwyk, using the Latin Vulgate, the last two psalms would have been numbered 135 and 146. An excellent example is the psalter commissioned by Abbot Geoffrey de Gorham around 1125 (Hildesheim, Dombibliothek MS St Godehard 1). Psalm 8 is online

at https://www.abdn.ac.uk/stalbanspsalter/english/commentary/
page083.shtml.

CHAPTER 2: THE RECKONING OF TIME

1. G. Chaucer, *The Canterbury Tales: Prologue to the Monk's Tale*,
   VII.1929–30, in *The Riverside Chaucer*, ed. L.D. Benson (Oxford,
   1987), 240.
2. J. Greatrex, *The English Benedictine Cathedral Priories: Rule and
   Practice, c.1270–c.1420* (Oxford, 2011), 67.
3. *Gesta Abbatum monasterii Sancti Albani (GASA)*, ed. H. Riley
   (London, 1867), 2:370, 3:393.
4. The arrangements for performance of the office varied slightly
   between monasteries but conformed to the general template laid
   down in the Rule (Chapters 8–18), and refined in the Monastic
   Constitutions (*c.*1077) of Archbishop Lanfranc. The pattern of the
   day is summarised in *The Monastic Constitutions of Lanfranc*, ed.
   D. Knowles, rev. edn by C.N.L. Brooke (Oxford, 2008), xx–xxv.
   For timings, see D. Knowles, *The Monastic Order in England: A
   History of Its Development from the Times of St. Dunstan to the
   Fourth Lateran Council, 940–1216* (Cambridge, 2nd edn 1963),
   450–51.
5. *GASA* (note 3), 2:428, 451.
6. *The Monastic Constitutions of Lanfranc* (note 4), ch. 87, pp. 122–7.
7. Oxford, Bodl. MS Bodley 38, ff. 19v–23v. Ed. G. Constable, Horolo-
   gium Stellare Monasticum, in *Consuetudines Benedictinae Variae*,
   Corpus Consuetudinum Monasticarum 6 (Siegburg, 1975), 1–18.
8. Gregory of Tours, *De Cursu Stellarum*, ed. B. Krusch, *Monumenta
   Germaniae Historica* I.2 (Hanover, 1969), 404–22; see S.C. McCluskey, 'Gregory of Tours, Monastic Timekeeping, and Early Christian
   Attitudes to Astronomy', *Isis* 81 (1990): 8–22. The earliest surviving
   manuscript, Bamberg Staatsbibliothek MS Patres 61 (8th century),
   ff. 75v–82v, has been digitised: see http://bit.ly/DeCursuStellarum.
   See 'Matters Arising', *Nature* 325 (1 Jan. 1987): 87–9.
9. B. Brady, D. Gunzburg and F. Silva, 'The Orientation of Cistercian
   Churches in Wales: A Cultural Astronomy Case Study', *Cîteaux –
   Commentarii cistercienses*, 67 (2016): 275–302.
10. J.L. Heilbron, *The Sun in the Church: Cathedrals as Solar Obser-
    vatories* (Cambridge, MA, 2001).

11. Lists of bedding and other personal items issued to monks at comparable houses are printed in Greatrex, *English Benedictine Cathedral Priories* (note 2), 58–60.

12. C.B. Drover, 'A Medieval Monastic Water-Clock', *Antiquarian Horology* 1 (1954): 54–8 & 63.

13. It is now at the Archivo de la Corona de Aragón (Barcelona), MS Ripoll 225, ff. 87r–93r, http://pares.mcu.es/ (Signatura: ACA, COLECCIONES, Manuscritos, Ripoll, 225). Text ed. J.M. Millás Vallicrosa, *Assaig d'història de les idees físiques i matemàtiques a la Catalunya medieval*, Estudis Universitaris Catalans: Sèrie monogràfica 1 (Barcelona, 1931), 316–18; tr. F. Maddison, B. Scott and A. Kent, 'An Early Medieval Water-Clock', *Antiquarian Horology* 3 (1962): 348–53.

14. See, e.g., the description of Good Friday 1333 by a Franciscan friar, in Cambridge University Library MS Hh.6.8, f. 124r; R. Bartlett, *The Hanged Man: A Story of Miracle, Memory, and Colonialism in the Middle Ages* (Princeton, 2006), 63–4.

15. 'De utilitatibus astrolabii', MS Ripoll 225 (note 13), ff. 13v–14r, ed. N. Bubnov in Gerbert, *Opera Mathematica* (Berlin, 1899), 114–47, at 129–30, though the Ripoll manuscript differs slightly from those Bubnov consulted. It is possible, but unlikely, that the author was Gerbert of Aurillac, later Pope Sylvester II: see C. Burnett, 'King Ptolemy and Alchandreus the Philosopher: The Earliest Texts on the Astrolabe and Arabic Astrology at Fleury, Micy and Chartres', *Annals of Science* 55 (1998): 329–68, at 330.

16. Burnett, 'King Ptolemy and Alchandreus' (note 15).

17. Cambridge, Corpus Christi College MS 111, pp. 47–8; ed. in J. Handschin, 'Hermannus Contractus-Legenden - nur Legenden?,' *Zeitschrift für deutsches Altertum und deutsche Literatur* 72 (1935): 1–7; tr. in L. Ellinwood, ed., *The Musica of Hermannus Contractus*, rev. J.L. Snyder (Rochester, 2015), 166–7; W. Berschin, 'Ego Herimannus: Drei Fragen zur Biographie des Hermannus Contractus', in *Hermann der Lahme: Reichenauer Mönch und Universalgelehrter des 11. Jahrhunderts*, ed. F. Heinzer and T.L. Zotz (Stuttgart, 2016), 19–24.

18. A twelfth-century copy is Durham Cathedral Library MS C.III.24, https://iiif.durham.ac.uk/index.html?manifest=t2mw0892992f.

19. D. Juste, 'Hermann der Lahme und das Astrolab im Spiegel der neuesten Forschung,' in *Hermann der Lahme: Reichenauer Mönch*

und Universalgelehrter des 11. Jahrhunderts, ed. F. Heinzer and T.L. Zotz, (Stuttgart, 2016), 273–84.

20. Oxford, Bodl. MS Ashmole 304, f. 2v, http://bit.ly/Ashmole304. See A. Iafrate, 'Of Stars and Men: Matthew Paris and the Illustrations of MS Ashmole 304', *Journal of the Warburg and Courtauld Institutes* 76 (2013): 139–77.

21. C. Eagleton, 'John Whethamsteade, Abbot of St. Albans, on the Discovery of the Liberal Arts and Their Tools: Or, Why were Astronomical Instruments in Late-Medieval Libraries?' *Mediaevalia* 29 (2008): 109–36.

22. M. Arnaldi and K. Schaldach, 'A Roman Cylinder Dial: Witness to a Forgotten Tradition', *Journal for the History of Astronomy* 28 (1997): 107–31; Juste, 'Hermann der Lahme und das Astrolab' (note 19), 278–82. See also C. Kren, 'The Traveler's Dial in the Late Middle Ages: The Chilinder', *Technology and Culture* 18 (1977): 419–35.

23. Chaucer, *The Canterbury Tales: The Shipman's Tale*, VII.201 6.

24. *The Rule of St Benedict*, ed. T. Fry as *RB 1980* (Collegeville, 1982), 41.1, p. 63.

25. Robertus Anglicus, commentary on Sacrobosco's *De Sphera*, XII. In L. Thorndike, ed., *The Sphere of Sacrobosco and Its Commentators* (Chicago, 1949), 185–6, 235–6.

26. A.A. Mills, 'Altitude Sundials for Seasonal and Equal Hours', *Annals of Science* 53 (1996): 75–84.

27. Palladius, *On Husbondrie* (Chapter 1, note 19), VI.225–8.

28. Chaucer, *The Canterbury Tales: The Nun's Priest's Tale*, VII.2853–7, 3187–99. See J.D. North, *Chaucer's Universe* (Oxford, 1988), 117–120.

29. *GASA* (note 3) 2:280–81, 385.

30. Paris, Bibliothèque Nationale de France, MS Français 19093, f. 5r, https://c.bnf.fr/xHR.

31. Robertus Anglicus, commentary on Sacrobosco's *De Sphera*, XI. In Thorndike, *The Sphere of Sacrobosco* (note 25), 180, 230. See also L. White, *Medieval Technology and Social Change* (Oxford, 1962), 132–3; A.J. Cárdenas, 'A Learned King Enthralls Himself: Escapement and the Clock Mechanisms in Alfonso X's *Libro del saber de astrologia*', in *Constructions of Time in the Late Middle Ages*, ed. C. Poster and R. Utz, *Disputatio* 2 (Evanston, 1997), 71–87.

32. Norwich Cathedral Priory, *Camera Prioris*, Roll no. 3, cited in J.D. North, *Richard of Wallingford: An Edition of His Writings*

(Oxford, 1976), 2:316. See also J. North, *God's Clockmaker: Richard of Wallingford and the Invention of Time* (London, 2005), 153; C.F.C Beeson, *English Church Clocks, 1280–1850* (London, 1971), 13–15.

33. Norfolk Record Office DCN 1/4/23 and 29; Beeson, *English Church Clocks* (note 32), 16–18 (fig. 2 reproduces the sacrist's roll for 1324–5), 104–5; *Victoria County History of Norfolk* (London, 1906), 2:318.

34. J. Needham, L. Wang and D.J. de S. Price, *Heavenly Clockwork: The Great Astronomical Clocks of Medieval China* (Cambridge, 2nd edn 1986).

35. North, *God's Clockmaker* (note 32), 175–81.

36. Cambridge, Gonville and Caius College MS 230 (116), ff. 116v, 11v–14v; D.J. Price, 'Two Medieval Texts on Astronomical Clocks', *Antiquarian Horology* 1 (1956), 156.

37. S.A. Bedini and F.R. Maddison, 'Mechanical Universe: The Astrarium of Giovanni de' Dondi', *Transactions of the American Philosophical Society* 56, no. 5 (1966): 1–69, at 8.

38. Oxford, Bodl. MS Ashmole 1796, ff. 130r, 160r. The list of St Albans monks in 1380 is in the Book of Benefactors, London, BL MS Cotton Nero D.VII, ff. 81v–83v; edited in W. Dugdale, *Monasticon Anglicanum*, new edn ed. J. Caley, H. Ellis and B. Bandinel (London, 1819), 2:209n. See E.M. Thompson (ed.), *Customary of the Benedictine Monasteries of St. Augustine, Canterbury and St. Peter, Westminster* (London, 1902), 1:117.

39. Leonardo da Vinci, Codex Madrid I, f. 12r, http://leonardo.bne.es/index.html; North, *God's Clockmaker* (note 32), 182–90.

40. A clock of comparable complexity, the Astrarium, was built in Padua (Italy) in the 1360s. See Giovanni Dondi dall'Orologio, *Tractatus astrarii*, ed. and tr. E. Poulle (Geneva, 2003).

41. L. Watson, K. McCann, and H. Horton, 'Big Ben: Why Has Westminster's Great Bell Been Silenced – and for How Long?', *Telegraph*, 21 Aug. 2017, https://www.telegraph.co.uk/news/2017/08/21/big-ben-row-everything-need-know-westminsters-great-bell-silenced/.

42. *Rule of Benedict* (note 24), 11.13, p. 41.

43. Knowles, *The Monastic Order in England* (note 4), 462–4, 455–6; *GASA* (note 3), 2:441–2, 1:194, 207–9; North, *Richard of Wallingford* (note 32), 2:532–8.

44. University of Aberdeen MS 123, f. 84r.

45. R.M. Kully, 'Cisiojanus: comment savoir le calendrier par cœur', in

Jeux de mémoire : aspects de la mnémotechnie médiévale (Montreal, 1985), 149–56.

46. Greatrex, *English Benedictine Cathedral Priories* (note 2), 66.
47. Hugh of St Victor, 'The Three Best Memory-Aids for Learning History', ed. W.M. Green, 'Hugo of St. Victor: De Tribus Maximis Circumstantiis Gestorum', *Speculum* 18 (1943): 484–93; tr. in M. Carruthers, *The Book of Memory: A Study of Memory in Medieval Culture* (Cambridge, 2nd edn 2008), 339–44. One language school explicitly using medieval techniques is Linguisticator (https://linguisticator.com).
48. Carruthers, *The Book of Memory* (note 47), 1–4.
49. J.G. Clark, *A Monastic Renaissance at St. Albans: Thomas Walsingham and His Circle, c.1350–1440* (Oxford, 2004), 54–5; British Library Lansdowne MS 763, ff. 97v–104r, http://www.bl.uk/manuscripts/Viewer.aspx?ref=lansdowne_ms_763_f098v; *GASA* (note 3), 2:106.
50. Alexander of Villedieu, *Massa Compoti*, in W.E. Van Wijk (ed.), *Le Nombre d'or: étude de chronologie technique suivie du texte de la Massa compoti d'Alexandre de Villedieu* (La Haye, 1936), 55. See also L. Means, ' "Ffor as Moche as Yche Man May Not Haue Þe Astrolabe": Popular Middle English Variations on the Computus', *Speculum* 67 (1992): 595–623, at 606; L. Thorndike, 'Unde Versus', *Traditio* 11 (1955): 163–93, at 168–71.
51. Oxford, Bodl. MS Ashmole 1522, f. 190r.
52. Bede, *The Reckoning of Time*, Ch. 40, ed. and tr. F. Wallis (Liverpool, 2004), 109–10; C.P.E. Nothaft, *Scandalous Error: Calendar Reform and Calendrical Astronomy in Medieval Europe* (Oxford, 2018), 25–6.
53. Nothaft, *Scandalous Error* (note 52), 57–8.
54. e.g. Oxford, St John's College MS 17 (Thorney, 12th century), http://digital.library.mcgill.ca/ms-17. British Library Royal MS 12 F II is a St Albans computus manuscript from the same period.
55. The key chronological table, produced by the Scythian monk Dionysius Exiguus (Dennis the Insignificant), was based on earlier Easter tables; C.P.E. Nothaft, *Dating the Passion: The Life of Jesus and the Emergence of Scientific Chronology (200–1600)* (Leiden, 2011), 75–6.
56. Bede, *The Reckoning of Time* (note 52), Ch. 43, p. 115; Nothaft, *Scandalous Error* (note 52), 61–4.
57. Hermann of Reichenau, *Abbreviatio compoti*, Ch. xxv, edited in N. Germann, *De temporum ratione: Quadrivium und Got-*

teserkenntnis am Beispiel Abbos von Fleury und Hermanns von Reichenau (Leiden, 2006), 326.

58. Herrad of Hohenburg, *Hortus Deliciarum* (destroyed 1870), ff. 318v–321v; reconstruction ed. R. Green et al. (London, 1979), 496–502. See O. Pedersen, 'The Ecclesiastical Calendar and the Life of the Church', in *Gregorian Reform of the Calendar*, ed. G.V. Coyne, M.A. Hoskin and O. Pedersen (Vatican City, 1983): 75–113, at 60–61.

59. L. White, 'Eilmer of Malmesbury, an Eleventh Century Aviator', *Technology and Culture* 2 (1961): 97–111; J. Paz, 'Human Flight in Early Medieval England: Reality, Reliability, and Mythmaking (or Science and Fiction)', *New Medieval Literatures* 15 (2013): 1–28; Leonardo da Vinci, Codex Madrid I, Biblioteca Nacional de España Ms. 8937, f. 64r, http://leonardo.bne.es/index.html.

60. Walcher of Malvern, *De lunationibus and De dracone*, ed. C.P.E. Nothaft (Turnhout, 2017).

61. Cambridge University Library MS Ii.6.11, f. 99r. See J. Tolan, *Petrus Alfonsi and His Medieval Readers* (Gainesville, 1993), 74–82, 182–204.

62. Walcher, *De dracone* 2.2 (note 60), 199.

63. See especially Oxford, Corpus Christi College MSS 157 (http://image.ox.ac.uk/show-all-openings?collection=corpus&manuscript=ms157) and 283, and Bodl. MS Auct. F.1.9. On the tables of al-Khwarizmi, see O. Neugebauer, ed., *The Astronomical Tables of Al-Khwārizmī* (Copenhagen, 1962); and R. Mercier, 'Astronomical Tables in the Twelfth Century', in *Adelard of Bath: An English Scientist and Arabist of the Early Twelfth Century*, ed. C. Burnett (London, 1987): 87–118.

64. Roger Bacon, *Opus Tertium*, Ch. 67, in *Opera Quaedam Hactenus Inedita*, ed. J.S. Brewer (London, 1859), 1:272; J.D. North, 'The Western Calendar – "Intolerabilis, Horribilis, et Derisibilis"; Four Centuries of Discontent', in *Gregorian Reform of the Calendar* (note 58): 75–113, repr. with addenda in North, *The Universal Frame: Historical Essays in Astronomy, Natural Philosophy and Scientific Method* (London, 1989): 39–77, at 46–8.

65. Chantilly, Musée Condé MS 65, ff. 1v–13r, http://bitly.com/TRHeures; a column marked 'Nombre d'or nouvel' contains the new numerals marked in gold; See C.P.E. Nothaft, 'The Astronomical Data in the Très Riches Heures and Their Fourteenth-Century Source', *Journal for the History of Astronomy* 46 (2015): 113–29;

C.P.E. Nothaft, 'Science at the Papal Palace: Clement VI and the Calendar Reform Project of 1344/45', *Viator* 46 (2015): 277–302.

66. François Rabelais, *Pantagruel*, Ch. 1 (Paris, 1988), 43–5.

67. *The Kalendarium of Nicholas of Lynn*, ed. S. Eisner (Athens, GA, 1980). Oxford, Bodl. MS Digby 41, ff. 57r–90v.

68. Bede, *The Reckoning of Time* (note 52), Ch. 55, pp. 137–9.

69. 'The Hand of Guido', in J.E. Murdoch, *Album of Science: Antiquity and the Middle Ages* (New York, 1984), 81. C. Burnett, 'The Instruments which are the Proper Delights of the Quadrivium: Rhythmomachy and Chess in the Teaching of Arithmetic in Twelfth-Century England', *Viator* 28 (1997): 175–201. A rhythmomachy board from the Franciscan convent at Coventry is in Cambridge, Trinity College MS R.15.16, f. 60r, http://trin-sites-pub.trin.cam.ac.uk/james/viewpage.php?index=1168.

70. *GASA* (note 3), 2:306, 302.

71. Described in Oxford, Bodl. MS Laud Misc. 697, ff. 27v–28r. Ed. in M.R. James, 'On the Glass in the Windows of the Library at St Albans Abbey', *Proceedings of the Cambridge Antiquarian Society* 8 (1895): 213–20.

72. *GASA* (note 3), 3:392–3; Clark, *Monastic Renaissance* (note 49), 84–99.

73. Greatrex, *English Benedictine Cathedral Priories* (note 2), 83–99.

74. *Rule of Benedict* (note 24), 58.17–20, pp. 79–80.

75. Register of Robert Braybrooke, London Metropolitan Archives DL/A/A/004/MS09531/003, ff. 2v, 5r, 7v. Scanned in *Registers of the Bishops of London, 1304–1660* (Brighton, 1984). Data collated in V. Davis, *Clergy in London in the Late Middle Ages: A Register of Clergy Ordained in the Diocese of London, Based on Episcopal Ordination Lists, 1361–1539* (London, 2000). I am grateful to Professor Davis for sharing a spreadsheet of the ordination lists with me.

76. *Incomprehensibilis* (5 Feb. 1156) and *Religiosam vitam elegentibus* (14 May 1157), papal bulls ed. in W. Holtzmann, *Papsturkunden in England* (Gottingen, 1952), 3:234–8 and 258–61, cited in M. Still, *The Abbot and the Rule* (Aldershot, 2002), 22–3.

CHAPTER 3: *UNIVERSITAS*

1. Register of R. Braybrooke, Metropolitan Archives DL/A/A/004/MS09531/003, f. 5r.

2. *Gesta Abbatum monasterii Sancti Albani (GASA)*, ed. H. Riley (London, 1867), 3:425, 486, 447; *Annales Monasterii S. Albani A.D. 1421–1440, A Johanne Amundesham, monacho, ut videtur, conscripti*, ed. H. Riley (London, 1870), 1:30; J.D. North, *Richard of Wallingford* (Oxford, 1976), 2:532–8; British Library Lansdowne MS 375, ff. 26v–27r.

3. J.I. Catto, 'Citizens, Scholars and Masters', in *The History of the University of Oxford, Volume I: The Early Oxford Schools*, ed. Catto (Oxford, 1984): 151–92, at 188.

4. British Library Harley MS 3775, ff. 129r–137r, 'De Altaribus, Monumentis, et locis Sepulcrorum, in Ecclesia Monasterii Sancti Albani', printed in *Amundesham* (note 2), 1:437.

5. J.G. Clark states that Westwyk was at Oxford ('University Monks in Late Medieval England', in *Medieval Monastic Education*, ed. G. Ferzoco and C. Muessig (London, 2000): 56–71, at 62). K.A. Rand is less sure ('The Authorship of The Equatorie of the Planetis Revisited', *Studia Neophilologica* 87 (2015): 15–35, at 20).

6. D. Wilkins, ed., *Concilia Magnae Britanniae et Hiberniae* (London, 1737), 2:595; J.G. Clark, *A Monastic Renaissance at St. Albans: Thomas Walsingham and His Circle, c.1350–1440* (Oxford, 2004), 65.

7. The completion rate for a bachelor's degree was probably below 5 per cent. See J. Greatrex, *The English Benedictine Cathedral Priories: Rule and Practice, c.1270–c.1420* (Oxford, 2011), 128–9; J. Greatrex, 'From Cathedral Cloister to Gloucester College', in *Benedictines in Oxford*, ed. H. Wansbrough and A. Marett-Crosby (London, 1997): 48–60, at 54–5; cf. J.G. Clark, *Monastic Renaissance* (note 6), 69.

8. G.E.M. Gasper et al., 'The Liberal Arts: Inheritances and Conceptual Frameworks', in *The Scientific Works of Robert Grosseteste, Volume 1: Knowing and Speaking*, ed. Gasper et al. (Oxford, 2019), 45–55.

9. W.H. Stahl and R. Johnson, *Martianus Capella and the Seven Liberal Arts* (New York, 1971).

10. *The Etymologies of Isidore of Seville*, tr. S.A. Barney et al. (Cambridge, 2006), 3. It is often stated online that Isidore was nominated as patron saint of the internet, perhaps by Pope John Paul II, but such a nomination has never been officially confirmed. If Isidore would be a fitting patron, it is perhaps equally fitting for the internet that his nomination should be an unsubstantiated rumour. L. Ante-

quera, '¿Pero es o no es, San Isidoro de Sevilla, el santo patrono de la red?', 27 April 2014, https://www.religionenlibertad.com/blog/35241/pero-es-o-no-es-san-isidoro-de-sevilla-el-santo.html

11. Petrus Alfonsi, *Die Disciplina clericalis des Petrus Alfonsi*, ed. A. Hilka and W. Söderhjelm (Heidelberg, 1911), 10–11; *The Scholar's Guide*, tr. J.R. Jones and J.E. Keller (Toronto, 1969).

12. Petrus Alfonsi, *Epistola ad peripateticos*, in J. Tolan, *Petrus Alfonsi and His Medieval Readers* (Gainesville, 1993), 166–7, 174–5; C. Burnett, 'Advertising the New Science of the Stars circa 1120–50,' in *Le XIIe Siècle: mutations et renouveau en France dans la première moitié du XIIe siècle*, ed. F. Gasparri (Paris, 1994): 147–57.

13. Hugh of St Victor, *De tribus diebus* 4, printed as book 7 of *Didascalicon* in Migne, *Patrologia latina* 176, col. 814B, http://pld.chadwyck.co.uk/; Psalm 92; G. Tanzella-Nitti, 'The Two Books Prior to the Scientific Revolution', *Annales Theologici* 18 (2004): 51–83.

14. W. Rüegg, 'Themes', in *A History of the University in Europe, Volume I: Universities in the Middle Ages*, ed. H. de Ridder-Symoens (Cambridge, 1992): 3–34, esp. 9–12.

15. R.W. Southern, 'From Schools to University', in *The University of Oxford, Vol. I*, ed. Catto (note 3): 1–36.

16. R.C. Schwinges, 'Admission', in *The University in Europe, Vol. I*, ed. Ridder-Symoens (note 14): 171–94, at 188.

17. N. Siraisi, 'The Faculty of Medicine', in *The University in Europe* (note 14): 360–87, at 364–5.

18. J.M. Fletcher, 'The Faculty of Arts', in *The University in Europe* (note 14): 369–99, at 370–72.

19. Petrus Alfonsi, *Epistola ad peripateticos* (note 12), 166–8, 174–6.

20. Biobibliography of Gerard of Cremona, attached to his translation of Galen's *Ars parva*, tr. M. McVaugh in *A Source Book in Medieval Science*, ed. E. Grant (Cambridge, MA, 1974), 35.

21. The crocodile's tongue is in *The Parts of Animals*, II.17 (660b); knowledge and the principles of scientific demonstration are treated most explicitly in the *Posterior Analytics*.

22. Calcidius, *On Plato's Timaeus*, tr. J. Magee (Cambridge, MA, 2016). A twelfth-century copy of Calcidius' translation (Oxford, Bodl. MS Digby 23, ff. 3r–54v), formerly at Osney Abbey near Oxford, is online at http://bit.ly/Digby23.

23. J.A. Weisheipl, 'Curriculum of the Faculty of Arts at Oxford in the Early Fourteenth Century', *Mediaeval Studies* 26 (1964): 143–85.

24. Robertus Anglicus, Commentary (1271) on the *Sphere*, *Lectio* I, in L. Thorndike, ed., *The Sphere of Sacrobosco and Its Commentators* (Chicago, 1949), 143; O. Pedersen, 'In Quest of Sacrobosco', *Journal for the History of Astronomy* 16 (1985): 175–220.

25. Greatrex, 'From Cathedral Cloister' (note 7), 55.

26. Ed. and tr. L. Thorndike, *Sphere of Sacrobosco* (note 24).

27. W. Irving, *A History of the Life and Voyages of Christopher Columbus*, vol. 1 (New York, 1828), 73–8; J.B. Russell, *Inventing the Flat Earth: Columbus and Modern Historians* (New York, 1991).

28. e.g. A.D. White, *A History of the Warfare of Science with Theology in Christendom* (New York, 1896), 1:89–109.

29. Thorndike, *Sphere of Sacrobosco* (note 24), 38–40; Grant (ed.), *Source Book in Medieval Science* (note 20), 630–39.

30. Sacrobosco, *The Sphere* (note 24), Ch. 1, 83; Aristotle, *On the Heavens*, II.14 (297b).

31. Aristotle, *On the Heavens*, II.14 (298a), tr. J. Barnes, *The Complete Works of Aristotle* (Princeton, 1984), 1:489.

32. Eratosthenes' solstice demonstration at Alexandria is recorded by Pliny the Elder (who does not name him): *Natural History* II.75, tr. H. Rackham (Cambridge, MA, 1967), 1:316–17; also Cleomedes, *De motu circulari corporum caelestium*, in T.L. Heath, ed., *Greek Astronomy* (Cambridge, 1932), 109–12. Other accounts locate the demonstration between Syene and Meroë, at the equinox rather than the solstice.

33. Ptolemy, *Geographia* 1.7.1, ed. K. Müller (Paris, 1883), 1:17; Columbus, marginal notes in the *Imago Mundi* of Pierre d'Ailly, printed in Cristóbal Colón, *Textos y documentos completos*, ed. C. Varela (Madrid, 1997), 10–11; G.E. Nunn, *The Geographical Conceptions of Columbus* (New York, 1924), 1–2, 9–10.

34. e.g. Carl Sagan, *Cosmos* (London, 1981), 14–15.

35. Aristotle, *Physics*, IV.1 (208b); *On the Heavens*, II.14 (297b), I.10–II.1 (279b–284a); *On Generation and Corruption*, II.10 (337a); *Meteorology* I.3 (339b–340a). Jean Buridan developed Aristotle's idea of constant change into a theory of dynamic equilibrium between land and sea: J. Kaye, *A History of Balance, 1250–1375: The Emergence of a New Model of Equilibrium and Its Impact on Thought* (Cambridge, 2014), 445–8.

36. K.A. Vogel, 'Sphaera terrae – das mittelalterliche Bild der Erde und

die kosmographische Revolution' (PhD thesis, University of Göttingen, 1995), 154.

37. Sacrobosco, *The Sphere* (note 24), Ch. 1, 78–9; Robertus Anglicus, Commentary (1271) on the *Sphere*, II, in Thorndike, *Sphere of Sacrobosco* (note 24), 150, 205; D. Wootton, *The Invention of Science: A New History of the Scientific Revolution* (London, 2015), 110–37 (esp. on the more intense fifteenth-century debates).

38. Aristotle, *On Generation and Corruption*, II.10 (337a); Commentary (*c*.1230) attributed to Michael Scot, III, in Thorndike, *Sphere of Sacrobosco* (note 24), 277; Robert Grosseteste, *De Sphera*, Ch. 4, in *Moti, virtù e motori celesti nella cosmologia di Roberto Grossatesta*, ed. C. Panti (Florence, 2001), 290.

39. Aristotle, *On the Heavens*, II.14 (297b); Genesis 1:9–10.

40. Exodus 12:35–6; Origen, *Letter to Gregory*; M. Pereira, 'From the Spoils of Egypt: An Analysis of Origen's *Letter to Gregory*', in *Origeniana Decima: Origen as Writer*, ed. S. Kaczmarek and H. Pietras (Leuven, 2011), 221–48. Augustine, *On Christian Doctrine* II.40.60; D.C. Lindberg, 'The Medieval Church Encounters the Classical Tradition: Saint Augustine, Roger Bacon, and the Handmaiden Metaphor', in *When Science & Christianity Meet*, ed. Lindberg and R.L. Numbers (Chicago, 2003), 7–32.

41. Dante, *The Divine Comedy* (*c*.1310–20), Inferno IV, 118–47.

42. H. Denifle and E. Chatelaine, eds., *Chartularium Universitatis Parisiensis* (Paris, 1889), 1:70.

43. G. Leff, 'The Trivium and the Three Philosophies', in *The University in Europe, Vol. I*, ed. Ridder-Symoens (note 14): 307–36, at 319–22.

44. Sources collected in F. Duncalf and A.C. Krey, *Parallel Source Problems in Medieval History* (New York, 1912), 137–74; Matthew Paris, *Chronica Majora*, ed. H. Luard (London, 1880), 3:166–9; N. Gorochov, 'The Great Dispersion of the University of Paris and the Rise of European Universities (1229–1231)', *CIAN-Revista de Historia de las Universidades* 21 (2018): 99–119.

45. C.H. Lawrence, 'The University in State and Church', in *The University of Oxford, Vol. I*, ed. Catto (note 3): 97–150, at 126–7, 139.

46. 'Parens scientiarum Parisius', in *Chartularium* (note 41), 1:136–9; I.P. Wei, *Intellectual Culture in Medieval Paris: Theologians and the University, c.1100–1330* (Cambridge, 2012), 109; Pedersen, 'In Quest of Sacrobosco' (note 24), 192.

47. *Chartularium* (note 42), 1:486–7.

48. B.C. Bazán, 'Siger of Brabant', in *A Companion to Philosophy in*

the Middle Ages, ed. J.J.E. Gracia and T.B. Noone (Malden, MA, 2006), 632–40.

49. *Chartularium* (note 42), 1:499–500, 543–55; E. Grant, 'The Condemnation of 1277, God's Absolute Power, and Physical Thought in the Late Middle Ages', *Viator* 10 (1979): 211–44.

50. B.C. Bazán, 'Boethius of Dacia', in *Philosophy in the Middle Ages* (note 48): 227–32; J.F. Wippel, 'The Parisian Condemnations of 1270 and 1277', in *Philosophy in the Middle Ages*: 65–73; J.M.M.H. Thijssen, 'What Really Happened on 7 March 1277? Bishop Tempier's Condemnation and Its Institutional Context', in *Texts and Contexts in Ancient and Medieval Science*, ed. E. Sylla and M. McVaugh (Leiden, 1997): 84–114; cf Thijssen, *Censure and Heresy at the University of Paris, 1200–1400* (Philadelphia, 1998), 52–6.

51. J.E. Murdoch, 'Pierre Duhem and the History of Late Medieval Science and Philosophy in the Latin West', in *Gli studi di filosofia medievale fra otto e novecento*, ed. R. Imbach and A. Maierù (Rome, 1991): 253–302; J.D. North, 'Eternity and Infinity in Late Medieval Thought', in *Infinity in Science*, ed. G. Toraldo di Francia (Rome, 1987): 245–56; repr. *Stars, Minds and Fate: Essays in Ancient and Medieval Cosmology* (London, 1989): 233–43.

52. Lawrence, 'University in State and Church' (note 45), 116–17; P.O. Lewry, 'Grammar, Logic and Rhetoric', in *The University of Oxford, Vol. I*, ed. Catto (note 3): 401–33, at 419–26.

53. M.W. Sheehan, 'The Religious Orders 1220–1370', in *The University of Oxford, Vol. I*, ed. Catto (note 3): 193–221, at 220; J. Campbell, 'Gloucester College', in *Benedictines in Oxford* (note 7): 37–47, at 37.

54. W.A. Pantin, ed., *Documents Illustrating the Activities of the General and Provincial Chapters of the English Black Monks, 1215–1540: Vol. 1*, Camden Third Series vol. 45 (London, 1931), 75.

55. Pantin, *Chapters* (note 54), 1:75; Greatrex, *English Benedictine Cathedral Priories* (note 7), 130.

56. I. Mortimer, *The Time Traveller's Guide to Medieval England: A Handbook for Visitors to the Fourteenth Century* (London, 2009), 100; *Amundesham* (note 2), 2:105–8; Clark, *Monastic Renaissance* (note 6), 71; M.R.V. Heale, 'Dependant Priories and the Closure of Monasteries in the Late Medieval England, 1400–1535', *English Historical Review* 119 (2004): 1–26, at 19–20.

57. Campbell, 'Gloucester College' (note 53), 40; Clark, *Monastic*

Renaissance (note 6), 68; Greatrex, *English Benedictine Cathedral Priories* (note 7), 129.

58. *GASA* (note 2), 2:182; J. North, *God's Clockmaker: Richard of Wallingford and the Invention of Time* (London, 2005), 51.

59. Sheehan, 'The Religious Orders' (note 53), p. 216.

60. Lambeth Palace Library MS 111; Clark, *Monastic Renaissance* (note 6), 67.

61. Richard Trevytlam, 'De laude universitatis Oxoniae', in a Glastonbury Abbey manuscript: Cambridge, Trinity College MS O.9.38, ff. 49v–54r, at 52v, http://trin-sites-pub.trin.cam.ac.uk/james/viewpage.php?index=985; printed in *Collectanea*, ed. M. Burrows (Oxford, 1896), 188–209, at 203; tr. A.G. Rigg, *A History of Anglo-Latin Literature, 1066–1422* (Cambridge, 1992), 274.

62. This abbot, Walter of Monington, clearly had a high opinion of John Sene, the drunken monk Trevytlam slandered in his poem. Sene frequently deputised for Monington on official business, including at the provincial chapter; A.G. Rigg, 'An Edition of a Fifteenth-Century Commonplace Book (Trinity College, Cambridge, MS O.9.38)' (D.Phil. thesis, University of Oxford, 1966), 337–8; Pantin, *Chapters* (note 54), vol. 3, Camden Third Series, vol. 54 (London, 1937), 29–30, 201–2.

63. Richard of Wallingford imposed a hunting ban on the monks staying at Redbourn; *GASA* (note 2), 204; tr. M. Still, *The Abbot and the Rule: Religious Life at St Alban's, 1290–1349* (Aldershot, 2002), 268–70.

64. Pantin, *Chapters* (note 54), 3:31–2, 53–4, 60; D. Knowles, *The Religious Orders in England* (Cambridge, 1957), 2:22–3.

65. K. Bennett, 'The Book Collections of Llanthony Priory from Foundation until Dissolution (c.1100–1538)' (PhD thesis, University of Kent, 2006), 182–7.

66. M.B. Parkes, 'The Provision of Books', in *The History of the University of Oxford, Volume II: Late Medieval Oxford*, ed. J.I. Catto and R. Evans (Oxford, 1992), 407–83, at 421–4.

67. A.J. Ray, 'The Pecia System and Its Use in the Cultural Milieu of Paris, c.1250–1330' (PhD thesis, UCL, 2015), 24–6, 225–7.

68. Greatrex, *English Benedictine Cathedral Priories* (note 7), 131; e.g. Oxford, Bodl. MS Selden Supra 24, f. 3v.

69. *The Philobiblon of Richard de Bury*, Ch. 17, ed. and tr. E.C. Thomas (London, 1888), 237–8.

70. Parkes, 'The Provision of Books' (note 66), 431–2, 449–51.

71. Albertus Magnus, *Book of Minerals*, tr. D. Wyckhoff (Oxford, 1967), III.i.1, p. 153. See also *Albertus Magnus and the Sciences: Commemorative Essays 1980*, ed. J.A. Weisheipl (Toronto, 1980), esp. essays by Kibre and Riddle/Mulholland.

72. Aristotle, *On Generation and Corruption*, II.2 (329b), II.8 (335a); Avicenna, *Avicennae de Congelatione et Conglutinatione Lapidum*, ed. E.J. Holmyard and D.C. Mandeville (Paris, 1927), 18–19, 26–8, 35–6; Albertus Magnus, *Book of Minerals* (note 71), I.i.2, pp. 13–14; G. Freudenthal, '(Al-)Chemical Foundations for Cosmological Ideas: Ibn Sînâ on the Geology of an Eternal World', in *Physics, Cosmology and Astronomy, 1300–1700: Tension and Accommodation*, ed. S. Unguru (Dordrecht, 1991): 47–73, repr. *Science in the Medieval Hebrew and Arabic Traditions* (Aldershot, 2005): XII.

73. E.A. Synan, 'Introduction' in *Albertus Magnus and the Sciences* (note 71), 1–12, at 2-4.

74. Roger Bacon, *Opus Minus* and *Opus Tertium*, in *Opera Quaedam Hactenus Inedita*, ed. J.S. Brewer (London, 1859), 30–31, 37–8, 327–38. Bacon never names Albertus, but most historians agree that he was the angry Franciscan's target; J.M.G. Hackett, 'The Attitude of Roger Bacon to the Scientia of Albertus Magnus', in *Albertus Magnus and the Sciences* (note 71), 53–72.

75. C. Burnett, 'Shareshill [Sareshel], Alfred of', *Oxford Dictionary of National Biography* (2004); D.A. Callus, 'Introduction of Aristotelian Learning to Oxford', *Proceedings of the British Academy* 29 (1943): 229–81, at 236–8.

76. Roger Bacon, *Compendium studii philosophiae*, in *Opera Quaedam Hactenus Inedita*, ed. Brewer (note 74), 469; R. Lemay, 'Roger Bacon's Attitude toward the Latin Translations and Translators of the Twelfth and Thirteenth Centuries', in *Roger Bacon and the Sciences: Commemorative Essays*, ed. J. Hackett (Leiden, 1997), 25–48.

77. Roger Bacon, *Opus Maius*, ed. J.H. Bridges (Oxford, 1897), 1:66–7; *Opus Tertium* (note 74), 33–4.

78. A.C. Dionisotti, 'On the Greek Studies of Robert Grosseteste', in *The Uses of Greek and Latin: Historical Essays*, ed. Dionisotti, A. Grafton and J. Kraye, (London, 1988), 19–39.

79. Robert Grosseteste, *De Sphera* (note 38), 289–319.

80. J.A. Weisheipl, 'Science in the Thirteenth Century', in *University of Oxford, Vol. I*, ed. Catto (note 3): 435–69, at 452; Robert Gros-

seteste, *De Luce*, in N. Lewis, 'Robert Grosseteste's On Light: An English Translation', in *Robert Grosseteste and His Intellectual Milieu*, ed. J. Flood, J.R. Ginther and J.W. Goering (Toronto, 2013), 239–47.

81. Robert Grosseteste, *De iride*, ed. L. Baur, *Die Philosophischen Werke des Robert Grosseteste, Bischofs von Lincoln* (Münster, 1912), 74; Bacon, *Opus Maius*, V.3.3.3–4 (note 77) 2:164–6, this part ed. and tr. D.C. Lindberg, *Roger Bacon and the Origins of Perspectiva in the Middle Ages* (Oxford, 1996), 330–35; V. Ilardi, *Renaissance Vision from Spectacles to Telescopes* (Philadelphia, 2007), 3–10.

82. R. French and A. Cunningham, *Before Science: The Invention of the Friars' Natural Philosophy* (Aldershot, 1996).

83. John 1:4–8, and Aquinas' commentary (https://aquinas.cc/188/190/~268); Lindberg, *Origins of Perspectiva* (note 81), lxviii, 355n163; *Roger Bacon's Philosophy of Nature: De Multiplicatione Specierum and De Speculis Comburentibus*, ed. D.C. Lindberg (Oxford, 1983), 2–5, 365n10; Weisheipl, 'Science in the Thirteenth Century' (note 80), 444.

84. 1 Corinthians 13:12; Psalm 119:130.

85. John Pecham, *Perspectiva communis*, ed. D.C. Lindberg, *John Pecham and the Science of Optics* (Madison, 1970); Bodl. MS Ashmole 341 (St Augustine's Canterbury), ff. 115r–120r.

86. Aristotle, *Posterior Analytics*, I.13 (78b); Robert Grosseteste, *Commentary on the Posterior Analytics* I.12, ed. P. Rossi, *Commentarius in posteriorum analyticorum libros* (Florence, 1981), 189–92; Weisheipl, 'Science in the Thirteenth Century' (note 80), 446–51; W.R. Laird, 'Robert Grosseteste on the Subalternate Sciences', *Traditio* 43 (1987): 147–69.

87. J. Hackett, 'Scientia Experimentalis: From Robert Grosseteste to Roger Bacon', in *Robert Grosseteste: New Perspectives on His Thought and Scholarship*, ed. J. McEvoy, (Turnhout, 1995): 89–119, at 103–7.

88. Bacon, *Opus Maius* (note 77), 2:167–222, esp. 172, 214, 221; Lindberg, *Roger Bacon and the Origins of Perspectiva* (note 81), lii–lvii; J. Hackett, 'Roger Bacon on Scientia Experimentalis', in *Roger Bacon and the Sciences* (note 76): 277–316; A. Power, *Roger Bacon and the Defence of Christendom* (Cambridge, 2013), 166–78. Bacon, 'Epistola de secretis operibus artis et naturae et de nullitate magiae', in *Opera Quaedam Hactenus Inedita*, ed. Brewer (note 74), 533.

89. Bacon, *Opus Maius* (note 77), 2:172–8; C.B. Boyer, *The Rainbow: From Myth to Mathematics* (Princeton, 2nd edn 1987), 88–119; D.C. Lindberg, 'Roger Bacon's Theory of the Rainbow: Progress or Regress?', *Isis* 57 (1966): 235–48.

90. Aristotle, *On Generation and Corruption*, II.3 (330a); D. Skabelund and P. Thomas, 'Walter of Odington's Mathematical Treatment of the Primary Qualities', *Isis* 60 (1969): 331–50; J.D. North, 'Natural Philosophy in Late Medieval Oxford', in *The University of Oxford, Vol. II*, ed. Catto and Evans (note 66): 65–102, at 74–7.

91. North, 'Natural Philosophy' (note 90), 82–5; M. McVaugh, 'Arnald of Villanova and Bradwardine's Law', *Isis* 58 (1967): 56–64.

92. Thomas Bradwardine, *Tractatus de Proportionibus*, ed. and tr. H.L. Crosby (Madison, 1955).

93. J.A. Weisheipl, 'Roger Swyneshed, O.S.B., Logician, Natural Philosopher, and Theologian', in *Oxford Studies Presented to Daniel Callus* (Oxford, 1964), 231–52; North, 'Natural Philosophy' (note 90), 80n44.

94. North, 'Natural Philosophy' (note 90), 85.

95. North, 'Natural Philosophy' (note 90), 92–3; J.E. Murdoch and E.D. Sylla, 'Swineshead, Richard', in *Complete Dictionary of Scientific Biography*, vol. 13 (Detroit, 2008), 184–213.

96. W.J. Courtenay, 'The Effect of the Black Death on English Higher Education', *Speculum* 55 (1980): 696–714; J.D. North, '1348 and All That: Oxford Science and the Black Death', *in Stars, Minds and Fate: Essays in Ancient and Medieval Cosmology* (London): 361–71.

97. Walter also wrote on astronomy and the calendar: North, *Richard of Wallingford* (note 2), 3:238–70.

98. Cambridge University Library MS Gg.6.3, f. 164v; P. Zutshi, 'An Urbanist Cardinal and His Books: The Library and Writings of Adam Easton', in *Der Papst und das Buch im Spätmittelalter (1350–1500)*, ed. R. Berndt (Münster, 2018): 24–46; S. Falk, 'A Merton College Equatorium: Text, Translation, Commentary', *SCIAMVS* 17 (2016): 121–59, at 130–31.

99. M.R. James, Unpublished catalogue description for Cambridge UL MS Gg.6.3.

100. W.R. Knorr, 'Two Medieval Monks and Their Astronomy Books: MSS. Bodley 464 and Rawlinson C.117', *Bodleian Library Record* 14 (1993): 269–84.

101. Oxford, Bodl. MS Digby 176, ff. 40r–41v, 50r–53v; K. Sned-

egar, 'John Ashenden and the Scientia Astrorum Mertonensis' (D.Phil. thesis, University of Oxford, 1988), 55–9, 265–70.

CHAPTER 4: ASTROLABE AND ALBION

1. *Gesta Abbatum monasterii Sancti Albani (GASA)*, ed. H. Riley (London, 1867), 2:182, 296.

2. J.G. Clark, *A Monastic Renaissance at St. Albans: Thomas Walsingham and His Circle, c.1350–1440* (Oxford, 2004), 71–2; J. Greatrex, *The English Benedictine Cathedral Priories: Rule and Practice, c.1270–c.1420* (Oxford, 2011), 145–7.

3. Clark, *Monastic Renaissance* (note 2), 111–23; *GASA* (note 1), 2:433.

4. *GASA* (note 1), 3:393; Greatrex, *English Benedictine Cathedral Priories* (note 2), 181.

5. J.D. North, *Richard of Wallingford* (Oxford, 1976), 2:127–30, 287; S. Falk, ' "I found this written in the other book": Learning Astronomy in Late Medieval Monasteries', in *Churches and Education*, ed. M. Ludlow, C. Methuen and A. Spicer, *Studies in Church History*, vol. 55 (2019): 129–44.

6. Ptolemy, *Almagest* V.1, tr. G.J. Toomer (London, 1984), 217–19. The spheres Ptolemy and his contemporaries used for observation were, it should be noted, far larger than those of the medieval class room.

7. It is unclear whether the Latin torquetum can be traced back to Jabir ibn Aflah, in twelfth-century Seville; see North, *Richard of Wallingford* (note 5), 2:297–300; R.P. Lorch, 'The Astronomical Instruments of Jābir Ibn Aflah and the Torquetum', *Centaurus* 20 (1976): 11–35; E. Poulle, '*Bernard de Verdun et Le Turquet*', *Isis* 55 (1964): 200–208; L. Thorndike, 'Franco de Polonia and the Turquet', *Isis* 36 (1945): 6–7.

8. Richard of Wallingford, 'Rectangulus' I, ed. North, *Richard of Wallingford* (note 5), 1:407.

9. Jean de Lignières, 'Quia nobilissima scientia astronomie', in Cambridge UL MS Gg.6.3, f. 217v. Here Jean's prologue was repurposed to introduce a treatise he had not written. Ed. and tr. S. Falk, 'A Merton College Equatorium: Text, Translation, Commentary', *SCIAMVS* 17 (2016): 121–59; C. Eagleton, 'John Whethamstede,

Abbot of St. Albans, on the Discovery of the Liberal Arts and Their Tools: Or, Why were Astronomical Instruments in Late-Medieval Libraries?', *Mediaevalia* 29 (2008): 109–36.

10. G. Chaucer, *The Canterbury Tales: The Miller's Tale*, I.3209, in *The Riverside Chaucer*, ed. L.D. Benson (Oxford, 1987), 68; J. Gower, *Confessio Amantis*, VI.1890, ed. R.A. Peck (Kalamazoo, 2004), 251; Paris, Bibliothèque Nationale de France, MS Latin 19093, f. 108r, https://gallica.bnf.fr/ark:/12148/btv1b8510021r/f223.item.

11. Chaucer, 'A Treatise on the Astrolabe', I.21, in *The Riverside Chaucer* (note 10), 667.

12. Cambridge, Peterhouse MS 75.I, f. 71r. Another almost identical list, from the monastery of Bury St Edmunds, is in Cambridge University Library MS Add. 6860, f. 70v–71r. The classic study of medieval star lists is P. Kunitzsch, *Typen von Sternverzeichnissen in astronomischen Handschriften des zehnten bis vierzehnten Jahrhunderts* (Wiesbaden, 1966). Wh.1264 matches Kunitzsch's Type VIII quite closely. See S. Falk, 'Sacred Astronomy? Beyond the Stars on a Whipple Astrolabe', in *The Whipple Museum of the History of Science: Instruments and Interpretations*, ed. L. Taub, J. Nall and F. Willmoth (Cambridge, 2019): 11–31; on this astrolabe, see also J. Davis and M. Lowne, 'An Early English Astrolabe at Gonville & Caius College, Cambridge, and Walter of Elveden's Kalendarium', *Journal for the History of Astronomy* 46 (2015): 257–90.

13. D.A. King, *In Synchrony with the Heavens: Studies in Astronomical Timekeeping and Instrumentation in Medieval Islamic Civilization. Volume Two: Instruments of Mass Calculation* (Leiden, 2005), XIIIe, 595.

14. O. Neugebauer, *A History of Ancient Mathematical Astronomy* (Berlin, 1975), 868–70.

15. Chaucer, 'Astrolabe' (note 11), I.21, 668.

16. For Chaucer's use of wax on the astrolabe, see 'Astrolabe' (note 11), II.40, 680.

17. Chaucer, 'Astrolabe' (note 11), II.1, 669.

18. Chaucer, 'Astrolabe' (note 11), II.3, 669–70.

19. Chaucer, 'Astrolabe' (note 11), II.3, 670.

20. Quoted in S.R. Sarma, *A Descriptive Catalogue of Indian Astronomical Instruments*, revised edn, 2019, www.srsarma.in, p. 17. See also Sarma, 'On the Life and Works of Rāmacandra Vājapeyin', in *Śrutimahatī: Glory of Sanskrit Tradition: Professor Ram Karan*

Sharma Felicitation Volume, ed. R. Tripathi (New Delhi, 2008), 2: 645–61.

21. Chaucer, 'Astrolabe' (note 11), Prol.1, 662.

22. Chaucer, *A Treatise on the Astrolabe*, ed. S. Eisner (Norman, OK, 2002), 12–15, 103; S. Horobin, 'The Scribe of Bodleian Library MS Bodley 619 and the Circulation of Chaucer's Treatise on the Astrolabe', *Studies in the Age of Chaucer* 31 (2009): 109–24.

23. Jalal ad-Din Rumi, *Masnavi* 1:110, http://www.masnavi.net; Sacrobosco, *The Sphere* I, in L. Thorndike, ed., *The Sphere of Sacrobosco and Its Commentators* (Chicago, 1949), 78; Chaucer, 'Astrolabe' (note 11), Prol.56–8, 662; S. Lerer, 'Chaucer's Sons', *University of Toronto Quarterly* 73 (2004): 906–15.

24. P.G. Schmidl, 'Using Astrolabes for Astrological Purposes: The Earliest Evidence Revisited', in *Heaven and Earth United: Instruments in Astrological Contexts*, ed. R. Dunn, S. Ackermann and G. Strano (Leiden, 2018), 4–23; S. Falk, 'What's on the Back of an Astrolabe? Astrolabes as Supports for Planetary Calculators', in *Heaven and Earth United*, 24–41.

25. Chaucer, 'Astrolabe' (note 11), I.3, 663; J.A. Mitchell, 'Transmedial Technics in Chaucer's Treatise on the Astrolabe: Translation, Instrumentation, and Scientific Imagination', *Studies in the Age of Chaucer* 40 (2018): 1–41.

26. A. Chapman, 'A Study of the Accuracy of Scale Graduations on a Group of European Astrolabes', *Annals of Science* 40 (1983): 473–88.

27. (pseudo-)Masha'allah, *De compositio astrolabii*, I.2, ed. and tr. R.T. Gunther, *Early Science in Oxford, Vol. 5: Chaucer and Messahalla on the Astrolabe* (Oxford, 1929), 197–8; History of Science Museum, Oxford 49359 (#4755 in the International Checklist of Astrolabes); Chaucer, 'Astrolabe' (note 11), Prol.73–6, 663.

28. D.J. Price, 'An International Checklist of Astrolabes', *Archives internationales d'histoire des sciences* 32/33 (1955): 243–63, 363–81.

29. London, BL MS Cotton Nero C.VI, ff. 147r–156v; A. Hiatt, 'The Reference Work in the Fifteenth Century: John Whethamstede's *Granarium*', in *Makers and Users of Medieval Books: Essays in Honour of A.S.G. Edwards*, ed. C.M. Meale and D. Pearsall (Woodbridge, 2014): 13–33; North, *Richard of Wallingford* (note 5), 3:112–14; C. Eagleton, 'Instruments in Context: Telling the Time in England, 1350–1500' (PhD thesis, University of Cambridge, 2004), 11–21, 241–60.

30. BL MS Cotton Nero C.VI, f. 154v; on medieval sources of Semiramis, see J. Parr, 'Chaucer's Semiramis', *Chaucer Review* 5 (1970): 57–61.

31. *Registra quorundam Abbatum Monasterii S. Albani, Vol. 1: Registrum Abbatiae Johannis Whethamstede*, ed. H.T. Riley, (London, 1872), 311–12; W. Dugdale, *Monasticon Anglicanum*, ed. J. Caley, H. Ellis and B. Bandinel (London, 1819), 2:209–10. Stickford was ordained deacon in 1368: Register of Robert Braybrooke, London Metropolitan Archives DL/A/A/004/MS09531/003, f. 5r.

32. BL MS Cotton Nero C VI, f. 149r; Richard of Wallingford, 'Tractatus Albionis' III, ed. North (note 5), 1:340.

33. *GASA* (note 1), 2:207.

34. Cambridge, Clare College MS 27. Bede notes that Britain was 'formerly known as Albion' in the first line of his *Ecclesiastical History of the English People*.

35. A. Bernau, 'Beginning with Albina: Remembering the Nation', *Exemplaria* 21 (2009): 247–73; J.J. Cohen, *Of Giants: Sex, Monsters, and the Middle Ages* (Minneapolis, 1999), 29–55; *Hardyng's Chronicle*, ed. J. Simpson and S. Peverley (Kalamazoo, 2015), I.1–322.

36. Genesis 6:1–7; Revelation 20:7–8. See also Ezekiel 38:2. Two St Albans MSS (Cambridge UL MS Dd.6.7 and Bodl. MS Rawlinson B.189) contain a Latin version of the Albina/Albion story as a prologue to the Brutus legend; L. Johnson, 'Return to Albion', and J.P. Carley and J. Crick, 'Constructing Albion's Past: An Annotated Edition of De Origine Gigantum', both in *Arthurian Literature*, XIII, ed. Carley and F. Riddy (Cambridge, 1995), 19–40 and 41–114. St Albans MSS featuring blue-skinned men, apparently giants, include Cambridge, Trinity College MS O.5.8, f. 148v, and Cambridge, Corpus Christi College MS 48, f. 263v.

37. 1 Samuel 17:4–9; *GASA* (note 1), 2:183–4.

38. *GASA* (note 1), 2:184–5. Richard himself was an elector, but it was not unusual for such panels to include likely candidates for the abbacy: D. Knowles, *The Religious Orders in England, Volume II: The End of the Middle Ages* (Cambridge, 1955), 248–50.

39. *GASA* (note 1), 2:127–9, 199; 3:367–8; North, *Richard of Wallingford* (note 5), 5–8.

40. *GASA* (note 1), 2:208; see 1344 statutes for one St Albans leper hospital, tr. in M. Still, *The Abbot and the Rule: Religious Life at St*

Alban's, 1290–1349 (Aldershot, 2002), 281–91; C. Rawcliffe, *Medicine & Society in Later Medieval England* (London, 1999), 14–17.

41. *GASA* (note 1), 2: 200–201, 281–2; Wallingford, 'Tractatus Albionis' (note 32), III, 1:340.

42. Attributed to Bernard of Chartres; C. Burnett, 'The Twelfth-Century Renaissance', in *The Cambridge History of Science, Vol. 2: Medieval Science*, ed. D.C. Lindberg and M.H. Shank (New York, 2013): 365–84, at 371.

43. D.A. King, 'On the Early History of the Universal Astrolabe in Islamic Astronomy, and the Origin of the Term "Shakkaziyya" in Medieval Scientific Arabic', *Journal for the History of Arabic Science* 3 (1979): 244–57; R. Puig, 'Concerning the Safīḥa Shakkāziyya', *Zeitschrift für Geschichte der arabisch–islamischen Wissenschaften* 2 (1985): 123–39.

44. Wallingford, 'Tractatus Albionis' (note 32), II.24, III.36, 1:332, 380; see also Appendix 28, 3:165–7. Trinity College Dublin MS 444, probably bought for the abbey by Richard, contains canons and tables composed by Arzachel.

45. Oxford, Bodl. MS Laud Misc. 657, f. 43r. This is as described in the 'Tractatus Albionis' (note 32), II.22, 1:328.

46. Bodl. MS Laud Misc. 657, f. 1v.

47. Bodl. MS Laud Misc. 657, f. 45r.

48. Cf 'Altayir' in Oxford, Corpus Christi College MS 144, f. 76v, with 'Altayn' in MS Laud Misc. 657, f. 37v. Westwyk also partially copied a diagram (f. 17r) which divides the limb of the first face of the second disc into 18; of all the surviving *Albion* copies, this feature only appears in Corpus Christi MS 144 (f. 59v). Badly copied table on MS Laud Misc. 657, ff. 44v–45r.

49. Bodl. MS Laud Misc. 657, f. 45r.

50. Bodl. MS Laud Misc. 657, f. 11r.

51. Bodl. MS Laud Misc. 657, ff. 43r–44r. 'The Abbot's Albion' is referred to in Corpus Christi MS 144, f. 59v, produced at St Albans a few decades earlier.

52. *GASA* (note 1), 2:237–57; J. North, *God's Clockmaker: Richard of Wallingford and the Invention of Time* (London, 2005), 115–36; M. Freeman, *St Albans: A History* (Lancaster, 2008), 95–100.

53. Thomas Walsingham, *The St Albans Chronicle: The Chronica Maiora of Thomas Walsingham*, ed. L. Watkiss, W.R. Childs and J. Taylor (Oxford, 2003), 1:456–9; *GASA* (note 1), 2:202.

54. *GASA* (note 1), 1:222–3, 255–8.

55. Walsingham, *Chronica Maiora* (note 53), 1:442–9, 456–9.
56. F.P. Mackie, 'The Clerical Population of the Province of York: An Edition of the Clerical Poll Tax Enrolments 1377–1381' (D.Phil. thesis, University of York, 1998), 1:9–14, 2:152–3; K.A. Rand, 'The Authorship of The Equatorie of the Planetis Revisited', *Studia Neophilologica* 87 (2015): 15–35, at 19–23; Book of Benefactors, London, BL MS Cotton Nero D.VII, ff. 81v–83r, ed. in Dugdale, *Monasticon Anglicanum* (note 31), 2:209-210.
57. Bodl. MS Laud Misc. 657, f. 1v.

CHAPTER 5: SATURN IN THE FIRST HOUSE

1. Orderic Vitalis, *The Ecclesiastical History*, VIII.23, ed. M. Chibnall (Oxford, 1973), 4:278.
2. The main sources for these events are the chronicles of Orderic Vitalis (note 1), Florence of Worcester and Simeon of Durham: both ed. and tr. J. Stevenson in *The Church Historians of England* (London, 1853–55), 2:317–18, 3:577 and 603. See H.H.E. Craster, *A History of Northumberland, Volume VIII: The Parish of Tynemouth* (Newcastle, 1907), 45–54, and F. Barlow, *William Rufus* (London, 1983).
3. National Archives, Patent Rolls, C66/329, Membrane 8 (13 Richard II), 23 Feb.1390. Ed. Craster, *Parish of Tynemouth* (note 2), 100n1; *Calendar of the Patent Rolls Preserved in the Public Record Office, 1388–92* (1902), 194; *Calendar of the Close Rolls [CCR] Preserved in the Public Record Office, 1389–92* (1922), 194–5, 401; CCR 1392–6 (1925), 31–2.
4. R. Westall, *A Time of Fire* (London, 1994), 30.
5. Bede, *Life and Miracles of St. Cuthbert*, ed. and tr. J.A. Giles, *Ecclesiastical History of the English Nation* (London, 1910), 286–349.
6. H.E. Savage, 'Abbess Hilda's First Religious House', *Archaeologia Aeliana*, 2nd series, 19 (1898): 47–75. Bede, *Ecclesiastical History of the English People*, III.14.
7. Matthew Paris, *Chronica Majora Volume 6: Additamenta*, ed. H.R. Luard (London, 1882), 372.
8. Clerical poll-tax accounts of Lincoln diocese, National Archives E179/35/16 (1381). Many abbots, including Richard of Wallingford and Thomas de la Mare, carried out visitations of the priory, usually passing by Belvoir; see, e.g., *Gesta Abbatum monasterii Sancti Albani (GASA)*, ed. H. Riley (London, 1867), 2:208, 394.

9. National Archives SC 8/144/7157 (1380). Ed. in Craster, *Parish of Tynemouth* (note 2), 97n2.

10. Richard of Wallingford, 'Tractatus Albionis' IV, ed. J.D. North, *Richard of Wallingford* (Oxford, 1976), 1:388.

11. Wallingford, 'Tractatus Albionis' (note 10) II.19, 1:324. Wallingford's word *clima* refers to a zone or band of latitude; he also uses the word to denote an astrolabe plate for a set latitude (III.37, at 1:382).

12. Wallingford, 'Tractatus Albionis' (note 10) IV.17, 1:400.

13. On the identification of the seventeen-hour parallel 'Mouths of Tanais' as the River Don, see G.J. Toomer, review of O. Pedersen, *A Survey of the Almagest*, *Archives internationales d'histoire des sciences* 27 (1977): 137–50, at 148.

14. The relationship between chord and sin is: crd α = 2 sin (α/2), or sin α = ½ crd 2α. On Ptolemy's table of chords (*Almagest*, I.11) and its dependence on his predecessors, such as Hipparchus, see G. Van Brummelen, *The Mathematics of the Heavens and the Earth: The Early History of Trigonometry* (Princeton, 2009), 33–93; O. Pedersen, *A Survey of the Almagest*, rev. edn by A. Jones (New York, 2011), 94–121.

15. Richard of Wallingford, 'Quadripartitum' and 'De sectore', ed. North, *Richard of Wallingford* (note 10), 1:21–178, 2:23–82. Cf H. Zepeda, 'The Medieval Latin Transmission of the Menelaus Theorem' (Ph.D. diss., University of Oklahoma, 2013), 160–81.

16. Van Brummelen, *Mathematics of the Heavens and the Earth* (note 14), 56–68; G. J. Toomer, ed. and trans., *Ptolemy's Almagest* (London, 1984), 69n84, 336, 338. There may well have been a copy of Menelaus' *Spherics* at St Albans: R. Sharpe, *English Benedictine Libraries: The Shorter Catalogues*, Corpus of British Medieval Library Catalogues, vol. 4 (London, 1996), B87:48, p. 561.

17. Cf Oxford, Corpus Christi College MS 144, f. 77v, '9;31' at λ = 5,8°, and Oxford, Bodl. MS Laud Misc. 657, f. 41r, '9;41'.

18. For example, an early-fourteenth-century manuscript of Oriel College, Oxford (now split between Bodl. MSS Digby 190 and 191), contained the *Almagest*'s descriptions of the parallels of latitude and table of oblique ascensions (II.6 and II.8) within a collection of mostly arithmetical texts, along with Robert Grosseteste's treatise on the rainbow. For Richard of Wallingford's use of the *Almagestum parvum* (aka. *Almagesti minor*) see North, *Richard of Wallingford* (note 10), 2:140; Zepeda, 'Medieval Latin Transmission' (note 15),

166–71, 444–92; H. Zepeda, *The First Latin Treatise on Ptolemy's Astronomy: The Almagesti minor (c.1200)* (Turnhout, 2018).

19. B. van Dalen, 'Ancient and Mediaeval Astronomical Tables: Mathematical Structure and Parameter Values' (PhD diss., University of Utrecht, 1993), 67, 185.

20. J. Chabás and B.R. Goldstein, *A Survey of European Astronomical Tables in the Late Middle Ages* (Leiden, 2012), 23; S. Falk, 'Copying and Computing Tables in Late Medieval Monasteries', in *Editing and Analysing Numerical Tables: Towards a Digital Information System for the History of Astral Sciences*, ed. M. Husson, C. Montelle and B. van Dalen (forthcoming).

21. *GASA* (note 8), 1:258.

22. *GASA* (note 8), 1:221–4.

23. Undated letter; fifteenth-century copy in Cambridge UL MS Ee.4.20. Ed. and summarised in Craster, *Parish of Tynemouth* (note 2), 71–3. Here the writer quotes the twelfth-century poet Hugh Primas (himself alluding to the liturgy for Compline), Poem 4, line 15, and Juvenal, *Satires*, VI; the remark about teeth set on edge is probably an allusion to Jeremiah 31:29–30. Among other allusions in the letter are Gregory the Great's homilies, and Ovid.

24. Edward I, *Close Rolls*, 31, m. 3 (19 Sept. 1303). In *Calendar of Close Rolls*, vol. 5 (1302–7); Wardrobe account of Edward II, in London, BL MS Stowe 553, pp. 45, 124–5, cited in M. Saaler, *Edward II* (London, 1997), 116.

25. *Annales Monasterii S. Albani a.d. 1421–1440, A Johanne Amundesham, monacho, ut videtur, conscripti*, ed. H. Riley (London, 1870), 1:214.

26. Cambridge, Pembroke College MS 82, f. 1r; K.A. Rand, 'The Authorship of The Equatorie of the Planetis Revisited', *Studia Neophilologica* 87 (2015): 15–35, at 34–5; Sharpe, *Shorter Catalogues* (note 16), B93 p. 588; cf B85, pp. 538–9; R.H. Rouse and M.A. Rouse, eds., *Registrum Anglie de libris doctorum et auctorum veterum* (London, 1991), 310–11.

27. Matthew 16:2; Pliny, *Natural History*, 18.78. Oxford, New College MS 274.

28. Aristotle, *On Generation and Corruption*, II.10 (336a–337a); *Meteorology*, I.2–3 (339a–341a); *Generation of Animals*, II.4 (738a); Ptolemy, *Tetrabiblos*, I.2.

29. Ptolemy, *Tetrabiblos*, I.1–2, tr. F.E. Robbins (Cambridge, MA, 1980), 3–15.

30. Aristotle, *Physics*, III.1 (200b).

31. R. Lemay, *Abu Ma'shar and Latin Aristotelianism in the Twelfth Century: The Recovery of Aristotle's Natural Philosophy through Arabic Astrology* (Beirut, 1962).

32. Richard of Wallingford, 'Exafrenon pronosticacionum temporis', ed. North, *Richard of Wallingford* (note 10), 1:182–243 (240–43 for the Thales story, which comes from Aristotle's *Politics*, 1:10), 2: 83–126. The Latin word *tempus* here means 'weather' rather than 'time' (cf French *temps*), but it can denote conditions rather wider than weather.

33. Ptolemy, *Tetrabiblos* (note 29), I.17.

34. Wallingford, 'Exafrenon' (note 32), 5, 1:232; A. Bouché-Leclercq, *L'Astrologie grecque* (Paris, 1899), 256–88.

35. e.g. Wallingford, 'Exafrenon' (note 32), 1, 1:190.

36. J.D. North, *Chaucer's Universe* (Oxford, 1988), 190; E.S. Kennedy, 'A Horoscope of Messehalla in the Chaucer Equatorium Manuscript', *Speculum* 34 (1959): 629–30.

37. Bodl. MS Laud Misc. 657, f. 30r. The method is exactly that described in (pseudo-)Masha'allah's astrolabe treatise, in a passage later translated with little modification by Geoffrey Chaucer (*Treatise on the Astrolabe*, II.36). Cf Wallingford, 'Tractatus Albionis' (note 10), III.39, 1:382–4.

38. Bodl. MS Laud Misc. 657, ff. 55bv 56v. This is an unusual presentation using midheaven as a starting point. It also has the additional feature of a time column, allowing the house divisions given for noon to be adjusted for any other time of day. On such presentations, see J.D. North, *Horoscopes and History* (London, 1986), esp. 126–30; Chabás and Goldstein, *European Astronomical Tables* (note 20), 207–11.

39. Abū Ma'šar, *The Abbreviation of The Introduction to Astrology*, ed. C. Burnett, K. Yamamoto and M. Yano (Leiden, 1994), 28–31, 61–3; *The Kalendarium of Nicholas of Lynn*, ed. S. Eisner (Athens, GA, 1980).

40. Robertus Anglicus, Commentary on the *Sphere*, *Lectio*, I, in L. Thorndike (ed.), *The Sphere of Sacrobosco and Its Commentators* (Chicago, 1949), 143–4.

41. John Gower, *Confessio Amantis*, VII.633–54, ed. R.A. Peck (Kalamazoo, 2004), 3:279.

42. Augustine, *The City of God against the Pagans*, V.6, tr. R.W. Dyson (Cambridge, 1998), 194–5; Zeno of Verona (d. 371), Tractatus, 1.38,

ed. B. Löfstedt, Corpus Christianorum 22 (Turnhout, 1971) 105–6; S.C. McCluskey, *Astronomies and Cultures in Early Medieval Europe* (Cambridge, 1998), 38–40.

43. Thomas Aquinas, *Summa Theologiae*, 2.2.95.5, 1.115.4, ed. Fundación Tomás de Aquino, http://www.corpusthomisticum.org. On Aquinas' careful consideration of this topic, see T. Litt, *Les Corps célestes dans l'univers de saint Thomas d'Aquin* (Louvain, 1963).

44. Nos. 206, 162, 6, in H. Denifle and E. Chatelaine, eds., *Chartularium Universitatis Parisiensis* (Paris, 1889), 1:544–55; C.P.E. Nothaft, 'Glorious Science or "Dead Dog"? Jean de Jandun and the Quarrel over Astrology in Fourteenth-Century Paris', *Vivarium* 57 (2019): 51–101; Nicole Oresme, 'Ad pauca respicientes', in *De proportionibus proportionum and Ad pauca respicientes*, ed. E. Grant (Madison, 1966), 382.

45. C.P.E. Nothaft, '*Vanitas Vanitatum et Super Omnia Vanitas*: The Astronomer Heinrich Selder and a Newly Discovered Fourteenth-Century Critique of Astrology', *Erudition and the Republic of Letters*, 1 (2016): 261–304, esp. 295–8. On conjoined twins, see Albertus Magnus, *De Animalibus*, 18.2.3.

46. G. Bos and C. Burnett, eds., *Scientific Weather Forecasting in the Middle Ages: The Writings of Al-Kindī* (London, 2000); Robert Grosseteste, *Hexaëmeron*, ed. R.C. Dales and S. Gieben (London, 1982), V.viii–xi, pp. 165, 170; Grosseteste, 'De natura locorum', ed. in *Die philosophischen Werke des Robert Grosseteste*, ed. L. Baur (Münster, 1912), 70.

47. *The Opus Maius of Roger Bacon*, ed. J.H. Bridges (Oxford, 1897), 1:138; J.D. North, 'Celestial Influence – the Major Premiss of Astrology', in '*Astrologi Hallucinati*': *Stars and the End of the World in Luther's Time*, ed. P. Zambelli (Berlin, 1986), repr. in North, *Stars, Minds and Fate: Essays in Ancient and Medieval Cosmology* (London, 1989): 243–98.

48. G. Chaucer, *Troilus and Criseyde*, III.624–8, in *The Riverside Chaucer*, ed. L.D. Benson (Oxford, 1987), 522.

49. Ptolemy, *Tetrabiblos*, I.2; Thomas Aquinas, *De caelo*, II.18, ed. Fundación Tomás de Aquino (note 43); see also *Summa contra Gentiles*, III.82; G. Bezza, 'Saturn–Jupiter Conjunctions and General Astrology: Ptolemy, Abu Ma'shar and Their Commentators', in *From Māshā'allāh to Kepler: Theory and Practice in Medieval and Renaissance Astrology*, ed. C. Burnett and D.G. Greenbaum (Ceredigion, 2015): 5–48. Notable prior predictions were by Jean

des Murs in Paris and John Ashenden in Oxford; See J-P. Boudet, 'Jean des Murs, Astrologer', *Erudition and the Republic of Letters* 4 (2019): 123–45; L. Thorndike, *A History of Magic and Experimental Science* (New York, 1934), 3:326–37.

50. John Ashenden, *Summa iudicialis de accidentibus mundi* (ed. Santritter, 1489), II.12.3, f. D(2).5v. Quoted from Oxford, Oriel College MS 23 f. 225v in Thorndike, *History of Magic* (note 49), 3:332n12.

51. Matthew Paris, in Bodl. MS Ashmole 304, f. 40v; quoted in A. Iafrate, 'The Workshop of Fortune: St Albans and the Sortes Manuscripts', *Scriptorium* 66 (2012): 55–87, at 82; K. Yamamoto and C. Burnett, eds., *Abū Ma'šar On Historical Astrology: The Book of Religions and Dynasties (on the Great Conjunctions)* (Leiden, 2000), 1.1.12–15 and 2.8.3, vol. 1, pp. 11, 123; on the cloister windows, see Chapter 2, note 71. Geared wheels are set into the inside front cover of Oxford, Bodl. MS Digby 46, probably produced at St Albans; Thorndike, *History of Magic* (note 49), 2: 110–18; C. Burnett, 'What is the *Experimentarius* of Bernardus Silvestris? A Preliminary Survey of the Material', *Archives d'histoire doctrinale et littéraire du Moyen Âge* 44 (1977): 79–125.

52. *Speculum astronomiae*, Ch. 11, ed. in P. Zambelli, *The Speculum astronomiae and Its Enigma: Astrology, Theology, and Science in Albertus Magnus and His Contemporaries* (Dordrecht, 1992), 208–73, at 240–51; R. Kieckhefer, 'Rethinking How to Define Magic' and C. Burnett, 'Arabic Magic. The Impetus for Translating Texts and Their Reception', both in *The Routledge History of Medieval Magic*, ed. S. Page and C. Rider (London, 2019): 15–25, 71–84; S. Page, *Magic in the Cloister: Pious Motives, Illicit Interests, and Occult Approaches to the Medieval Universe* (University Park, PA, 2013), 43–5.

53. Gower, *Confessio Amantis* (note 41), VII.1296-1318, 3:293-4.

54. See, e.g., British Library Harley MS 1612. Gower's source, the *Tractatus Enoch*, survives in at least eleven manuscripts, ed. L. Delatte in *Textes latins et vieux français relatifs aux Cyranides* (Paris, 1942), 276–89. P. Lucentini and V. Perrone Compagni list the manuscripts in *I testi e i codici di Ermete nel Medioevo* (Florence, 2001), no. 14, pp. 47–8; S. Falk, 'Natural Sciences', in *Historians on John Gower: Society, Religion and Politics*, ed. S.H. Rigby (Woodbridge, 2019); Page, *Magic in the Cloister* (note 52), 112–29 (on the magical-liturgical *Ars notoria*), 1–4.

55. British Library Harley MS 4664 (The Coldingham Breviary), f. 125v;

R.W. Pfaff, *The Liturgy in Medieval England: A History* (Cambridge, 2009), 223–4. A fuller version of the (Norman-inflected) French instructions is in British Library Arundel MS 220, f. 314v; C.P.E. Nothaft, *Scandalous Error: Calendar Reform and Calendrical Astronomy in Medieval Europe* (Oxford, 2018), 169.

56. I.B. Cowan and D.E. Easson, *Medieval Religious Houses: Scotland*, 2nd edn (London, 1976), 55–6.
57. J. Hsy, 'Lingua Franca: Overseas Travel and Language Contact in The Book of Margery Kempe', in *The Sea and Englishness in the Middle Ages*, ed. S.I. Sobecki (Cambridge, 2011): 159–78, at 173–4.

CHAPTER 6: THE BISHOP'S CRUSADE

1. *The St Albans Chronicle: The* Chronica Maiora *of Thomas Walsingham*, ed. J. Taylor, W.R. Childs and L. Watkiss (Oxford, 2003), 636.
2. J. Sumption, *The Hundred Years War, Volume III: Divided Houses* (London, 2009), 452–60.
3. *St Albans Chronicle* (note 1), 492–4.
4. *The Westminster Chronicle, 1381–1394*, ed. L.C. Hector and B.F. Harvey (Oxford, 1982), 33.
5. *Knighton's Chronicle 1337–1396*, ed. G.H. Martin (Oxford, 1995), 324.
6. Parliamentary Rolls, 6–7 Richard II (Feb./Oct.1383), iii.147–8, 153–4, http://www.sd-editions.com/PROME/home.html.
7. National Archives C76/67 m.18-16, https://www.medievalsoldier.org/; *Knighton's Chronicle* (note 5), 332.
8. *Eulogium (historiarum sive temporis): Chronicon ab orbe condito usque ad MCCCLXVI*, ed. F.S. Haydon (London: 1863), continuation, 3:357; *Gesta Abbatum monasterii Sancti Albani (GASA)*, ed. H. Riley (London, 1867), 2:416. Marginal addition (not in the printed edition) in British Library Cotton MS Claudius E.IV, f. 239v.
9. Parl. Rolls, 6 Richard II (Feb. 1383), iii.148 (note 6); *Westminster Chronicle* (note 4), 39; *St Albans Chronicle* (note 1), 670.
10. British Library Cotton MS Nero D.VII, f. 83r; ed. in W. Dugdale, *Monasticon Anglicanum*, new edn ed. J. Caley, H. Ellis and B. Bandinel (London, 1819), 2:209n.
11. Robert the Monk, *Historia Hierosolymitana*, in *Recueil des historiens des croisades: historiens occidentaux* (Paris, 1866), 3:729.

12. R. Sharpe, ed., *English Benedictine Libraries: The Shorter Catalogues* (London, 1996), B107, 627–9.

13. Oxford, Bodl. MS Ashmole 1796, ff. 58r–59r. Also e.g. Bodl. MS Laud Misc. 674, ff. 73r–74ar, Cambridge UL MS Hh.6.8, f. 184r. Cambridge, Corpus Christi MS 16I, f. ivv; J.B. Mitchell, 'The Matthew Paris Maps', *Geographical Journal* 81 (1933): 27–34.

14. Paris, Bibliothèque Nationale de France MS esp. 30; facsimile in *Mapamundi, the Catalan Atlas of the Year 1375*, ed. G. Grosjean (Dietikon-Zurich, 1978); review by T. Campbell in *Imago Mundi* 33 (1981), 115–16; K. Kogman-Appel, 'The Geographical Concept of the Catalan *mappamundi*', in *Knowledge in Translation: Global Patterns of Scientific Exchange, 1000–1800 CE*, ed. P. Manning and A. Owen (Pittsburgh, 2018), 19–40.

15. T. Campbell, 'Portolan Charts from the Late Thirteenth Century to 1500', in *The History of Cartography, Volume I: Cartography in Prehistoric, Ancient and Medieval Europe and the Mediterranean*, ed. J.B. Harley and D. Woodward (Chicago, 1987), 371–463, at 384–5; M.R. Cohen and I.E. Drabkin, *A Source Book in Greek Science* (New York, 1948), 310–14, incl. excerpts by Pliny, Plato, Lucretius and Galen; Augustine, *The City of God against the Pagans*, XXI.4, tr. R.W. Dyson (Cambridge, 1998), 1051; J. Needham, *Science and Civilisation in China*, vol. 4.1 (Cambridge, 1962), 229–334, esp. 249–50.

16. J.A. Smith, 'Precursors to Peregrinus: The Early History of Magnetism and the Mariner's Compass in Europe', *Journal of Medieval History* 18 (1992): 21–74, at 25–7.

17. London, College of Arms MS Arundel 6, f. 135v, cited in R.W. Hunt, *The Schools and the Cloister: The Life and Writings of Alexander Nequam (1157–1217)* (Oxford, 1984), 1.

18. Alexander Neckam, *De nominibus utensilium*, ed. T. Wright, *A Volume of Vocabularies* (1857), 96–119, at 114.

19. Alexander Neckam, *De naturis rerum*, ed. T. Wright (Cambridge, 1863), 1–3; II.16–20: 136–41; Hunt, *Schools and the Cloister* (note 17), 67–82.

20. Neckam, *De naturis rerum* (note 19), II.98: 182–3.

21. Jacques de Vitry, *Historia Orientalis seu Hierosolymitana*, Ch. 91, cited in Smith, 'Precursors' (note 16), 41; Isidore, *Etymologies*, 16.21.3, tr. S.A. Barney et al. (Cambridge, 2006), 331. Isidore does not identify the statue; in Pliny's version (*Natural History*, 34:42) it is of Arsinoë II, co-ruler of Egypt with Ptolemy II Philadelphus.

22. Jean de St-Amand, commentary on the *Antidotarium Nicolai*, ed. in Pseudo-Mesuë, *Opera omnia cum expositione mondini super canones vniuersales* (Venice, 1502), 294–331, at 330v–331r; some parts (without the snake or poison) ed. and tr. in L. Thorndike, 'John of St. Amand on the Magnet', *Isis* 36 (1946): 156–7; J. Rubin, 'The Use of the "Jericho Tyrus" in Theriac', *Medium Aevum* 83 (2014): 234–53.

23. Petrus Peregrinus de Maricourt, 'Epistula de Magnete', ed. L. Sturlese in *Opera* (Pisa, 1995), 63–89, at 65–6; Bodl. MS Ashmole 1522, ff. 181v–187v.

24. Geoffrey Chaucer, *The Canterbury Tales*, I.388–410, in *The Riverside Chaucer*, ed. L.D. Benson (Oxford, 1987), 29–30.

25. Neckam, *De naturis rerum* (note 19), II.17: 138; British Library Cotton MS Julius D.7, f. 45v; ed. in *Rara Mathematica*, ed. J.O. Halliwell (London, 1839), 55; E.G.R. Taylor, *The Haven-Finding Art: A History of Navigation from Odysseus to Captain Cook* (London, 1956), 136.

26. *De viis maris*, ed. P. Gautier Dalché, *Du Yorkshire à l'Inde: une 'géographie' urbaine et maritime de la fin du XIIe siècle* (Geneva, 2005), 173–229, at 177, 184, 215.

27. I have modernised his Middle English. Edited in G.A. Lester, 'The Earliest English Sailing Directions', in *Popular and Practical Science of Medieval England*, ed. L.M. Matheson (East Lansing, 1994), 331–67, at 342–3.

28. John Tapp, *The Seamans Kalender* (London, 1622), B7r; BNF MS esp. 30 (note 14), sheet 1a; M.E. Schotte, *Sailing School: Navigating Science and Skill, 1550–1800* (Baltimore, 2019), 35–7.

29. Avicenna, *Canon of Medicine*, 1.3.5.8 (Venice, 1489), f. g1v; tr. O.C. Gruner, *A Treatise on the Canon of Medicine* (London, 1930), §903, p. 456; P. Horden, 'Travel Sickness: Medicine and Mobility in the Mediterranean from Antiquity to the Renaissance', in *Rethinking the Mediterranean*, ed. W.V. Harris (Oxford, 2005): 179–99, at 193–4.

30. Sumption, *Divided Houses* (note 2), 131–40, 582–5, 890n28.

31. *St Albans Chronicle* (note 1), 672, 676.

32. *St Albans Chronicle* (note 1), 678–82.

33. *St Albans Chronicle* (note 1), 687, 693; *GASA* (note 8), 2:416.

34. *Eulogium* (note 8), 3:356–7.

35. Luke 13:2-5; John 9:1–3.

36. Jean of Joinville, *Life of Saint Louis*, 306, 310, tr. M.R.B. Shaw

in *Chronicles of the Crusades* (Harmondsworth, 1963), 240–41; J. Phillips, 'The Experience of Sickness and Health during Crusader Campaigns to the Eastern Mediterranean, 1095–1274' (PhD thesis, University of Leeds, 2017), 241–2.

37. P.D. Mitchell, E. Stern and Yotam Tepper, 'Dysentery in the Crusader Kingdom of Jerusalem: An ELISA Analysis of Two Medieval Latrines in the City of Acre (Israel)', *Journal of Archaeological Science* 35 (2008): 1849–53; L. Demaitre, *Medieval Medicine: The Art of Healing, from Head to Toe* (Santa Barbara, 2013), 261–5.

38. Chaucer, *The Canterbury Tales*, I.411–40, in *The Riverside Chaucer* (note 24), 30.

39. Chaucer, *The Canterbury Tales*, I.429–34, in *The Riverside Chaucer* (note 24), 30.

40. e.g. British Library Harley MS 3698, f. 69r (Bernard of Gordon, *Lilium medicine*).

41. Windows: see Chapter 2 (note 71); *GASA* (note 8), 1:194–212; Dugdale, *Monasticon* (note 10), 3:355–6.

42. *GASA* (note 8), 1:217, 246–9.

43. Ralph of Coggeshall, *Radulphi de Coggeshall chronicon anglicanum*, ed. J. Stevenson (Rolls Series, 1875), 183–4; Matthew Paris, *Chronica maiora*, ed. H.R. Luard (Rolls Series, 1874), 2:667-8; K. Park, 'The Life of the Corpse: Division and Dissection in Late Medieval Europe', *Journal of the History of Medicine* 50 (1995): 111–32.

44. Park, 'Life of the Corpse', (note 43), 111

45. K. Park, 'Medical Practice', in *The Cambridge History of Science, Vol. 2: Medieval Science*, ed. D.C. Lindberg and M.H. Shank (Cambridge, 2013), 611–29, at 617–20.

46. Cambridge, Jesus College MS 60 (Q.G.12), f. 45r; P.M. Jones, 'Mediating Collective Experience: the *Tabula Medicine* (1416–1425) as a Handbook for Medical Practice', in *Between Text and Patient: The Medical Enterprise in Medieval & Early Modern Europe*, ed. F.E. Glaze and B.K. Nance (Florence, 2011), 279–307

47. Muhammad al-Idrisi, *Géographie d'Édrisi*, French tr. by P.A. Jaubert (Paris, 1836), 1.51, 182.

48. British Library Arundel MS 22, f. 202r.

49. L.E. Demaitre, *Doctor Bernard de Gordon: Professor and Practitioner* (Toronto, 1980), 51–4.

50. Bernard of Gordon, *Lilium Medicine*, V.14 (Lyon, 1574), 476–80; corrected with British Library Harley MS 3698, ff. 69v–70r; Demaitre, *Medieval Medicine* (note 37), 261.

51. Chaucer, 'The Miller's Tale', I.3690–911; Isidore, *Etymologies* (note 21), XVI.ii.6, 318; G.E.M. Gasper and F. Wallis, 'Salsamenta Pictavensium: Gastronomy and Medicine in Twelfth-Century England', *English Historical Review* 131 (2016): 1353–85, at 1377–8.

52. Bernard, *Lilium Medicine* (note 50), V.14, 480 (70r); I.14, 59 (8r); Rubin, 'Jericho Tyrus' (note 22); Demaitre, *Medieval Medicine* (note 37), 72; M. McVaugh, 'Theriac at Montpellier', *Sudhoffs Archiv* 56 (1972): 113–44.

53. Chaucer, *The Canterbury Tales*, I.423–44, in *The Riverside Chaucer* (note 24), 30.

54. C. Rawcliffe, 'The Doctor of Physic', in *Historians on Chaucer: The 'General Prologue' to the Canterbury Tales*, ed. S.H. Rigby and A.J. Minnis (Oxford, 2014), 297–318, at 316.

55. John Gower, *Mirour de l'Omme* 25621–32, in *The Complete Works*, ed. G.C. Macaulay (Oxford, 1899), 3.283–4.

56. Paris, BNF MS Lat. 7443, ff. 184r–211v, at 186v. The case is summarised in L. Thorndike, *A History of Magic and Experimental Science* (New York, 1934), 4.139–42.

57. Paris, Archives Nationales LL//85, ed. in L. Mirot, 'Le Procès de Maître Jean Fusoris', *Mémoires de La Société de l'histoire de Paris et de l'Île-de-France* 27 (1900): 137–287, at 213. See also E. Poulle, *Un Constructeur d'instruments astronomiques au XVe siècle: Jean Fusoris* (Paris, 1963); J.H. Wylie, *The Reign of Henry the Fifth* (Cambridge, 1914–29) 1:498-510, 2:42-3.

58. Mirot, 'Le Procès' (note 57), 245–6, 236.

59. Mirot, 'Le Procès' (note 57), 223.

60. *St Albans Chronicle* (note 1), 702.

61. John Gower, *Vox Clamantis*, III.6.343, in *The Complete Works* (note 55), 4.116; John Wyclif, 'Of Prelates' 9, and ' 'The Office of Curates' 16, both in *The English Works of Wyclif*, ed. F.D. Matthew, EETS 74 (London, 1880), 73, 152.

62. Parl. Rolls (note 6), 7 Richard II (Oct. 1383), iii.153–8.

63. *GASA* (note 8), 2:416.

CHAPTER 7: COMPUTER OF THE PLANETS

1. Cambridge, Peterhouse MS 75.I, f. 72r, https://cudl.lib.cam.ac.uk/view/MS-PETERHOUSE-00075-00001.

2. C.P. Christianson, *A Directory of London Stationers and Book Artisans, 1300–1500* (New York, 1990), 31.

3. C.M. Barron, *London in the Later Middle Ages: Government and People, 1200–1500* (Oxford, 2004), 241; London Metropolitan Archives, Hustings Rolls 1373-4, CLA/023/DW/01/100 (160).

4. Peterhouse MS 75.I (note 1), f. 64r.

5. Geoffrey Chaucer, *The Canterbury Tales*, V.1274-6, in *The Riverside Chaucer*, ed. L.D. Benson (Oxford, 1987), 185.

6. Jean de Lignières, *Tabule magne*, dedicated to Robert of Florence, Dean of Glasgow. Cambridge, Gonville & Caius College MS 110 (179), 1.

7. Peterhouse MS 75.I (note 1), ff. 38v–44v.

8. Peterhouse MS 75.I (note 1), f. 72r; Chaucer, *The Canterbury Tales*, V.1281-2, in *The Riverside Chaucer* (note 5), 185.

9. *Alfontii regis castelle illustrissimi celestium motuum tabule* (Venice, 1483), ff. c4r–d1r.

10. Oxford, Bodl. MS Ashmole 1796, 58v–59r (5°). See also Bodl. MS Laud Misc. 674, ff. 73v–74ar (5;30° and 8;26° – tables attributed to the Oxford astronomers John Maudith, John Walter and William Worcester), CUL MS Hh.6.8, f. 184r (4°), Ptolemy, *Cosmographia*, II (Ulm, 1482), b4r, b7r (10°).

11. Full fiddly details in S. Falk, 'Learning Medieval Astronomy through Tables: The Case of the Equatorie of the Planetis', *Centaurus* 58 (2016): 6–25.

12. Peterhouse MS 75.I (note 1), f. 14r; D.J. Price, *The Equatorie of the Planetis* (Cambridge, 1955), 182-7.

13. Peterhouse MS 75.I (note 1), f. 5v; J.D. North, *Chaucer's Universe* (Oxford, 1988), 173-4.

14. Peterhouse MS 75.I (note 1), ff. 64r, 70r, 63v; on 'R.B.' for Roger Bacon, see L.E. Voigts, 'The "Sloane Group": Related Scientific and Medical Manuscripts from the Fifteenth Century in the Sloane Collection', *British Library Journal* 16 (1990): 26–57.

15. Thomas Hoccleve, *La Male regle*, 143-4, in *Chaucer to Spenser: An Anthology*, ed. D. Pearsall (Oxford, 1999), 320.

16. Thomas Hoccleve, *The Regiment of Princes*, ed. C.R Blyth (Kalamazoo, 1999), ll. 1-7.

17. John Gower, *Confessio Amantis*, VII.38-9, ed. R.A. Peck (Kalamazoo, 2004), 3:265; S.C. McCluskey, *Astronomies and Cultures in Early Medieval Europe* (Cambridge, 1998), 131-4.

18. P. Kibre, 'Lewis of Caerleon, Doctor of Medicine, Astronomer, and

Mathematician', *Isis* 43 (1952): 100–108; Cambridge, St John's College MS B.19 (41); British Library Royal MS 12 G I; manuscript in private collection. See also BL Arundel MS 66; H.M. Carey, 'Henry VII's Book of Astrology and the Tudor Renaissance', *Renaissance Quarterly* 65 (2012): 661–710.

19. E.K. Rand, 'Editor's Preface', *Speculum* 1 (1926): 3–4, at 4; R. Bradley, 'Backgrounds of the Title Speculum in Mediaeval Literature', *Speculum* 29 (1954): 100–115; Alain de Lille, hymn *De incarnatione Christi*, in Migne, *Patrologia latina*, 210, col. 579A-B, http://pld. chadwyck.co.uk/.

20. British Library Harley MS 3244, ff. 40r–v. Some medieval sources on the beaver are collected at D. Badke, *The Medieval Bestiary*, http://bestiary.ca/beasts/beast152.htm. P. Canvane, *A Dissertation on the Oleum Palmae Christi, Sive Oleum Ricini; Or, (as it is commonly call'd) Castor Oil* (Bath, 1764), 4.

21. Hoccleve, *Regiment of Princes* (note 16), ll. 1962, 1964, 2087–8; J.D. North, *God's Clockmaker: Richard of Wallingford and the Invention of Time* (London, 2005), 131–2; *Gesta Abbatum monasterii Sancti Albani (GASA)*, ed. H. Riley (London, 1867), 3:274–5; J.G. Clark, *A Monastic Renaissance at St. Albans: Thomas Walsingham and His Circle, c.1350–1440* (Oxford, 2004), 40.

22. *GASA* (note 21), 1:289, 322, 471–2; 2:281; *Annales Monasterii S. Albani a.d. 1421–1440, A Johanne Amundesham, monacho, ut videtur, conscripti*, ed. H. Riley (London, 1870), 1:47; C.M. Barron, 'Centres of Conspicuous Consumption: The Aristocratic Town House in London, 1200–1550', *London Journal* 20 (1995): 1–16.

23. C.M. Barron, 'The Expansion of Education in Fifteenth-Century London', in *The Cloister and the World: Essays in Medieval History in Honour of Barbara Harvey*, ed. J. Blair and B. Golding (Oxford, 1996), 219–45.

24. Peterhouse MS 75.I (note 1), f. 71r. Astronomical position data from Stellarium (www.stellarium.org).

25. Aristotle, *Metaphysics*, XII.8, 1073b–74a.

26. Plato, *Timaeus*, 34; Aristotle, *On the Heavens*, II.6 (288a–289a); *Ptolemy's Almagest*, IX.2, tr. G.J. Toomer (London, 1984), 422. The equant is introduced in IX.5.

27. *Campanus of Novara and Medieval Planetary Theory: Theorica Planetarum*, ed. F.S. Benjamin and G.J. Toomer (Madison, 1971); Roger Bacon, *Compendium studii philosophiae*, VIII and

Opus Tertium, XI, both in *Opera quaedam hactenus inedita*, ed. J.S. Brewer (London, 1859), 1:465, 35.

28. Jean de Lignières, 'Quia nobilissima scientia astronomie', Oxford, Bodl. MS Digby 168, f. 64v, ed. Price, *Equatorie of the Planetis* (note 12), 188–96.

29. Peterhouse MS 75.I (note 1), f. 71v. The text *On the Construction and Operation of the Astrolabe*, which was – along with Sacrobosco's *Sphere* – Chaucer's principal source, was commonly attributed to the eighth-century astrologer Masha'allah ibn Athari (known to the Latins as Messahalla); ed. and tr. in R. T. Gunther, *Early Science in Oxford, Vol. 5: Chaucer and Messahalla on the Astrolabe* (Oxford, 1929), 137–231.

30. BL MS Arundel 292, f. 72v, ed. in T. Wright and J.O. Halliwell, *Reliquae antiquae* (London, 1845), 1:240.

31. John Buridan, *Questions on the Eight Books of the Physics of Aristotle*, IV.8, tr. in E. Grant, *A Source Book of Medieval Science* (Cambridge, MA, 1974), 326; I.8, ed. M. Streijger and P.J.J.M. Bakker, *Quaestiones super octo libros Physicorum Aristotelis, Books I–II* (Leiden, 2015), 87–9.

32. Peterhouse MS 75.I (note 1), ff. 74r, 73v.

33. Peterhouse MS 75.I (note 1), ff. 76r, 72v, 79r.

34. L.E. Voigts, 'What's the Word? Bilingualism in Late-Medieval England', *Speculum* (1996): 813–26; P. Pahta and I. Taavitsainen, 'Vernacularisation of Scientific and Medical Writing in Its Sociohistorical Context', in *Medical and Scientific Writing in Late Medieval English*, ed. Taavitsainen and Pahta (Cambridge, 2004), 1–22; *The Fifty Earliest English Wills*, ed. F.J. Furnivall, EETS 78 (London, 1882), 3; Geoffrey Chaucer, 'A Treatise on the Astrolabe', prologue, in *The Riverside Chaucer* (note 5), 662; Barron, 'Expansion of Education' (note 23), 221–2.

35. Peterhouse MS 75.I (note 1), f. 73r; S. Partridge, 'The Vocabulary of the Equatorie of the Planetis and the Question of Authorship', in *English Manuscript Studies 1100–1700*, vol. 3, ed. P. Beal and J. Griffiths (London, 1992), 29–37.

36. S. Falk, 'Vernacular Craft and Science in *The Equatorie of the Planetis*', *Medium Ævum*, 88 (2019), 329–60, at 350–51.

37. Campanus of Novara, 'Theorica planetarum' VI, ed. Benjamin and Toomer (note 27), 347, 353.

38. E. Poulle, *Les Instruments de la théorie des planètes selon Ptolémée: équatoires et horlogerie planétaire du XIIIe au XVIe siècle* (Geneva,

1980), 158, 204; Oxford, Merton College SC/OB/AST/2; S. Falk, 'A Merton College Equatorium: Text, Translation, Commentary', *SCIAMVS* 17 (2016): 121–59.

39. Peterhouse MS 75.I (note 1), ff. 73r, 75r.

40. 'Peterhouse: Equatorie of the Planetis', University of Cambridge Digital Library, https://cudl.lib.cam.ac.uk/view/MS-PETERHOUSE-00075-00001.

41. Peterhouse MS 75.I (note 1), ff. 77r, 75v, 71v, 76r, 77r.

42. D.W. Robertson, *Chaucer's London* (New York, 1968), 59; Barron, *London in the Later Middle Ages* (note 3), 254.

43. *On Crowds*, VII.8, in Seneca, *Epistles, Volume I: Epistles 1–65*, tr. R.M. Gummere (Cambridge, MA, 1917), 34; G.G. Wilson, ' "Amonges Othere Wordes Wyse": The Medieval Seneca and the "Canterbury Tales" ', *Chaucer Review* 28 (1993): 135–45.

44. A significant later equatorium is that of Jamshid al-Kashi, which could compute planetary latitudes; E.S. Kennedy, *The Planetary Equatorium of Jamshīd Ghīyāth al-Dīn al-Kāshī* (Princeton, 1960); H. Bohloul, 'Kāshānī's Equatorium: Employing Different Plates for Determining Planetary Longitudes', in *Scientific Instruments between East and West*, ed. N. Brown, S. Ackermann and F. Günergun (Leiden, 2019): 122–41. Johannes Schöner produced a cut-out-and-assemble 'Æquatorium astronomicum' in 1521; printed in *Opera mathematica* (Nuremberg, 1561); see J. Evans, *The History and Practice of Ancient Astronomy* (Oxford, 1998), 405–10. The most sumptuous example is Petrus Apianus' *Astronomicum Caesareum* (1540), produced for Emperor Charles V; O. Gingerich, 'Apianus's *Astronomicum Caesareum* and Its Leipzig Facsimile', *Journal for the History of Astronomy* 2 (1971): 168–77.

45. Moses Maimonides, *The Guide of the Perplexed*, II.24, tr. S. Pines (Chicago, 1963), 325–6.

46. Aristotle, *On the Heavens*, II.10, 291a–b; Ptolemy, *Planetary Hypotheses*, 1.2.3, ed. B.R. Goldstein, 'The Arabic Version of Ptolemy's Planetary Hypotheses', *Transactions of the American Philosophical Society* 57:4 (1967): 3–55, at 7; L. Taub, *Ptolemy's Universe* (Chicago, 1993), 111–12.

47. A unique, fascinating example of monumental, painted astronomy in the age of print is a large, hinged compendium (polyptych) produced by the Cracow master Marcus Schinnagel (who also edited printed almanacs) in 1489: R.L. Kremer, 'Marcus Schinnagel's Winged Polyptych of 1489: Astronomical Computation in a Litur-

gical Format', *Journal for the History of Astronomy* 43 (2012): 321–45; Landesmuseum Württemberg 1995-323, https://bawue.museum-digital.de/index.php?t=objekt&oges=2758.

48. Nicolaus Copernicus, *De revolutionibus orbium caelestium* (Nuremberg, 1543), f. iiiv.

49. Jean Buridan, *Quaestiones super libris quattuor De caelo et mundo*, II.22, ed. E.A. Moody (Cambridge, MA, 1942), 226–33; Copernicus, *De revolutionibus* (note 48), I.10, f. 8v.

50. F.J. Ragep, 'Ibn Al-Haytham and Eudoxus: The Revival of Homocentric Modeling in Islam', in *Studies in the History of the Exact Sciences in Honour of David Pingree*, ed. C. Burnett et al. (Leiden, 2004), 786–809.

51. Nasir al-Din Tusi, introduction to *Ilkhani Zij*, tr. A.J. Arberry, *Classical Persian Literature* (London, 1958), 182, quoted in F.J. Ragep, ed., *Naṣīr al-Dīn al-Ṭūsī's Memoir on Astronomy* (New York, 1993), 10.

52. A. Sayılı, *The Observatory in Islam and Its Place in the General History of the Observatory* (Ankara, 1960), 189–223.

53. Nizam al-Din al-Isfahani, 'Fi madh Nasir al-Din al-Tusi wa fi wasf al-rasd' ('In Praise of Nasir al-Din al-Tusi and an Ekphrasis on Astronomy'), ed. A. Sayılı, 'Khwaja Nasir-i Tusi wa Rasadkhana-i Maragha', *Ankara Üniversitesi Dil ve Tarih-Cografya Fakültesi Dergisi* 14 (1956), 1–13, at 13; translated by Matthew Keegan with advice from James Montgomery, Jamil Ragep and Geert Jan van Gelder. 'Bright stars like dice' is an allusion to the Arabic game of astronomical chess, and hints at Tusi's power to influence even the stars.

تَقَرُّبُ الأَلْحاظِ والنَّفْسُ نُبَهَّجُ	—	بِناءٌ لَعُمْرِي مِثْلُ بانِيهِ مُعْجِزٌ
يُناغِي كِعابَ الزُّهْرِ مِنها تَبَرُّجُ	—	سَيَبْلُغُ أَسْبابَ السَّماءِ بِصَرْحِهِ
وشَيَّدَ قَصْرًا لم يَشِدْهُ مَتَوَّجُ	—	أَقولُ وقَدْ شادَ البِناءُ بِذِكْرِهِ

54. F.J. Ragep, 'Alī Qushjī and Regiomontanus: Eccentric Transformations and Copernican Revolutions', *Journal for the History of Astronomy* 36 (2005): 359–71; J. North, *Cosmos: An Illustrated History of Astronomy and Cosmology* (Chicago, 2008), 204–9.

55. Roger Bacon, *Opus Maius*, IV.4.16, ed. J.H. Bridges (Oxford, 1897), 1:399–400; R. Morrison, 'A Scholarly Intermediary between the Ottoman Empire and Renaissance Europe', *Isis* 105 (2014): 32–57.

56. N.M. Swerdlow and O. Neugebauer, *Mathematical Astronomy in Copernicus's De Revolutionibus* (New York, 1984), 295; D.M. Nicol, *Byzantium and Venice: A Study in Diplomatic and Cultural Relations* (Cambridge, 1992), 419.

EPILOGUE

1. A.R. Hall, 'The First Decade of the Whipple Museum', in *The Whipple Museum of the History of Science: Instruments and Interpretations*, ed. L. Taub and F. Willmoth (Cambridge, 2006).

2. L. Bragg, 'Physicists After the War', *Nature* 150 (1942): 75–80, at 78. 'Radio Newsreel', *BBC Light Programme* [now BBC Radio 3], 1 March 1952; transcript courtesy of the Price family. The production of Price's replica is documented in S. Falk, 'The Scholar as Craftsman: Derek de Solla Price and the Reconstruction of a Medieval Instrument', *Notes and Records of the Royal Society* 68 (2014): 111–34. The replica's inventory number is Wh.3271.

3. R.K. Merton, *On the Shoulders of Giants: A Shandean Postscript* (New York, 1965).

4. D.M. Miller, 'The Thirty Years War and the Galileo Affair', *History of Science* 46 (2008): 49–74.

5. M. Rees, *Before the Beginning: Our Universe and Others* (London, 1997), 100.

6. *Galileo on the World Systems: A New Abridged Translation and Guide*, tr. M.A. Finocchiaro (Berkeley, 1997), 90; see, e.g., Alain de Lille, *De Planctu Naturae*, in Migne, *Patrologia latina*, 210, col. 444A, http://pld.chadwyck.co.uk/; D.R. Danielson, 'The Great Copernican Cliché', *American Journal of Physics* 69 (2001): 1029–35.

7. N. deG. Tyson, 'The Cosmic Perspective', *Natural History* (April 2007), https://www.naturalhistorymag.com/universe/201367/cosmic-perspective.

8. Archivum Secretum Vaticanum, Lateran Regesta 45 (1396-7), f. 174v; W.H. Bliss and J.A. Twemlow, eds., *Calendar of Papal Registers Relating to Great Britain and Ireland, Vol. 5: 1398–1404* (London, 1904), https://www.british-history.ac.uk/cal-papal-registers/britie/vol5/pp23-64. Several monastic recipients of these indults (who tended to live on for some years afterwards) are recorded in J. Greatrex, *Biographical Register of the English Cathedral Priories of the*

Province of Canterbury, c.1066 to 1540 (Oxford, 1997), e.g. 806, 843, 900.

9. *Gesta Abbatum monasterii Sancti Albani (GASA)*, ed. H. Riley (London, 1867), 3:425–6, 480–81.

10. J. Greatrex, *The English Benedictine Cathedral Priories: Rule and Practice, c.1270–c.1420* (Oxford, 2011), 302.

11. J. Greatrex, personal correspondence, 10 Aug. 2019.

12. *The Philobiblon of Richard de Bury*, Ch. 1, ed. E.C. Thomas (London, 1888), 11.

13. L.C. Taub, *Ptolemy's Universe: The Natural Philosophical and Ethical Foundations of Ptolemy's Astronomy* (Chicago, 1993), 31–7.

Acknowledgements

If medieval science was – as I have tried to show – an intensely collaborative endeavour, the production of a book today is no less so. It is a great privilege and pleasure to be able to record my gratitude to all of the following people who have contributed to this one. (Shamefaced apologies go to anyone I may inadvertently neglect to name.)

The bulk of this book was written in the tower of Girton College, Cambridge. At Girton I was lucky to find not only the space and opportunity to research and write, but also a wonderful community, with interested colleagues who encouraged my efforts. However, my debts stretch back much further. This book grew – sometimes in unexpected ways – from my PhD at Cambridge, and I am pleased to acknowledge the support of the Arts and Humanities Research Council, which funded my project. I benefited hugely from being a student at Peterhouse (the current custodian of John Westwyk's *Equatorie* manuscript); I am thankful for the many ways in which the college has supported my work. My thanks go to all at the Department of History and Philosophy of Science in Cambridge, which has been a fabulous place to work and study. Still further back, I am glad to record my gratitude to my history teachers. Robert 'Hendy' Henderson first aroused my love of the subject. Ian Clark showed me that intellectual history could be exciting: it was in his classroom that I first heard the phrase 'Twelfth-century Renaissance'. Colin Pendrill taught me to take notes, not just copy from books. At Oxford I benefited from several inspiring teachers, but Katya Andreyev and Giles Gasper stand out for their enthusiasm and pedagogical care. At Cambridge, Leon Rocha encouraged me to dig deep into archives. Andrew Cunningham taught me to fear the phrase 'medieval science'; he may

ACKNOWLEDGEMENTS

disagree with some of this book, but he helped give me the confidence to write it. My polymathic doctoral advisor Nick Jardine was an inspiration to me; my supervisor Liba Taub first introduced me to the *Equatorie*, and has supported me in countless ways ever since.

Like any history book, this one leans heavily on the work of past scholars. The extent of my debts should be clear from the endnotes, and I hope interested readers will consult the Further Reading section, where I have highlighted the books and articles I found particularly valuable. I have also received enormous assistance from generous colleagues, family and friends around the world. The following people were kind enough to advise me on specific research questions, to assist with tricky translations, to help me improve my writing, or to support me in other ways, from image advice to accommodation: Seb Allen, Debby Banham, Caroline Barron, Winston Black, Jenny Blackhurst, Ben Blundell, Bernadette Brady, Paul Brand, Leah Broad, Peter Brown, Charles Burnett, Jason Bye, Hilary Carey, Martha Carlin, José Chabás, Karine Chemla, Rajat Chowdhury, James Clark, Paul Cobb, Katie Cooper, Lisa Cooper, Simon Cunningham, Jacob Currie, Richard Dance, John Davis, Virginia Davis, Andrew Dunning, Catherine Eagleton, Bella Falk, Margaret Gaida, John Gallagher, Samuel Gessner, Sarah Gilbert, Christopher Graney, Joan Greatrex, Monica Green, Sarah Griffin, Matthieu Husson, Boris Jardine, Peter Jones, Peter Joubert, David Juste, Matthew Keegan, Richard Kremer, Scott Mandelbrote, Iona McCleery, Stephen McCluskey, Laure Miolo, Clemency Montelle, James Montgomery, Nigel Morgan, Robert Morrison, Adam Mosley, Stephennie Mulder, Christopher Norton, Philipp Nothaft, Lea Olsan, Richard Oosterhoff, James Paz, Josie Pearson, Joanna Phillips, Jamil Ragep, Jennifer Rampling, Kari Anne Rand, Alison Ray, Stephen Rigby, Levi Roach, Petra Schmidl, Nathan Sivin, Jacqueline Smith, Keith Snedegar, Sigbjørn Sønnesyn, Neil Stratford, Tess Tavormina, Mark Thakkar, Richard Thomason, Rod Thomson, Glen Van Brummelen, Benno van Dalen, Geert Jan van Gelder, Linda Ehrsam Voigts, Daniel Wagner, Faith Wallis, Immo Warntjes, Tessa Webber, Seán Williams, Henry Zepeda. Joanne Edge suggested the title and provided frequent encouragement and dog chat. Stefan Bojanowski-Bubb asked consistently thought-provoking questions, while James Duffy and

Sam Brooks helped prevent the answers from becoming too tedious. I also benefited from the advice of the contributors to various online scholarly forums, notably HASTRO, MEDMED, and Rete. And I learned a huge amount from participating in collaborative projects, particularly the ALFA and TAMAS projects led by Matthieu Husson, the Astronomical Diagrams project led by Nick Jardine and Sachiko Kusukawa, and the Ordered Universe project led by Giles Gasper, Tom McLeish and Hannah Smithson.

Invidious as it is to single out any individual from the above list, I must pay tribute to Kari Anne Rand, without whose ground-breaking work this book could not exist. Throughout its development I have valued her encouragement and patient advice. The children of Derek de Solla Price – Jeffrey, Linda and Mark – also provided support at an early stage: they not only gave me an insight into their father's character, but allowed me access to a wealth of family memorabilia. I also benefited enormously from access to more formal archives and libraries. Among them, I must thank the staff of Cambridge University Library, the Whipple Library, and the libraries of Corpus Christi, Girton, Gonville & Caius, Pembroke and Trinity Colleges; the Bodleian Library, Oxford, and the libraries of Corpus Christi and Merton Colleges (especially Julia Walworth); the libraries of the universities of Aberdeen and Salamanca; the Bamberg State Library; the Adler Planetarium, Chicago; the British Library, British Museum and Royal Institution, London (especially Jane Harrison); the Vatican Secret Archives; and the Museum of the History of Science, Oxford (especially Silke Ackermann and Stephen Johnston). I have been immensely fortunate to have had unfettered access to the collections of the Whipple Museum of the History of Science in Cambridge, and am grateful to its staff, particularly Steve Kruse, Josh Nall and Claire Wallace.

In recent months I have been blessed with constructive feedback from a panel of expert readers. The following all kindly read at least one chapter in draft form, suggesting improvements large and small, and saving me from innumerable errors: Charles Burnett, John Davis, Joanne Edge, Margaret Gaida, Joan Greatrex, Peter Jones, Richard Kremer, Tom McLeish, Philipp Nothaft, Kari Anne Rand, Liba Taub, Glen Van Brummelen, and the members of the Cambridge Medieval ECR Work-in-Progress group convened by Emily Ward. With immense generosity and sensitivity, Harriet Campbell and Thony Christie both read and commented on the entire book. Mistakes will inevitably remain: if you notice one, please let me know via Twitter (@Seb_Falk) or sebfalk.com. As John Westwyk himself recognised, even as we teach we are all still learning.

Many people have worked to turn an idea and text into the book you now hold. The sage advice and support of my agent Andrew Gordon got this project off the ground. I am grateful to him and the whole team at David Higham Books, as well as to his co-agent Michelle Tessler. I have been supported by a marvellous team at Penguin, led by Casiana Ionita. Her belief in the project from the beginning inspired and motivated me, and her deft touches on the tiller have helped navigate it to a successful destination. Also at Penguin I must thank Isabel Blake, Thi Dinh, Richard Duguid, Sam Johnson, Ingrid Matts, Julie Woon, and the tireless master of detail Edward Kirke. Sarah Day was a thoughtful and meticulous copy-editor. Francis Young compiled the index with great skill and efficiency. I have felt warmly supported and encouraged by the team at W.W. Norton, especially Quynh Do, Drew Weitman, Erin Sinesky Lovett and Steve Colca.

In listing the names of so many friends, colleagues, mentors and idols, I am uncomfortably aware that the same credit is rarely offered to medieval scholars – and still less to the craftsmen (and women), to the farmers and physicians, the shipwrights and monks in the background of this book. Some I have tried to rescue from anonymity, but there will always be more whose identities we can never know. I was also wary of weighing down the story with too many inessential names. All I can do here, then, is apologise to those historical figures whose names I chose to omit, from Ascelin of Augsburg and Abu Rayhan al-Biruni to William of Ockham and Guy de Chauliac, Levi ben Gerson and Michael Mentmore. In the words of Ben Sirach (translated for the book of Ecclesiasticus in the King James Bible): 'There be of them, that have left a name behind them, that their praises might be reported. And some there be, which have no memorial; who are perished, as though they had never been . . . But these were merciful men, whose righteousness hath not been forgotten . . .

Their glory shall not be blotted out. Their bodies are buried in peace; but their name liveth for evermore.'

On every day from beginning to end of writing this book I have benefited from the support and advice of my inspiration as a historical writer, my wife Susannah. She read the first draft of every page of this book, and has always been ready with suggestions, a patient ear, or an energising 'Yeah!', when each was required. Amos and Oisín too have made their own contributions, playing with (or chewing) astrolabes and providing welcome distraction. Ridley has always been close by, offering calm and an opportunity for a reflective walk. They have all brought me daily joy and reminded me what is most important.

This book is dedicated to my parents, to whom I owe more than I will ever know; more than they can imagine.

SLDF
Good Friday, 2020
Clime of 16 hours, 46 minutes
4;10° east of Toledo

Index

Muslims 32, 74, 86, 121, 127, 155,
    173, 185, 202, 228, 242, 271,
    285, 286, 287
  instruments 147–8, 150
  calendar 19, 244, 318n8
myrobalan fruit 228, 230, 233

natural philosophy 95, 112, 120,
    183, 193, 213, 306
  distinct from science? 8
  importance of Aristotle 87–8,
    99, 185
  in universities 88, 99–100, 117
nature, book of (metaphor) 85,
    147, 257
navigation 13, 206, 210, 212–13,
    215–18
Neckam, Alexander 211–12,
    215–16, 225, 257
necromancy see under magic
Nectanabus 198
Newark, Nottinghamshire 226
Newcastle-upon-Tyne 16, 164,
    166, 208
Newton, Isaac 8, 292, 316n14
Nicolas of Benevento 295
Nicholas of Lynn 75–6, 191
Nicholas of Redclif (Radcliffe),
    archdeacon of St Albans
    34, 35
non-naturals 230
Norman Conquest 4, 44, 65, 73, 164
North, John 58–9, 254–5
North Pole 25, 75, 168
  celestial 25, 27, 62, 76, 125–6,
    133, 134, 135, 138–9, 170,
    174–5, 213
North Shields, Northumberland
    see Tynemouth Priory,
    Northumberland

Northampton,
    Northamptonshire 85
Norwich, Norfolk 16, 56, 131,
    134, 238
  Cathedral Priory 55, 120
novelty, attitudes to 150
Number Battle (boardgame) 78,
    327n69
numerals, Babylonian 33, 35
numerals, Hindu-Arabic 31–2, 34,
    39, 41, 76, 86, 292
  development of 32–3, 319n31
  in calendars 71, 76
numerals, Roman 33–5, 76
  calculating with 31, 35, 39–40
  in calendars 70–71
  and place-value notation 33, 35
Nuremberg, Germany 282
nutrition 224, 230

obliquity see ecliptic
observatories 12, 47, 285–7
optics 87, 112–15, 195, 214
Oresme, Nicole 119, 194, 195
Orford, Suffolk 216
Orkney Isles 207
Oswin, St, king of Northumbria
    162, 167
Ottoman Empire 288
Ouse, River 216
Ovid 89, 344n23
Oxford, Oxfordshire 16, 55, 80,
    91, 201
  latitude of 76, 95, 170, 178,
    191
Oxford, University of 45, 58, 81,
    94, 111, 117, 203, 211, 235,
    297, 306
  astronomy at 120–21, 124, 197
  Black Death 119